Experimental Physics

Experimental Physics

Modern Methods

R. A. Dunlap

Dalhousie University
Halifax, Nova Scotia

New York Oxford
OXFORD UNIVERSITY PRESS
1988

Oxford University Press

Oxford New York Toronto
Delhi Bombay Calcutta Madras Karachi
Petaling Jaya Singapore Hong Kong Tokyo
Nairobi Dar es Salaam Cape Town
Melbourne Auckland

and associated companies in
Berlin Ibadan

Published by Oxford University Press, Inc.
200 Madison Avenue, New York, New York 10016

Library of Congress Cataloging-in-Publication Data

Dunlap, R. A.
Experimental physics.

Bibliography: p.
Includes index.
1. Physics—Experiments—Methodology.
2. Physical measurements. I. Title.
QC33.D86 1988 530 87-34750
ISBN 0-19-504949-7

10 9 8 7 6 5 4 3 2

Printed in the United States of America
on acid-free paper

Preface

This text is the result of teaching a full-year, junior-level course in experimental physics over the past six years. The course consists of three hours of lecture per week and a complementary laboratory, and so this book serves as both source material for the lecture and reference for the laboratory. In cases where a lecture course on experimental physics is taught independently of laboratory work, this text should provide enough material for two semesters. In cases where a laboratory course is taught without accompanying lectures, this book should serve as a useful reference. However, it is not intended as a laboratory manual of physics experiments. The purpose of the text is to provide an overview of the physical principles of experimental apparatus and measurement techniques and, for the most part, does not make reference to specific experiments. The text assumes the student has a knowledge of introductory mechanics, electricity, and magnetism, including some background in wave mechanics. Some familiarity with basic quantum mechanics would also be helpful. Most of the discussions on optics and on nuclear, atomic, and solid-state physics are developed from basic principles. Typically, physics majors at the junior level or higher should have sufficient background to follow the presentation.

Experimental physics encompasses a vast number of different areas. It is not possible to provide even a brief description of experimental techniques from all these areas within a single text; nor would this necessarily be desirable. Since this text is designated for use by advanced undergraduate students, it deals with those subjects which these students are most likely to encounter from an experimental point of view. For this reason, the book concentrates on three particular areas of modern physics: solid-state physics, optics, and nuclear physics, though the apparatus and techniques described here are frequently applicable to other areas of physics as well. To understand the operation of modern laboratory apparatus, some knowledge of electronics is necessary. And since it has been assumed that many physics majors do not have any formal background in electronics, the necessary introductory material is also presented. This treatment of electronics emphasizes the physics of electronic devices rather than principles of circuit design.

Specifically, Chapter 1 is an introduction to the band theory of solids. This is necessary for an understanding of semiconducting devices, which first appear in Chapters 2 and 3, as well as for the discussions on the physics of sensors, which appear throughout the later chapters. Chapter 4 presents an overview of electronic techniques for signal processing and data accumulation. Chapters 5 to 7 deal with producing, controlling, and measuring temperature and pressure.

The remaining material can be divided into three main groups. Chapters 8 to 10, which deal with optics, Chapters 11 and 12, which deal with nuclear physics, and Chapters 13 and 14, which deal with solid-state properties. These three groups may be covered in any order with one exception. The section on photomultiplier tubes, which appears in section 10.1, is important for the discussion of scintillation detectors, which begins in section 12.1.

Finally, the use of nomenclature and units requires some comment. This book deals with topics from several diverse fields of physics. The nomenclature and units commonly used in the different fields are not always consistent. As much as possible, the nomenclature and symbols used for various quantities have been made uniform throughout the book. Also, wherever practical, the SI system of units has been used, although in some specific cases the more customary unit has been retained.

February 1988
Halifax, Nova Scotia R. A. D.

Contents

Experimental Physics

1
The electrical properties of solids

1.1 The free-electron model

The electrical conductivity of solids covers one of the largest ranges of values of any physical parameter known—more than 25 orders of magnitude. This range is illustrated in Figure 1.1. Materials are grouped roughly into three categories according to electrical properties: conductors, semiconductors, and insulators. To understand the reasons for these distinctions, it is necessary to gain some insight into the behavior of electrons.

Although in some cases the electron behaves like a particle, it can be shown (e.g., by diffraction experiments) to possess characteristics of a wave as well. The wave-like nature of the electron is described by a wave equation, just as are the displacement of a vibrating string and the density fluctuations in air caused by sound waves. For electrons, the relevant entity is the electron wavefunction. This is a function of the spatial coordinates, $\psi(\mathbf{r})$, and its square, $|\psi(\mathbf{r})|^2 = |\psi(\mathbf{r})\psi(\mathbf{r})^*|$, gives the probability of finding the electron at some particular location at a given time. For simplicity, we consider the one-dimensional case. The derivation of the three-dimensional case is somewhat more complicated but follows along the same lines. The wave equation for an electron is Schrödinger's equation and in one dimension it has the form

$$\frac{\hbar^2}{2m}\frac{\partial^2\psi}{\partial x^2} - V\psi = -\mathscr{E}\psi \tag{1.1}$$

where $\hbar$ is Planck's constant, m is the mass of the electron, $\mathscr{E}$ is its total energy, and V is the potential energy. As the simplest case, we consider a free electron. In this case, the electron does not interact with other electrons or with atoms or ions. This means that, at least for the time being, we assume that there are no perturbing fields (e.g., electric or magnetic fields). Hence $V = 0$, and the energy of the electron is merely the kinetic energy, $\mathscr{E} = \frac{1}{2}mv^2$. (We have ignored relativistic effects here and it will turn out that, in dealing with electrons in solids, this omission is justifiable.)

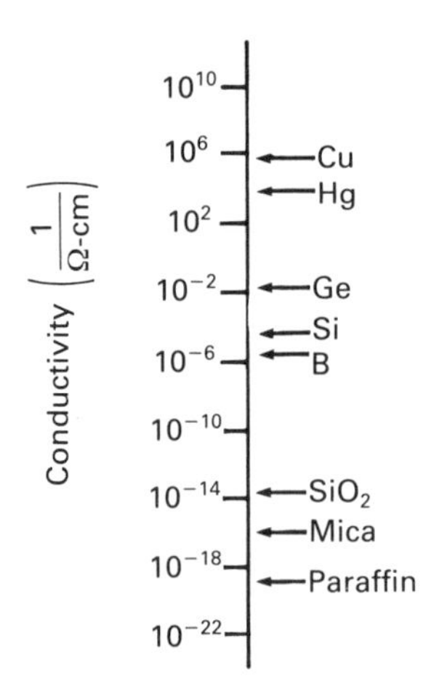

Figure 1.1. Electrical conductivity of some common materials.

Hence, substituting for $\mathscr{E}$ and rearranging, equation (1.1) becomes

$$\frac{\partial^2 \psi}{\partial x^2} = -\frac{m^2 v^2}{\hbar^2}\psi \tag{1.2}$$

We will substitute the nonrelativistic momentum $p = mv$ into subsequent steps of the derivation. Partial differential equations of the form of (1.2) can be solved most easily by guessing at an appropriate function $\psi(x)$, usually with some insight into what the answer should be. In this case, our insight provides the answer

$$\psi(x) = A \sin(px/\hbar) + B \cos(px/\hbar) \tag{1.3}$$

An important observation we can make from equation (1.3) is that the wavefunction of an electron has a wavelength λ that is inversely proportional to the electron momentum by

$$\lambda = \frac{2\pi\hbar}{p} \tag{1.4}$$

that is, the more energetic the electron, the shorter is its wavelength—we recall that the same is true for photons.

Now let us consider the behavior of an electron in a solid. We continue to make the assumption that the electron does not interact with the atoms or with other electrons, so equation (1.2) is still valid inside the material. At room temperature, the electron cannot normally escape from the solid, and the reason for this is simple. If a negatively charged electron were to escape from the surface of the material, it would immediately be attracted back into the material by the positive image charge created by the electron's absence. If an electron gains sufficient energy, as it can in materials at higher temperatures, it can escape from the attraction of its own image

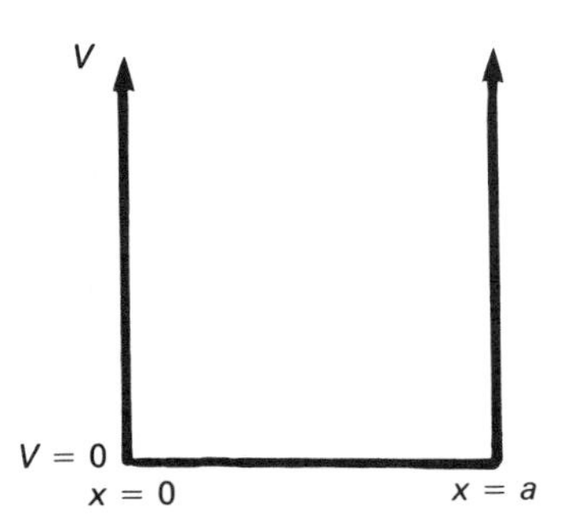

Figure 1.2. Potential energy diagram for an electron in a piece of material in one dimension.

charge. This phenomenon is called thermionic emission, and later we will learn about some of its characteristics. Considering the room-temperature situation, in which this does not occur, we see that there is, in effect, an infinite potential well at the edge of the solid. For a piece of material of length a, the potential energy is described by Figure 1.2. Hence, outside of the material, the wave equation for the electron has the form of equation (1.1) with $V = \infty$. This has only the trivial solution $\psi = 0$. Since ψ must be continuous, we can apply boundary conditions to equation (1.3) at $x = 0$ and $x = a$. The $\psi(0) = 0$ condition requires that $B = 0$, but places no restrictions on A. For $\psi(a) = 0$ we have $pa/\hbar = n\pi$, where n is an integer. Thus, the momentum, and hence the energy, is quantized and this quantization results from the application of the boundary conditions. So we have the momenta

$$p = \frac{n\pi\hbar}{a} \tag{1.5}$$

and the corresponding energy levels

$$\mathscr{E} = \frac{p^2}{2m} \tag{1.6}$$

which are

$$\mathscr{E} = \frac{n^2\pi^2\hbar^2}{2a^2m} \tag{1.7}$$

Thus, the wavefunction for the electron in one dimension has modes of oscillation in the material analogous to the modes of oscillation in a string with both ends fixed. This is shown in Figure 1.3. The square of ψ, which is always a positive quantity, gives the probability of finding the electron at a particular location. Since $\psi = 0$ outside of the material (i.e., for $x < 0$ and $x > a$), the probability of finding the electron outside of the material is, as we expect, zero.

The relationship between $\mathscr{E}$ and p is known as a dispersion relation and

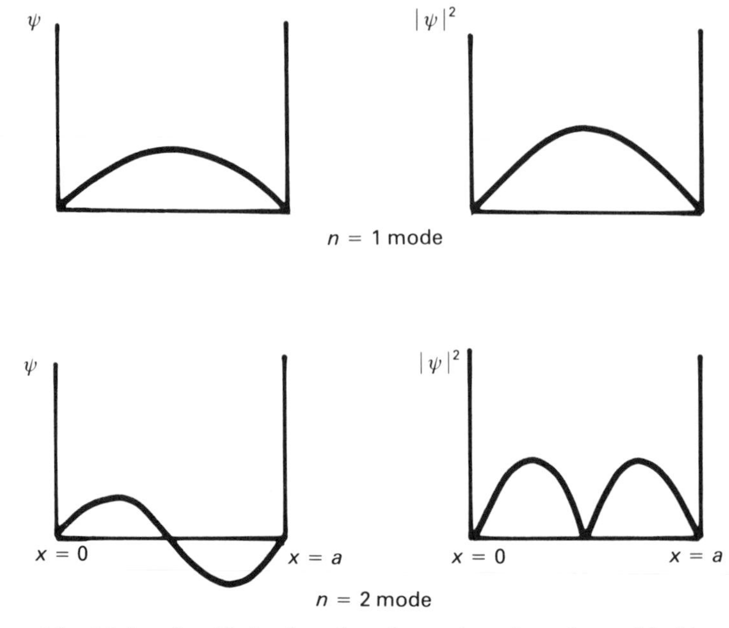

Figure 1.3. Modes of oscillation for a free electron in a piece of material of length a.

for the free electron it is seen from equation (1.6) to be parabolic (see Figure 1.4). The allowed energy (or momentum) states, as given by equation (1.7), are shown in the figure. We see from equation (1.6) that the dispersion relation for all free electrons is the same, but it is clear from equation (1.7) that the spacing in energy between the allowable states depends on a, the size of the piece of material. The number of states per unit energy is referred to as the "density of states." The larger the sample, the greater the density of states.

Let us now consider which states will be occupied. When the temperature is very low and there are no perturbing forces, the system will be in its ground state. For electrons, which are fermions, there are restrictions on how the energy states can be filled. This restriction is called the Pauli exclusion principle. Basically, it states that any given energy level can contain *at most* two electrons, one with spin up and one with spin down.

Let us look again at the dispersion relation in Figure 1.4. We see that each energy has *two* corresponding momenta, one positive and one negative. This is because energy is a scalar quantity and momentum is a vector. In one dimension the electron can have the same energy by travelling at a particular speed in one direction or the other. These two cases correspond to momenta of opposite signs. An electron can reverse its direction and hence the sign of its momentum without a change in energy by scattering elastically from the potential wall at either end of the piece of material. These two momenta [(+) and (−)] for a given energy are *not*

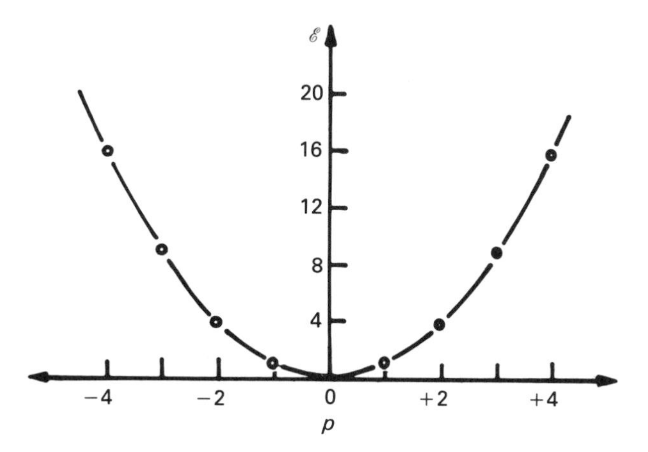

Figure 1.4. Dispersion relation for free electrons showing allowed energy states. The energy is given in units of $\pi^2\hbar^2/2ma^2$ and the momentum is given in units of $\pi\hbar/a$.

analogous to the two spin states! For simplicity, we redraw the dispersion relation for $|p|$ rather than for p. This is illustrated in Figure 1.5.

Now we can begin to fill up the available states, placing up to two electrons in each, as indicated in Figure 1.5. Ultimately what we would like to determine is the *maximum* energy an electron can have in the ground state. To find this qualitatively, we consider a material with a density n_0 of electrons. In one dimension this is the number of electrons per unit length. A length a of material then contains $N = n_0 a$ electrons. Since there can be two electrons per state, we require $an_0/2$ states to accommodate all of the electrons. The energy of the highest filled state is obtained from equation

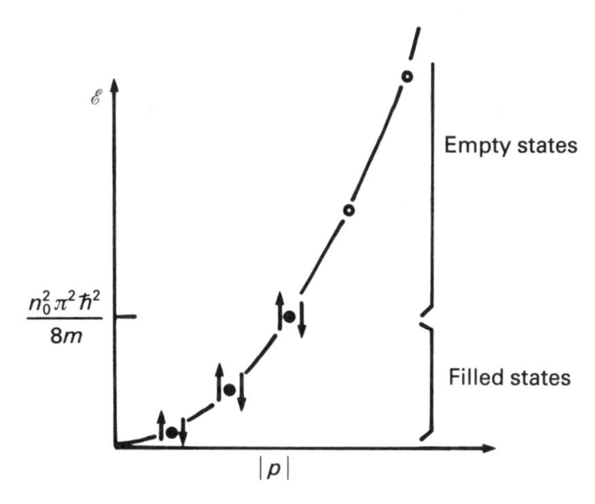

Figure 1.5. Population of the energy levels of a one-dimensional system of fermions in the ground state.

(1.7) by substituting $n = an_0/2$ and is called the Fermi energy. This is found to be

$$\mathscr{E}_F = \frac{n_0^2\pi^2\hbar^2}{8m} \tag{1.8}$$

We note that this energy depends on the *density* of the electrons in the material but not on the overall dimensions of the piece of material. We should emphasize that the *number* of energy states below $\mathscr{E}_F$ does depend on the size of the material since, the larger this is, the more electrons must be accommodated. Figure 1.5 shows how the electrons are arranged in the energy levels, one with spin up and one with spin down to each level. If we had proceeded with this derivation in three dimensions, we would have obtained the Fermi energy as

$$\mathscr{E}_F = \frac{\hbar^2}{2m}(3\pi^2 n_0)^{2/3} \tag{1.9}$$

where n_0 is the electron density per unit volume.

The above calculation applies at zero temperature. At finite temperature, the probability that an energy state is occupied is a function of energy and of temperature. This probability function is the Fermi–Dirac distribution and it is illustrated in Figure 1.6. These curves should, in fact, be a series of dots equally spaced in momentum, as the allowed states are quantized. If the sample is large and contains a large number of electrons, the dots are closely spaced and, as in the figure, appear as a solid line. The mathematical form of the Fermi–Dirac distribution is given by

$$f(\mathscr{E}) = \frac{2}{e^{(\mathscr{E}-\mathscr{E}_F)/k_BT} + 1} \tag{1.10}$$

where k_B is Boltzmann's constant. The 2 in the numerator comes from the

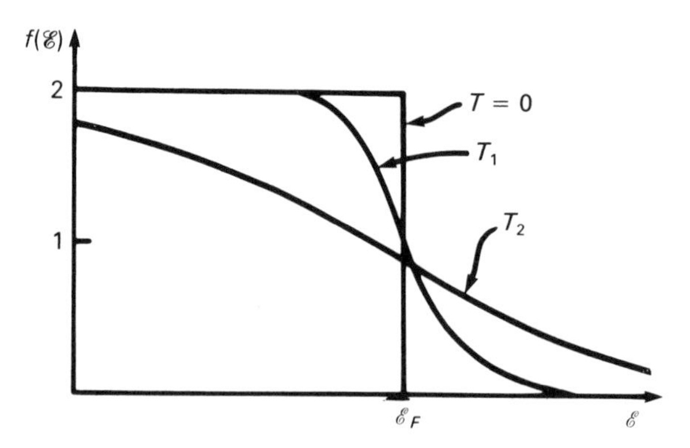

Figure 1.6. Fermi–Dirac distribution for electrons ($0 < T_1 < T_2$).

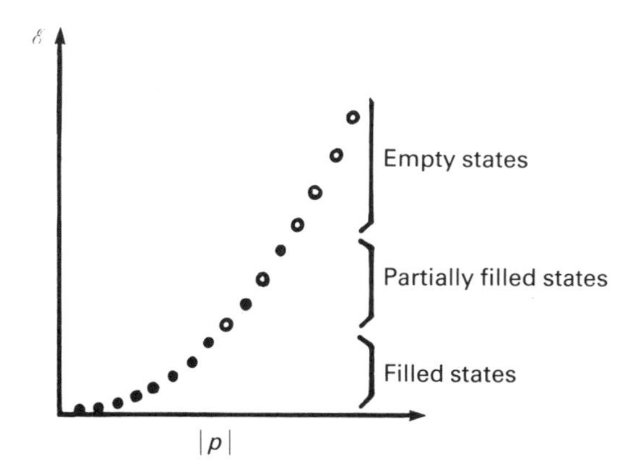

Figure 1.7. Dispersion relation for free electrons at $T > 0$.

two allowed spin states per energy state. By inspection of equation (1.10) we can see that $f(\mathscr{E})$ for $T = 0$ is a step function with a value of 2 for $\mathscr{E} > \mathscr{E}_F$ and a value of 0 for $\mathscr{E} > \mathscr{E}_F$. This represents the ground-state situation we have previously described. As T increases, the curve smears out around $\mathscr{E}_F$. This corresponds to the electrons near the Fermi energy jumping up into unoccupied higher-energy states. The lower-energy electrons cannot gain any additional energy unless there is a unoccupied energy state above them and they have sufficient energy to jump up into that empty state. At some intermediate temperature, the situation is as shown in Figure 1.7. Hence, the states can be divided into three regions; filled, partially filled, and vacant. The integral of $f(\mathscr{E})$ must be independent of temperature, as the number of electrons is conserved. You can see this for yourself by integrating equation (1.10) over energy.

Materials contain a lot of electrons, many of which are unimportant (more or less) in determining the electronic properties. It turns out that the electrons that are of importance are the valence electrons. The core electrons remain localized around particular atoms and, at least to first order, are not important to the electronic properties. In the general sense, the valence electrons are those that have "orbits" that encompass more than one atom. So, when we look back at equations like (1.8), it is only the valence electrons that we have to consider in determining n_0.

Now let us use what we have learned to calculate some measurable properties of a material. Consider a sample of cross section A and of length l. We attach leads to each end and supply a voltage V. Free electrons can gain energy via interaction of their charge with the electric field E produced by the applied voltage. We perform the experiment as shown in Figure 1.8. At some time (defined as $t = 0$) we close the switch and apply the electric field. The motion of an electron at the Fermi energy can be considered as that of an entirely free electron, because there are unlimited empty energy states above it. Hence we can write down the equation of motion for this

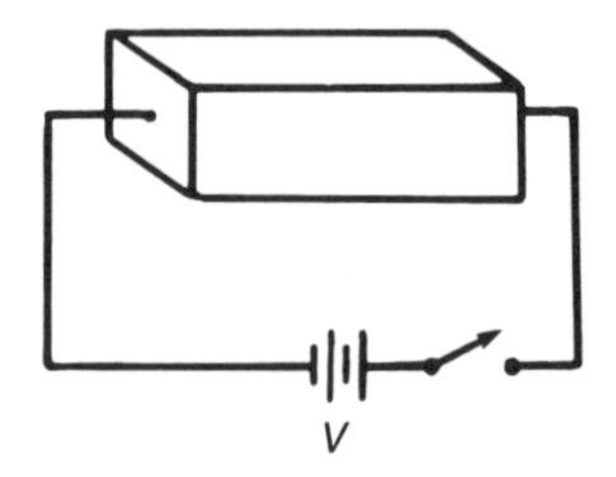

Figure 1.8. Experiment for measuring the electrical conductivity of a sample.

electron in terms of the Lorentz force:

$$\mathbf{F} = -e(\mathbf{E} + \mathbf{v} \times \mathbf{B}) \tag{1.11}$$

where e is the electron charge, $\mathbf{v}$ is the electron velocity, $\mathbf{E}$ is the applied electric field, and $\mathbf{B}$ is the applied magnetic flux density. Since, in our experiment, $\mathbf{B} = 0$, equation (1.11) is immediately separable and we can integrate it to obtain

$$\int_{p_0}^{p} dp = -eE \int_0^t dt \tag{1.12}$$

or

$$p = p_0 - eEt \tag{1.13}$$

where p_0 is the momentum of the electron before the electric field was applied. Since p_0 is small, the (nonrelativistic) energy of the electron at time t after the switch is closed is

$$\mathscr{E} = \frac{e^2 E^2 t^2}{2m} \tag{1.14}$$

To find the conductivity of a material, let us write the well-known form of Ohm's law

$$V = IR \tag{1.15}$$

Dividing each side of this equation by the length of the material to give

$$\frac{V}{l} = \frac{I}{A} \frac{RA}{l} \tag{1.16}$$

we see that the left-hand side is just the electric field inside the material. The first factor on the right-hand side is the current density j, and the second factor on the right-hand side is the resistivity ρ, or the inverse of the conductivity σ. Thus Ohm's law may be written in the form

$$E = \frac{j}{\sigma} \tag{1.17}$$

By definition, the current density is the charge per unit area flowing per unit time in the material. We can write this in terms of the electronic charge e, the number n_0 of electrons per unit volume, and the velocity of the electrons as

$$j = -en_0v \tag{1.18}$$

From equation (1.14) the velocity of the electrons after time t is given by

$$v = -\frac{eEt}{m} \tag{1.19}$$

We take the negative square root because of the negative sign of the charge on the electron. Substituting (1.18) and (1.19) into (1.17) and solving for σ we obtain

$$\sigma = \frac{n_0e^2t}{m} \tag{1.20}$$

This tells us that the conductivity increases with increasing density of valence electrons. This makes sense. It also tells us that the conductivity increases as a function of time after we close the switch. This does *not* make sense! Clearly, when we measure the conductivity (or resistivity) of a sample, it is independent of time.

The failure of this model to provide reasonable results goes back to our basic assumption that the valence electrons do not interact with the atoms in the material. This is not true—they do interact. In order to do the problem properly, it is necessary to go back to the beginning and modify Schrödinger's equation to account for the interactions. We will do this in the next section, at least in a nonrigorous manner. For now, let us try to save the free-electron model by fixing up equation (1.20) to take into account the electron–atom interactions.

What actually happens is that the electron travels along in the material, generally unaffected by the atoms, until it undergoes some major interaction with an atom. In the simplest case, this is an elastic collision. In the one-dimensional case, this corresponds to a change in the direction of $\mathbf{p}$. In the context of the dispersion relation, electrons, by their interaction with the applied field, acquire momenta that are progressively more positive (or negative if the direction of the electric field is reversed). Thus, electrons at the Fermi energy jump into vacant states above them and lower-energy electrons jump into the newly vacated states. This process is illustrated in Figure 1.9*a*, which shows that the occupied states on the dispersion curve tend to "slide" to more positive p. Figure 1.9*b* shows the case in which the electron travels forward for some time and is then elastically scattered by an interaction with an atom. The electron would thus "circulate" about this closed cycle in the momentum–energy diagram.

If we assume some mean time τ between interactions, the conductivity we

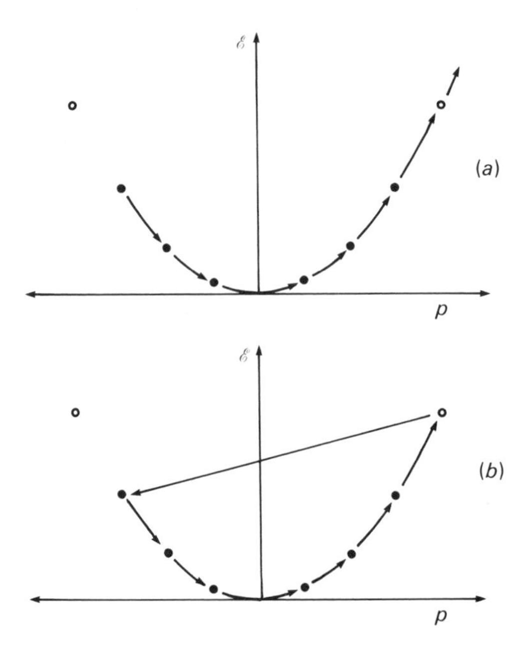

Figure 1.9. Motion of electrons in an applied electric field (*a*) without ionic interactions and (*b*) with ionic interactions.

would measure would be

$$\sigma = \frac{n_0 e^2 \tau}{m} \tag{1.21}$$

For pure copper it has been found that τ at room temperature is about 2×10^{-14} s.

An interesting aspect of equation (1.21) is its temperature dependence. At low temperature, the atoms in a solid are fixed fairly rigidly in their locations. An electron normally interacts with an atom only when it gets within a certain distance of it. At higher temperatures, the atom vibrates around its equilibrium position as a result of its thermal energy. Thus it represents a fuzzy volume in space that is somewhat larger than the actual size of the atom. This means that the electron is more likely to interact with any particular atom, and as a result the time τ becomes shorter. By comparison with the room-temperature value of τ for copper, the value can be 2×10^{-9} s at 4.2 K. Thus we would expect a conductivity σ that decreased with increasing temperature. Figure 1.10 shows that this is the case for copper and other transition metals. Obviously it is not true for germanium or silicon.

The free-electron model cannot explain the temperature dependence of the conductivity of semiconductors, nor can it explain the very low conductivities of semiconductors and insulators. To do this, the effects of

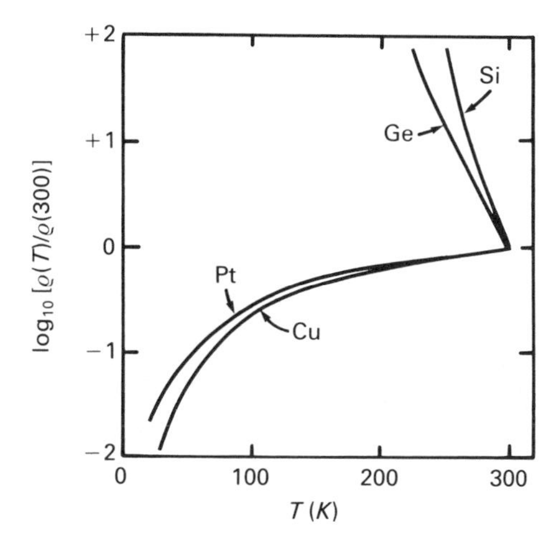

Figure 1.10. Resistivity ($\rho = 1/\sigma$) of some common materials as a function of temperature.

the atoms on the electron wavefunctions must be taken into account in a more rigorous manner. A straightforward method of doing this is described in the next section.

1.2 The band theory of solids

a. Nearly-free-electron model

In order to deal properly with electron–atom interactions we must go back to the wave equation. In one dimension, we write

$$\frac{\hbar^2}{2m}\frac{\partial^2 \psi}{\partial x^2} - V(x)\psi = -\mathscr{E}\psi \tag{1.22}$$

The potential $V(x)$ is a position-dependent function that describes the interaction between electrons and ions. In general, a quantitative mathematical description of $V(x)$, even in one dimension, is not a simple matter. Nor is the solution of the resulting differential equation necessarily straightforward. We can, of course, add a constant V (independent of x) to $\mathscr{E}$ and merely shift the energy levels up or down without affecting the general features of the electron wavefunction, but when V becomes x-dependent, this is no longer the case.

Since the ions (atoms with their valence electrons removed) are positively charged, they represent deep, attractive potential wells for the valence electrons. So, qualitatively speaking, the shape of the potential $V(x)$ for a solid shows a deep well wherever there is an atom. Generally, in a solid (certainly a crystalline solid), the atoms are evenly spaced. The distance between the atoms is called the lattice parameter a_0. In our one-dimensional

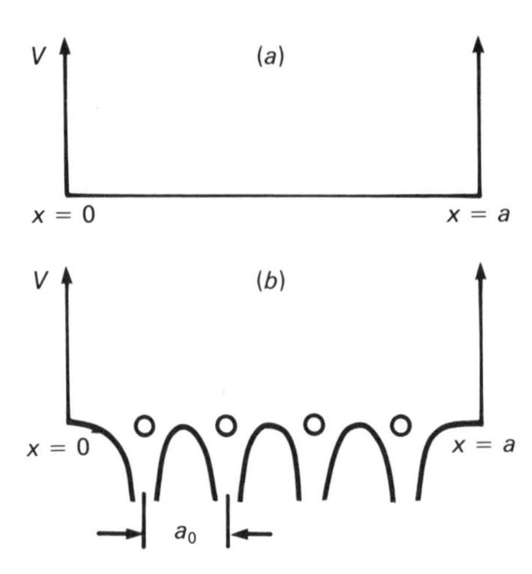

Figure 1.11. Potential energy for (*a*) the free-electron model and (*b*) the nearly-free-electron model

model there is only a single lattice parameter. A qualitative picture of $V(x)$ is shown in Figure 1.11. We are interested in determining how the addition of this potential affects the electron wavefunction. The electrons with wavefunctions of very long wavelength λ are not affected at all. For this same reason, an electron microscope using very short-wavelength electrons can be used to look at objects that cannot be seen with longer-wavelength optical photons. From equation (1.4), we see that the long-wavelength electrons are those with lowest energy (or momentum). As the wavelength of the electrons becomes comparable to the distance between the ions, the shape of the dispersion relation can be affected. Actually, when the wavelength of the electrons is such that

$$\frac{n\lambda}{2} = a_0 \tag{1.23}$$

unusual things happen. We can see in Figure 1.12 that $|\psi|^2$ has a zero at the

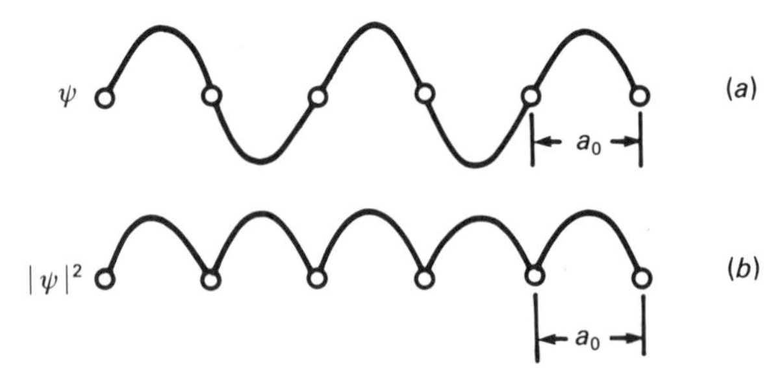

Figure 1.12. (*a*) Relationship of ψ for wavelength $\lambda/2 = a_0$ and the location of the ions and (*b*) relationship of $|\psi|^2$ and the location of the ions.

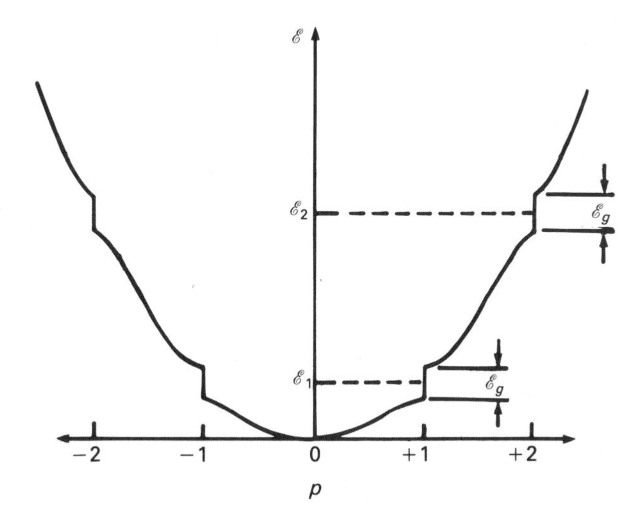

Figure 1.13. Dispersion relation for the nearly-free-electron model. The momentum is given in units of $\pi\hbar/a_0$.

location of each atom. This results in a sort of resonant condition, and standing waves are formed. So for values of p from equation (1.5) such that

$$p = \frac{\pi\hbar n}{a_0} \tag{1.24}$$

the dispersion relation becomes distorted as a result of the presence of the atoms. The result is shown in Figure 1.13. We see that although the momentum of the electrons is a continuous function, the energy has discontinuities; these are known as energy gaps and there are *no* allowed states with energies in these regions. The regions of allowed energy states are known as bands. The exact shape of the dispersion relation near $p = n\pi\hbar/a_0$ and the size of the energy gaps $\mathscr{E}_g$ depend on the details of the potential function $V(x)$. However, as long as $V(x)$ has the periodicity a_0, something generally like the situation of Figure 1.13 will occur.

We can show intuitively that the curve shown in Figure 1.13 makes sense. Consider the derivative of the dispersion relation $\partial\mathscr{E}/\partial p$. For simplicity, we look at the free-electron case,

$$\frac{\partial\mathscr{E}}{\partial p} = \frac{\partial(mv^2/2)}{\partial(mv)} = v \tag{1.25}$$

The slope of the dispersion curve is a velocity. It turns our that this is the group velocity of the electron waves. We recall that near $p = n\pi\hbar/a_0$ the waves are standing waves, because of the interaction of the electrons with the ions. The group velocity of a standing wave is zero, since it is a superposition of two waves travelling in opposite directions. Thus $\partial\mathscr{E}/\partial p$

should go to zero at $p = n\pi\hbar/a_0$. The bending over of the dispersion relation at these values of p, so that the curve intercepts the $p = n\pi\hbar/a_0$ axis horizontally, satisfies this requirement for the group velocity.

The location of the energy gap (see $\mathscr{E}_1$ in Figure 1.13) does not depend on the size of the sample. However, the number of states in each band does. We can easily calculate this number. The energy of the nth state is given by equation (1.7). The energy at the middle of the first energy gap is given by the energy from the free-electron model for $p = \pi\hbar/a_0$:

$$\mathscr{E}_1 + \tfrac{1}{2}\mathscr{E}_g = \frac{\pi^2\hbar^2}{2ma_0^2} \tag{1.26}$$

Equating this to (1.7), we solve for n, the number of energy levels, as

$$n = \frac{a}{a_0} \tag{1.27}$$

This is merely the number of atoms in the sample! So each band contains one energy level for each atom in the sample. Since each energy level can accommodate two electrons (of different spins), we can make the following observations for $T = 0$:

1. For monovalent atoms, the first band will be half full and all others empty.
2. For divalent atoms, the first band will be full, all others empty.

Similar statements can be made concerning materials of other valences. It is important to note that the valence we talk about here is the average valence and does not have to be an integer; even if we have only one type of atom present in our sample, we can still speak of an *average* valence of, say, 1.7. This means that the bands do not have to be either empty, half full, or full, but can be in any state between. We will see that this is very important in determining the electronic properties of solids.

b. Band structure and the conductivity of materials

In the simple model we have developed, there are three things that are important in determining the electronic properties of a material. These are (1) the temperature, (2) the size of the energy gap $\mathscr{E}_g$, and (3) the relative locations of the Fermi energy and the energy gap. This last factor is determined primarily by the average valence. For simplicity we concern ourselves only with materials with valences between 0 and 4. This means that only the lowest two bands are of interest, as the others will be empty. Later, we will discuss materials with larger valences but we can always (more or less) disregard the lowest filled band for higher-valence materials, since the electrons here are "trapped" and cannot contribute to the conduction. Rather than talking too much right now about valences, we will concentrate on the Fermi level and worry about what it means later. You can see from equation (1.9) how the Fermi energy is related to n_0 and hence to the valence.

Let us look at some different kinds of materials, as shown in Figure 1.14. For convenience we call the first band the *valence band* and the second band the *conduction band*. We begin by assuming that the temperature is low.

The material shown in Figure 1.14*a* is a metal. The Fermi energy is partway (but not very far) up the conduction band. This means that the valence band is completely filled and the conduction band is somewhat less than half-filled. When we apply an electric field, the electrons near the

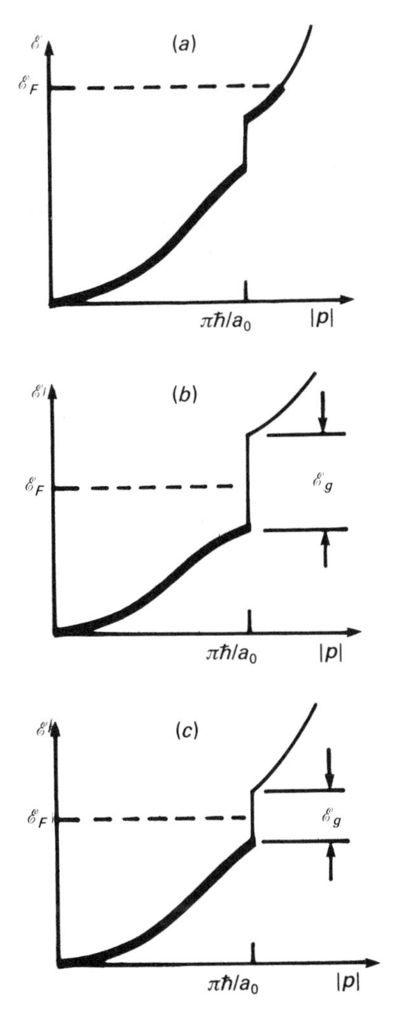

Figure 1.14. Energy band diagrams (*a*) metal (conductor), (*b*) insulator and (*c*) semiconductor at $T = 0$. States are occupied up to the Fermi level and are indicated by the heavy lines.

Fermi energy have many unoccupied states above them and they can gain quite a lot of energy before they reach the top of the conduction band. These electrons are therefore fairly similar to those we described in the free-electron model. In principle, all of the electrons in the conduction band are much like free electrons. Those in the valence band are not. These are in a filled band. Electrons near the bottom of the band cannot change state, because the states above them are filled, and those at the top of the band cannot gain energy, because there is the forbidden energy gap right above them. So we can describe the situation here by arguments that are essentially the same as those for the free-electron model, except that the electron density n_0 that is used should include only the valence electrons that are in the conduction band and not those in the valence band. Here, too, the conductivity is limited by the mean free time τ, and the conductivity as a function of temperature looks the same as it does for the free-electron model, which is desirable because this agrees with experimental evidence.

Now let us look at the band structure shown in Figure 1.14*b*. This represents an insulator. What happens to the electrons in this material when an electric field is applied? Simply speaking: nothing. Here the Fermi energy is just in the center of the energy gap. This means that all of the states in the valence band are filled and all of the states in the conduction band are empty. You have probably guessed (correctly) that this is a valence-2 material. In this case, none of the electrons can gain any energy. Those in the lower part of the band have occupied states above them and those at the top of the band have the forbidden energy gap right above them. So, when we apply the electric field, none of the electrons can move and (ideally) no conduction can take place. Hence we have an insulator.

In the above discussion we have taken the Fermi energy to be in the center of the energy gap when the valence band is filled and the conduction band is empty. Let us try to justify that assumption. The Fermi energy given by equation (1.9) is the free-electron Fermi energy, since it is obtained on the basis of the dispersion relation for the free-electron model. In the nearly-free-electron model or other more realistic treatments of this problem, the definition of the Fermi energy is a much more complicated problem. An "arm-waving" explanation can be based on the free-electron model.

Suppose we had a situation in which, according to the free-electron model, momentum states were filled up to $p = \pi\hbar/a_0$. This situation is shown in Figure 1.15. When ionic interactions are included, the dispersion relation bends over near $p = \pi\hbar/a_0$, so that the group velocity goes to zero. This is done in such a way as to place the free-electron Fermi energy in the center of the gap. The band below $\mathscr{E}_g$ still contains the same number of momentum states as for the free-electron case—that is, the same number of momentum states between $p = 0$ and $p = \hbar\pi/a_0$. Since each momentum state corresponds to a unique energy state, the same number of energy

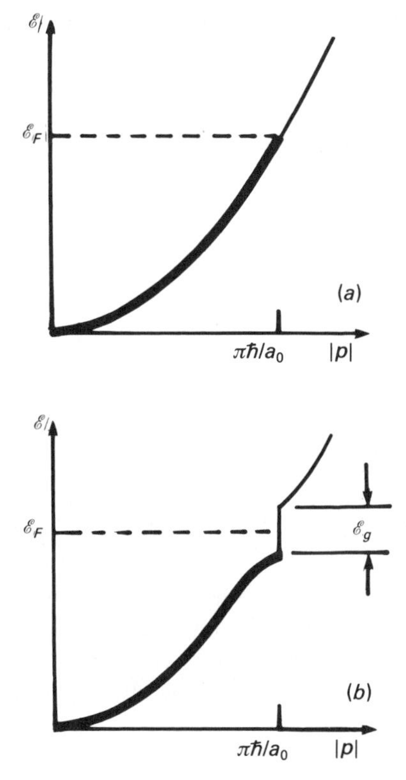

Figure 1.15. Occupied states in a semiconductor at $T = 0$ for (a) the free electron model and (b) the nearly-free-electron model.

states occur (and obviously the same number of electrons can be accommodated) in the valence band in Figure 1.15*b* as in Figure 1.15*a* below $p = \hbar\pi/a_0$. So the number electrons below $\mathscr{E}_F$ is the same in both Figures 1.15*a* and 1.15*b*. The reason for leaving $\mathscr{E}_F$ at the free-electron value rather than lowering it to the top of the valence band we can accept for now as merely a matter of convention. Also, as we will see, it turns out to explain nicely the physics of pure semiconductors.

The semiconductor shown in Figure 1.14*c* is very similar to the insulator. The Fermi energy is again at the center of the energy gap, but in this case the energy gap is a little smaller. This will become important later, but we can ignore it for the time being. The argument for the conductivity of the semiconductor now proceeds as for that of the insulator, and we come to the conclusion that the semiconductor is an insulator.

Remember, however, that the drawings in Figure 1.14 and the discussion on the last two pages referred to very low temperature. What happens at higher temperatures? Looking back at Figure 1.6 for the Fermi–Dirac (F–D) distribution, we see that at $T \approx 0$ this is a step function, the step occurring at $\mathscr{E}_F$. At the temperature increases, however, the function smears out. Let us see how much smearing occurs. To do this we have to estimate the temperature equivalent of the Fermi energy. Using equation (1.9) we assume two valence electrons per atom and a distance between atoms of 3 Å (this is about average). Therefore n_0 is given by

$$n_0 = 2/(3 \times 10^{-10})^3 = 7 \times 10^{22}\ \mathrm{m}^{-3} \tag{1.28}$$

and from (1.7)

$$\mathscr{E}_F = 9.9 \times 10^{-19}\ \mathrm{J} = 6\ \mathrm{eV} \tag{1.29}$$

Using Boltzmann's constant and $\mathscr{E} = k_B T$, the equivalent temperature here is

$$T_F = 7 \times 10^4\ \mathrm{K} \tag{1.30}$$

Remember that this is only the *equivalent* temperature. So the smearing of the F–D function as a result of raising the temperature from 0 to 300 K is not very much. Nonetheless, it is an important consideration for semiconductors.

How does this affect the situations shown in Figure 1.14? For the conductor there is essentially no change. Some of the electrons near $\mathscr{E}_F$ gain a little thermal energy, but, as these are all essentially free electrons, the description of the temperature dependence of the conductivity given for the free-electron model still holds.

The insulator and the semiconductor are very similar and can be dealt with together. Figure 1.16 shows the average number of electrons per unit energy near $\mathscr{E}_F$. This is just the F–D function. At $T = 0$, this is a step function at $\mathscr{E}_F$ so that all states in the valence band have two electrons (one spin up and one spin down), while all states in the conduction band are empty.

We can see in Figure 1.16*b* that at room temperature the F–D function has smeared out a little and there is no longer a zero probability of electrons occupying states above $\mathscr{E}_F + \frac{1}{2}\mathscr{E}_g$ (i.e., in the conduction band), nor is there a probability of unity for electrons to occupy states below $\mathscr{E}_F - \frac{1}{2}\mathscr{E}_g$ (in the valence band). This is indicated by the solid areas in Figure 1.16*b*. Simply speaking, we can describe this physically by saying that some of the electrons at the top of the valence band have gained enough energy as a result of thermal excitations to jump over the energy gaps and into the conduction band.

The amount of smearing of $f(\mathscr{E})$ of course depends on temperature, but it is the size of the energy gap that is crucial in determining how many electrons jump up into the conduction band at a given temperature. It is the size of the energy gap that is the important difference between an insulator and a semiconductor (see Figure 1.14). Diamond, one of the best insulators, has $\mathscr{E}_g = 5.4$ eV while germanium, a well-known semiconductor, has $\mathscr{E}_g = 0.7$ eV. This means that the solid areas in a figure like Figure 1.16*b* for

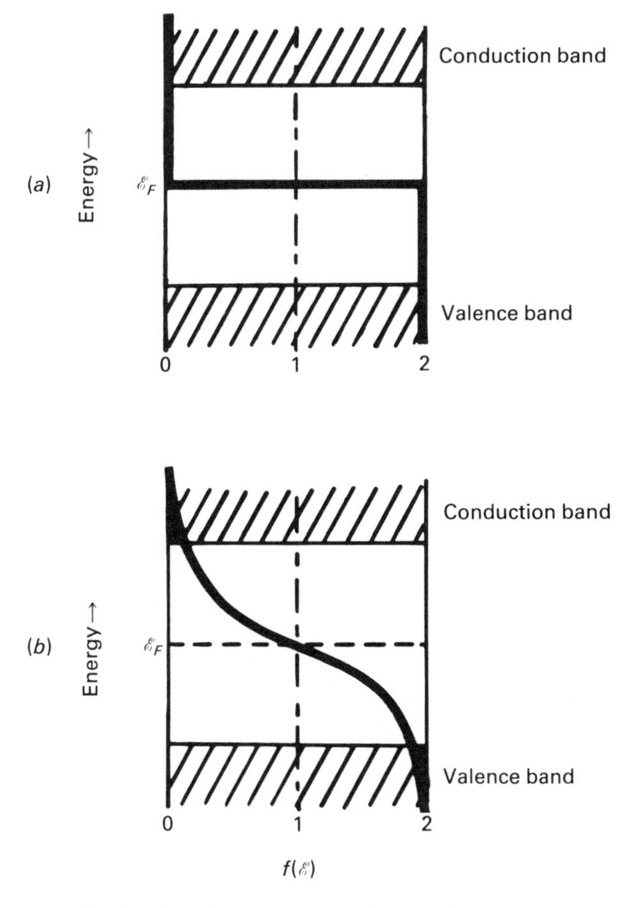

Figure 1.16. Energy distribution of electrons near the Fermi energy (a) at $T = 0$ and (b) at room temperature.

diamond are essentially nonexistent. For germanium they are small but important.

You can see that an insulator at room temperature is essentially the same (electrically) as it is at $T \approx 0$. From our previous discussion of Figure 1.14, we known why it does not conduct and the smearing of the F–D distribution at room temperature is not sufficient to alter the situation.

The semiconductor, however, is more interesting and deserves further investigation. The electrons that jump up into the conduction band of a semiconductor are like the electrons in the conduction band of a conductor—they are free to gain energy because they have unoccupied levels above them. So, in a limited sense, a semiconductor at room temperature is like a conductor, in that it does conduct to an extent. But there are three important differences (see Figure 1.17).

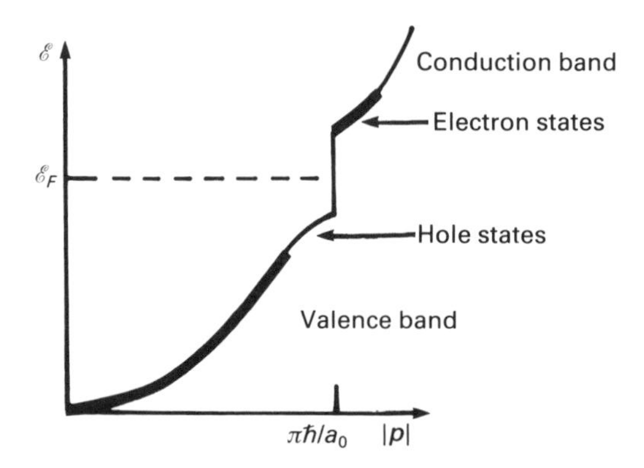

Figure 1.17. Electron and hole states in a semiconductor at room temperature.

1. The number of electrons in the conduction band is very small (because of the small solid areas in Figure 1.16*b*). So n_0 in equation (1.21) is much smaller than it is for a conductor.
2. n_0 is temperature dependent, because the shape of the F–D function, and hence the size of the solid areas in Figure 1.16*b*, is a function of temperature.
3. When an electron jumps up into the conduction band it leaves a vacant state or "hole" at the top of the valence band.

Item (1) has the obvious consequence that the conductivity of a semiconductor should be lower than that of a conductor.

Item (2) has important consequences. If we speak of a conductivity governed by equation (1.21), then we find that the temperature dependence of n_0 is much stronger than the temperature dependence of τ. As T increases, the smearing of $f(\mathscr{E})$ increases and the number of electrons hopping up into the conduction band increases. Thus σ increases; the behavior predicted for a conductor is just the opposite, but is consistent with the experimental evidence (see Figure 1.10).

Consider item (3). The hole produced at the top of the valence band when an electron jumps up into the conduction band behaves like a *positive* charge and thus, when an electric field is applied, can contribute to the conduction by interchanging energy states with an electron in the valence band. This of tremendous importance when we come to discuss the operation of semiconducting devices.

We can review the important features of the nearly-free-electron model as follows.

1. The temperature dependence of the conductivity of a conductor is governed by the mean free time τ of the electrons. Thus τ decreases with increasing T.
2. Insulators have energy gaps that are too large to allow them to conduct at any reasonable temperatures.

3. The temperature dependence of σ for a semiconductor is governed by the temperature dependence of n_0, and σ increases with increasing T. As $T \rightarrow 0$, the semiconductor is an insulator.

c. *Temperature dependence of σ for semiconductors*

We have developed a model that gives the essential features of $\sigma(T)$ for semiconductors. Let us see if we can make the model somewhat more quantitative. The conductivity due to a particular species (electrons or holes) can be written in the form of equation (1.21). A more convenient way of writing this is

$$\sigma_i = n_i e_i \mu_i \tag{1.31}$$

where n_i, e_i, and μ_i are the concentration, charge, and mobility of the species i. Comparison with equation (1.21) tells us that μ_i has a great deal to do with the mean free time τ. This makes sense. If both electrons (e) and holes (h) are present, we can write

$$\sigma = \sigma_{\text{total}} = n_e e \mu_e + n_h e \mu_h \tag{1.32}$$

We add the contributions because, although the species have opposite charges, they also move in opposite directions. In intrinsic or pure semiconducting materials, n_e and n_h will be the same, which helps to simplify matters.

To find n_e it is necessary to integrate the F–D function in Figure 1.16 over the solid area in the conduction band.

$$n_e(T) = \int_{\mathscr{E}_F + \mathscr{E}_g/2}^{\infty} n_e(\mathscr{E}, T)\, d\mathscr{E} = \int_{\mathscr{E}_F + \mathscr{E}_g/2}^{\infty} \eta(\mathscr{E}) f(\mathscr{E}, T)\, d\mathscr{E} \tag{1.33}$$

The integral on the right-hand side is an integral over energy from the botton of the conduction band up to ∞ (actually only to the top of the conduction band). Here $\eta(\mathscr{E})$ represents the density of electron states—that is, the number of available states per unit energy per unit volume available in the conduction band. Thus $f(\mathscr{E}, T)$ is the F–D function.

The evaluation of equation (1.33) is a messy process because $f(\mathscr{E}, T)$ is not an especially tractable function; because we do not really know what $\mathscr{E}_g$ is; and because we do not know *a priori* what $\eta(\mathscr{E})$ is. Rather than get involved in a rigorous and complex derivation of these quantities, we state some results and see whether we can understand what they mean. From equation (1.33) we find

$$n_e = 2UT^{3/2} \exp\left(-\frac{\mathscr{E}_g}{2k_B T}\right) \tag{1.34}$$

where U is a constant that comes from $\eta(\mathscr{E})$. From equation (1.32) we can now write (remember $n_h = n_e$)

$$\sigma = 2eU(\mu_e + \mu_h)T^{3/2} \exp\left(-\frac{\mathscr{E}_g}{2k_B T}\right) \tag{1.35}$$

The important factors appear as the temperature dependence on the right-hand side. We see two factors: a $T^{3/2}$ factor and an exponential factor. The effect of the $T^{3/2}$ factor is extremely small compared with that of the exponential factor. A plot of this function (i.e., plot $\log \sigma$ against $1/T$), shows essentially a straight line. Often (for practical purposes) you will see written

$$\sigma = (\text{constant}) \exp\left(-\frac{\mathscr{E}_g}{2k_B T}\right) \tag{1.36}$$

This has the general form that we expect from Figure 1.10, and experimental results of σ for very pure semiconductors show that this model gives good quantitative results as well.

d. Doping

The semiconductors described so far are referred to as *intrinsic* semiconductors. This means that they are made up of atoms that are all of the same kind and all have the same valence. An example is pure silicon or pure germanium.

When a semiconducting device is made, it is nearly always made from a semiconductor (usually silicon or germanium) that contains a small amount (usually a *very* small amount) of some impurity. These are doped semiconductors, and we will see in this section why they are important.

The semiconductors silicon and germanium are normally in the valence-4 state, so that the first two bands are valence bands and the third band is the conduction band. For convenience, we call the second band *the* valence band and the third band *the* conduction band. The conductivity of these materials in the pure state is really quite small. Each charge carrier has to be generated by thermal excitation, and we should reiterate that each electron in the conduction band has a corresponding hole in the valence band. So the number of (−) electron carriers is *exactly* equal to the number of (+) hole carriers.

Consider silicon as an example and assume that we add a small number of arsenic impurity atoms. Arsenic normally has valence-5 (one more than silicon); you can see this from looking at the periodic table. The band structure is determined by the silicon, since nearly all the atoms are silicon atoms. The one extra valence electron provided by the occasional arsenic atom *must* go into the conduction band. As a result, compared to what would result from thermally exciting the silicon electrons, there are many electrons in the conduction band when there are only a small number of arsenic impurity atoms present. An impurity with a greater valence than the host is called a *donor,* since it provides an extra electron. By controlling the amount of arsenic in the silicon crystal, we can adjust the conductivity to whatever value we want over a fairly wide range. More important, we have obtained electrons in the conduction band *without* forming corresponding holes in the valence band. So the equality of n_h and n_e no longer holds; rather, $n_e > n_h$. There are more electron carriers than there are hole carriers and this material is called an n-type material (n for negative, for obvious reasons).

The electrons are called *majority* carriers (since there are more of them) and the holes are called the *minority* carriers (since there are fewer of them). To give you an idea of how important impurities are in semiconductors, we note that 10 parts per million of arsenic in "pure" silicon increases σ at room temperature by a factor of about 1000.

If we now add to a semiconductor an impurity that has a smaller valence than the host, this impurity is called an *acceptor*. An example is boron added to germanium. The boron atom "binds up" one of the electrons that was a germanium valence electron. This leaves a "hole" or vacant state and we are therefore able to produce hole carriers without creating corresponding electron carriers. Thus this is a p-type material (p for positive), and $n_e < n_h$. The situation is now just the opposite of that for donor impurities. Table 1.1 gives a summary of these considerations.

This picture of doping is all rather phenomenological and we can, in fact, quite easily consider the problem in more detail. We will look at donor doping as an example. What we have done by adding the higher-valence impurity is to increase the average number of valence electrons per unit volume in the material, but only slightly. In the free-electron sense, this will raise the Fermi level slightly. A rigorous derivation would show that this is the case for the nearly-free-electron model as well. This means that the Fermi energy is no longer at the center of the energy gap but has been raised by some amount $\mathscr{E}_d$. The exact size of $\mathscr{E}_d$ is unimportant for the following discussion—so long as it is less than $\frac{1}{2}\mathscr{E}_g$. The situation at room temperature is shown in Figure 1.18. We see that the result is not only to *increase* n_e but also at the same time to *decrease* n_h. Calculating n_e and n_h is a straightforward matter. From equation (1.34) we see that all we have to

Table 1.1. Carrier types in doped semiconductors

Valence of host	Impurity valence	Impurity type	Semiconductor type	Carriers	
				Majority	Minority
4	5	Donor	n	Electrons	Holes
4	3	Acceptor	p	Holes	Electrons

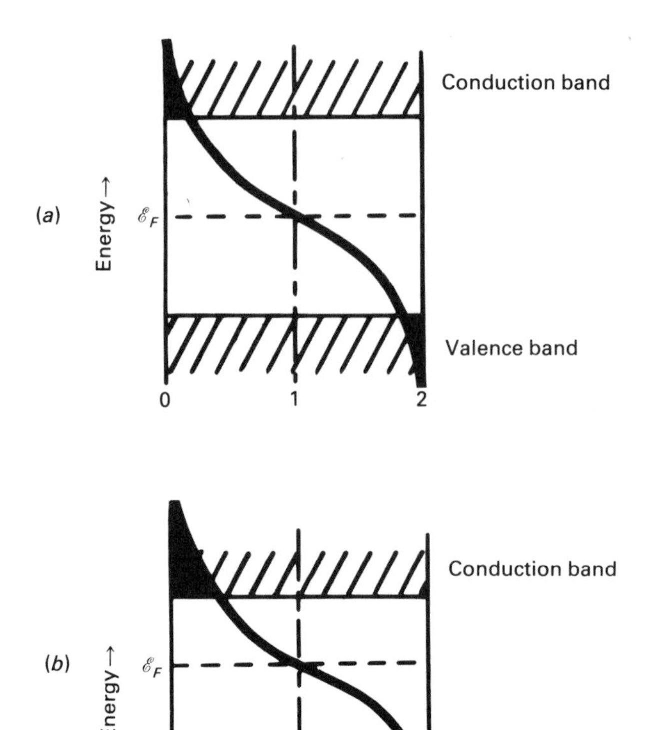

Figure 1.18. F–D distribution at room temperature for (*a*) pure semiconductor and (*b*) doped (*n*-type) semiconductor.

do is to shift the energy term in the exponent. For electrons this means

$$n_e = 2UT^{3/2} \exp\left[-\frac{\mathscr{E}_g - 2\mathscr{E}_d}{2k_B T}\right] \tag{1.37}$$

and for holes it is

$$n_h = 2UT^{3/2} \exp\left[-\frac{\mathscr{E}_g + 2\mathscr{E}_d}{2k_B T}\right] \tag{1.38}$$

The term $\mathscr{E}_d$ of course depends on the doping level, but the interesting thing that we find from this is that the product $n_e n_h$ is independent of $\mathscr{E}_d$—that is,

$$n_e n_h = 4U^2 T^3 \exp\left(-\frac{\mathscr{E}_g}{k_B T}\right) \tag{1.39}$$

This is true for p-type materials as well and is often referred to as the "constancy of the p–n product." Later we will see why it is important.

Thus we are able not only to control the conductivity of semiconductors but also to control the ratio of (+) to (−) charge carriers in the material. Both n- and p-type materials are crucial for the construction of semiconductor devices, as we shall see.

2
Discrete semiconductor devices

2.1 Diodes

a. The p–n junction

The simplest useful semiconducting device is the diode. This is a semiconductor that is n-type on one end and p-type on the other end. There is a highly developed technology for producing the p-type and n-type materials as well as for making the junction.

We assume the junction is in the form of a plane defined by $x = 0$. Let us draw the energy levels for the electrons in the device. We start out with the situation in which there is no applied electric field; this is illustrated in Figure 2.1. We know that in a p-type material the Fermi energy lies closer to the valence band than to the conduction band and that in a n-type material the Fermi energy lies closer to the conduction band than to the valence band. What happens when we form a junction of the two different materials? The two materials are in electrical contact at $x = 0$ as well as in physical contact. The implication of this is that the Fermi energy in one material must be the same as the Fermi energy in the other material. (This will no longer be true if there is an external electric field or when there is a thermal gradient across the material, but it is true for the case we are considering.) To satisfy the conditions stated above for the locations of the band edges relative to the Fermi energy, it is necessary for the band gap of one material to be displaced relative to the other. The result is illustrated in Figure 2.1.

This shift does not result in a discontinuity of the energy bands at the junction but occurs over a small region around $x = 0$, so that the change is smoothed out somewhat. The figure shows the junction of two pieces of the same host material (so that $\mathscr{E}_g$ is the same on both sides of $x = 0$) with equal concentrations (but different kinds) of impurities. This means that $\mathscr{E}_n + \mathscr{E}_p = \mathscr{E}_g$. The relative shift of the energy bands is given by $\mathscr{E}_0 = \mathscr{E}_n - \mathscr{E}_p$ and is called the *contact potential*.

To quantify the operation of this device, we need to consider the motions

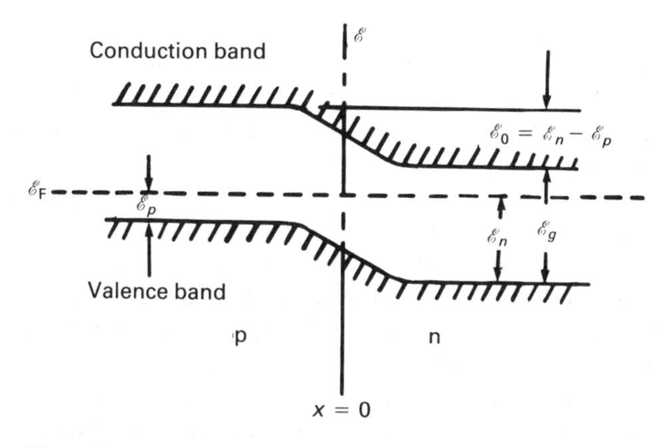

Figure 2.1. Energy level diagram of the p–n junction region.

of the charges. We consider the following types of charges:

1. Mobile electrons
2. Mobile holes
3. Impurity ions

The charge of types (1) and (2) can move; in fact, they can diffuse across the p–n junction, while those of type (3) are fixed in place and cannot move, at least not significantly at room temperature.

Figure 2.2 shows in detail what happens to the charges near the junction. The p-type material normally has more holes than electrons; that is, the holes are majority carriers. In the n-type material just the opposite applies. Before being brought into contact, however, the materials are themselves neutral. For example, in the p-type material the excess of holes is cancelled electrically by an excess of negative ions. When the materials are brought

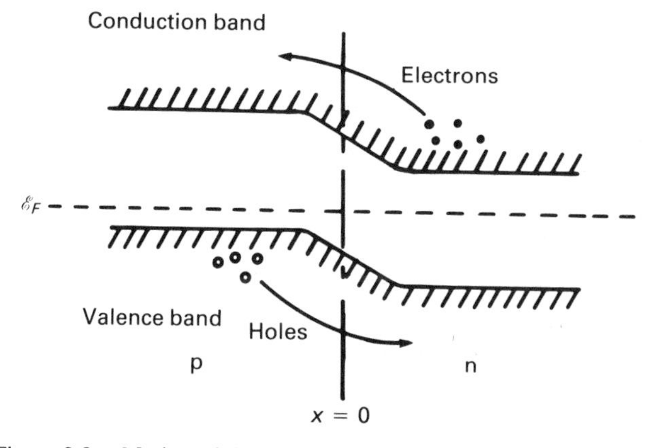

Figure 2.2. Motion of the majority carriers across the p–n junction.

into electrical contact, the holes in the valence band of the p-type material (where they are the majority carriers) experience a lower hole concentration on the other side of the junction, in the n-type material. This hole gradient across the junction causes a diffusion of holes from the p side to the n side of the junction. Similarly, there is a diffusion of electrons from the conduction band of the n side to the p side. While the two pieces of material taken together are electrically neutral, there is an imbalance of charges on each side of the junction. This situation resembles a charged parallel-plate capacitor, with the area around the junction as the dielectric. The effect of this is to leave a ($-$) charge on the p side (since it has lost holes) and a ($+$) charge on the n side. This charge separation inhibits the further diffusion of carriers across the junction; that is, there is an electric field across the junction that opposes the diffusion. What happens ultimately is that some equilibrium is reached at which the carrier concentration gradient is, in a sense, cancelled by the electric field set up by the charge imbalance. Some diffusion takes place, but not enough to equalize the carrier concentrations on the two sides of the junction. This equilibrium electric field is a result of the differences in energies of the bands on the two sides of the junction (the contact potential) and is exactly sufficient to make the Fermi energy constant across the junction. Electrons (which were majority carriers on the n side) that have migrated across the junction became minority carriers on the p side, and similarly for the holes. There are also minority carriers on both sides that could migrate across the junction and become majority carriers. However, this is a fairly small effect and we will deal with it later.

In the equilibrium situation, we need to look at what happens to the mobile majority carriers in the two halves of the device. Refer to Figure 2.3. The p region contains many acceptor atoms, most of which are ions because they have accepted an extra electron and have become negatively charged. Similarly, the n region contains many positive donor ions. The mobile hole carriers in the p region do not want to be near the junction because of the positive ions immediately on the other side, so there is a region near the

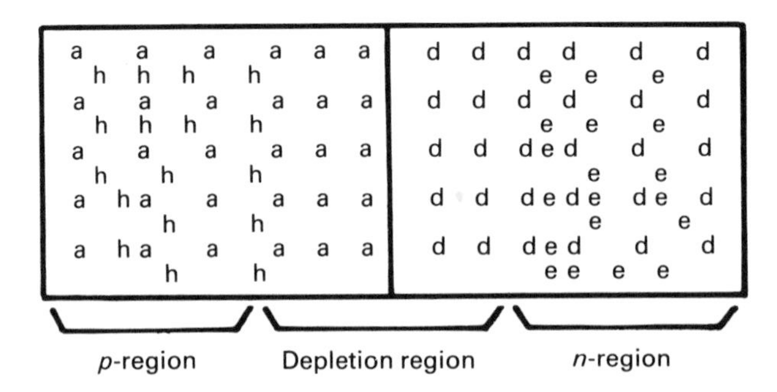

Figure 2.3. The p–n junction. a = acceptor impurity, d = donor impurity, e = electron and h = hole. The host atoms and the minority carriers are not shown.

junction on the p side where there are no majority hole carriers. Similarly, the region near the junction on the n side is void of majority electron carriers. Hence, in the unbiased p–n junction (no external electric field present) not only is there established an equilibrium concentration of carriers on the two sides of the junction but a region is formed on both sides near the junction where the electric fields in the material "sweep" away all of the mobile charges. This region is called the *depletion layer* or *depletion region.*

If we were able somehow to measure the resistance of this junction in the unbiased state (i.e., without applying an electric field), we would find that the regions of the p-type and n-type material farther away from the junction would have a reasonably high conductivity, because many majority carriers are present. The depletion region, however, has a very low conductivity, because no carriers are present.

Finally, let us look briefly at the problem of the minority carriers. Minority electrons can exist in the p region in some equilibrium concentration. Similarly, minority holes can exist in the n region. What happens when a minority electron from the p region drifts into the depletion region? It is immediately "swept" across the depletion layer (and across the junction) and into the n region by the electric field in the depletion region. Here it becomes a majority carrier. Similarly, minority holes from the n region that drift into the depletion region are swept over to the p region, where they become majority carriers.

We see that three things happen when we bring the p-type and n-type materials into contact: (1) diffusion of majority carriers, (2) diffusion of minority carriers, and (3) establishment of the depletion region and the accompanying electric fields. Some equilibrium is reached between the majority and minority carrier concentrations in the p-type and n-type materials and the size of the depletion layer. These concentrations depend on many things, most noticeably on the type of host atoms (this determines the size of the energy gaps) and the type and concentration of impurity atoms. A typical situation is shown in Figure 2.4. We note that:

1. The carrier levels are different on the p and n sides.
2. The product $n_e n_h$ is the same for the two sides.
3. The carrier concentrations in the depletion region are zero.

One other thing to notice here is that the depletion region is not symmetric about the junction. This follows from the fact that the majority (as well as the minority) carriers are not equal in the two regions.

The currents across the junction are referred to as drift currents and we can define four of these. Table 2.1 gives the details. Since there is no net flow of current out of the device, the equilibrium carrier concentrations do not change in time. We see that we can write

$$I_1 = -I_2 \qquad I_3 = -I_4 \tag{2.1}$$

The minus sign results from the convention that the current is positive when

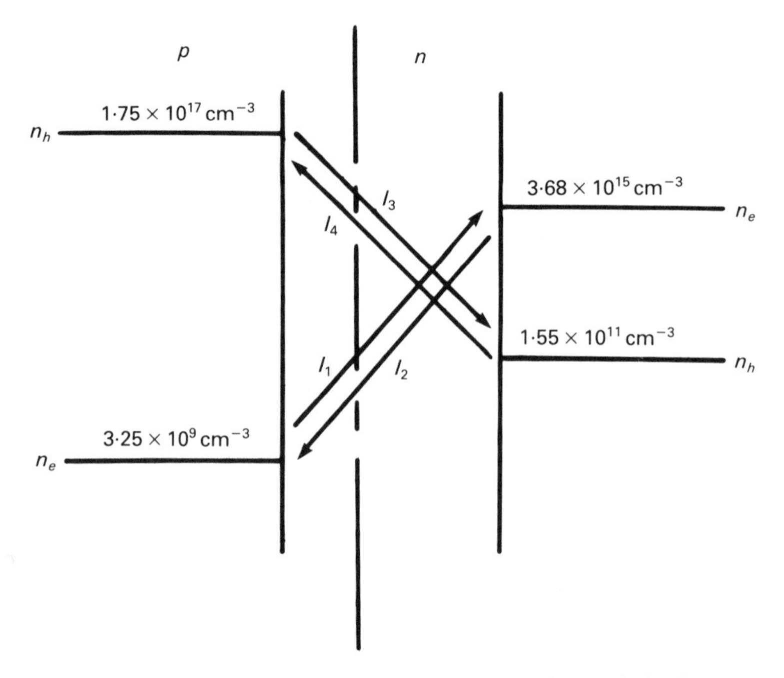

Figure 2.4. Carrier concentrations near the p–n junction; typical values.

the carriers move to more positive values of x and the current is negative when carriers move to more negative values of x. In the next section, we will see what happens when we apply an external electric field to the junction device.

b. *The biased junction*

The p–n junction may be biased in two ways, *forward*-biased or *reverse* biased, depending on whether the applied electric field *opposes* or *adds to* the natural field across the depletion layer.

Let us look at the forward-bias condition first. This is defined as a bias such that a positive potential is applied to the p side of the junction and a negative potential is applied to the n side of the junction. This produces an

Table 2.1. Drift currents across the p–n junction

Current	Carrier type	Direction
I_1	e	$p \rightarrow n$
I_2	e	$n \rightarrow p$
I_3	h	$p \rightarrow n$
I_4	h	$n \rightarrow p$

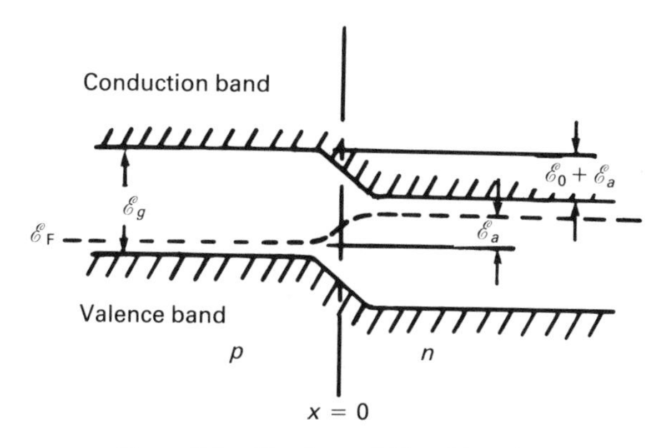

Figure 2.5. The forward-biased junction.

electric field that opposes that set up across the depletion region by the impurity ions. Referring to Figure 2.5, the application of the forward bias of (say) V_a volts (corresponding to an energy $\varepsilon_a = eV_a$) decreases the potential across the junction and effectively raises potential energies on the n side relative to the p side. It also raises the Fermi energy. This is illustrated in Figure 2.5. Note that for now we assume $\mathscr{E}_a < \mathscr{E}_g$.

The effect of this is to allow more majority electrons to cross the junction from the n side to become minority carriers on the p side. This is because there are now more electrons in the tail of the F–D distribution with sufficient energy to cross the junction. Their number has, in fact, increased by a factor

$$I_2^+ = I_2 \exp\left(\frac{eV_a}{k_B T}\right) \tag{2.2}$$

where the + indicates the forward-biased condition. Similarly, the current of holes is given by

$$I_3^+ = I_3 \exp\left(\frac{eV_a}{k_B T}\right) \tag{2.3}$$

However, the biasing has no effect on the minority currents, so that

$$I_1^+ = I_1 \qquad I_4^+ = I_4 \tag{2.4}$$

When the diode is reverse biased, the situation is as shown in Figure 2.6. In this case the potential across the junction is increased and it becomes more difficult for charges to get across. We can write the reverse-biased currents that result from majority carriers as

$$I_2^- = I_2 \exp\left(\frac{-eV_a}{k_B T}\right) \qquad I_3^- = I_3 \exp\left(\frac{-eV_a}{k_B T}\right) \tag{2.5}$$

and the minority currents are

$$I_1^- = I_1 \qquad I_4^- = I_4 \tag{2.6}$$

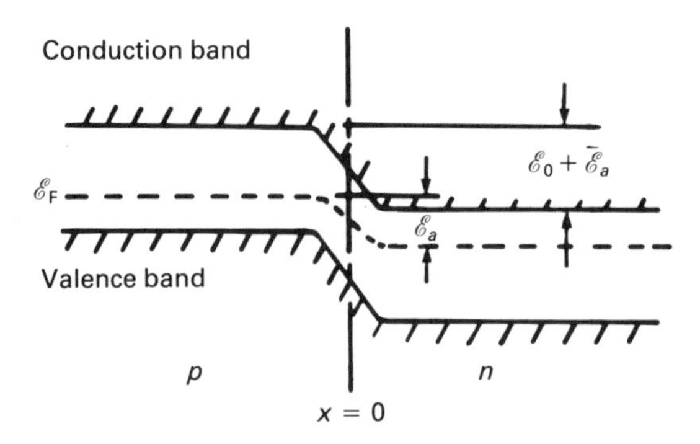

Figure 2.6. The reverse-biased junction.

It is interesting to note that the smaller potential drop across the junction that is caused by the forward biasing means that the spatial extent of the depletion region is less than for the unbiased junction. Similarly, for the reverse-biased junction the depletion region has become larger.

We can write down the total current I_T for either bias condition as

$$I_T = I_1 + I_2 + I_3 + I_4 \tag{2.7}$$

For the unbiased diode this is clearly zero. In the biased case we write

$$I_T = I_1 + I_2 \exp\left(\pm\frac{eV_a}{k_B T}\right) + I_3 \exp\left(\pm\frac{eV_a}{k_B T}\right) + I_4 \tag{2.8}$$

and from equations (2.1)

$$I_T = I_1\left[1 - \exp\left(\pm\frac{eV_a}{k_B T}\right)\right] + I_4\left[1 - \exp\left(\pm\frac{eV_a}{k_B T}\right)\right] \tag{2.9}$$

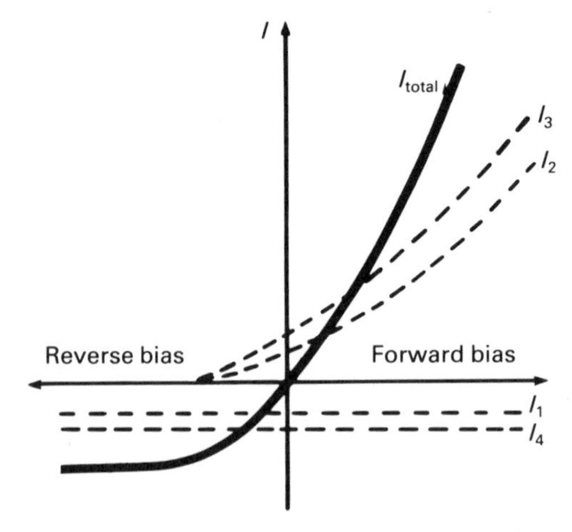

Figure 2.7. The total current in a junction diode is shown as the result of four separate components.

or

$$I_T = (I_1 + I_4)\left[1 - \exp\left(\pm\frac{eV_a}{k_B T}\right)\right] \tag{2.10}$$

where the $\pm$ represents the forward/reverse bias condition. Equation (2.10) is known as the *rectifier equation.* Figure 2.7 shows the contributions to the total current in a diode as a function of bias voltage. Note that this continues to increase with increasing forward-bias condition. Clearly, this device is non-ohmic and as a result has some interesting and useful applications.

c. *Diode applications*

Rectifiers. One of the most common uses of diodes is in the construction of *rectifiers.* Let us begin by looking at some simple DC diode circuits. The electrical symbol for a diode, and how it relates to the physical construction of the device, are shown in Figure 2.8. Simple forward- and reverse-biased circuits are illustrated in Figure 2.9.

For the forward-biased case, the diode conducts quite well and has a fairly low effective resistance (see Figure 2.7). Thus, the voltage drop across the diode is small and the voltage drop across the load is large. For the reverse-biased case, the diode does not conduct very well and as a result the voltage drop is large across the diode and small across the load.

The characteristics shown in Figure 2.7 are for a "real" diode and many diodes behave very much as the figure would indicate. We often speak of "ideal" diodes, which have an I–V curve as shown in Figure 2.10. We can define the resistance of a device at any particular bias voltage as $R = (dI/dV)^{-1}V_a$. This is referred to as the *dynamic resistance.* For the ideal diode, the dynamic resistance is zero in the forward-bias condition and infinity for the reverse-bias condition. For the ideal diode the entire voltage drop will be across either the diode or the load. For a real diode described by Figure 2.7, the voltage drop across the device will never be quite 0 or V. In the following discussion, we make the approximation that the diodes are ideal.

A useful circuit is the *full-wave rectifier,* shown in Figure 2.11*a*. Table 2.2

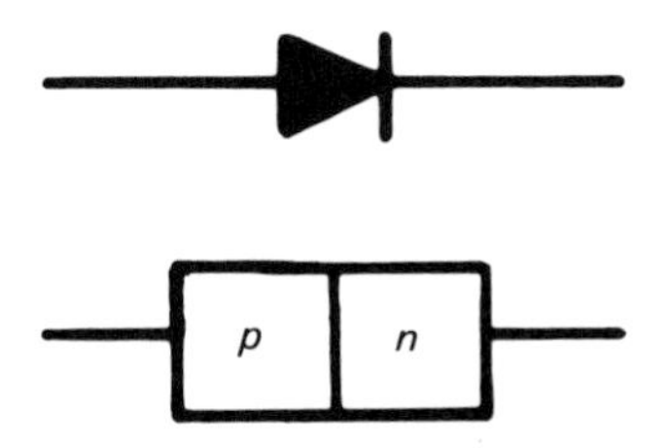

Figure 2.8. The diode and its schematic symbol.

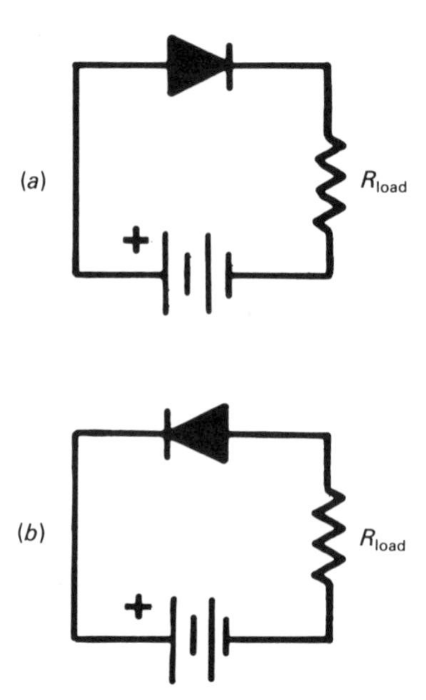

Figure 2.9. Circuits for (*a*) forward-biased and (*b*) reverse-biased diodes. Note the change in the power supply polarity.

gives the bias condition for the different diodes at different times in the cycle of the source voltage. Recalling that forward-biased diodes conduct and reverse-biased diodes do not conduct, we can easily determine the voltage drop across the load. The result is shown in Figure 2.11*b*. We see that the negative part of the source voltage has been inverted. We can now use appropriate filters and voltage regulators to obtain a constant DC output voltage.

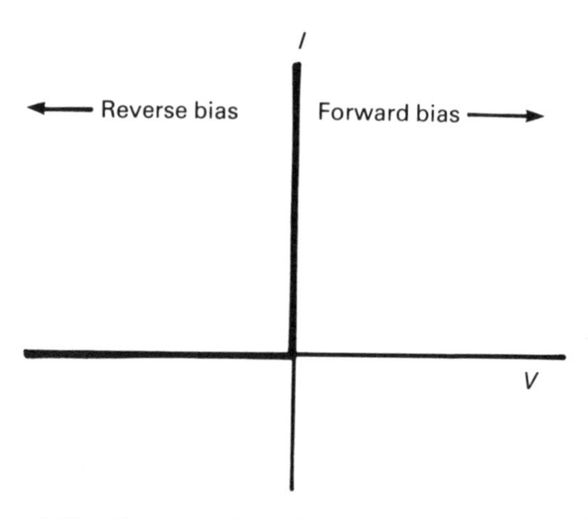

Figure 2.10. Current–voltage characteristics for the ideal diode.

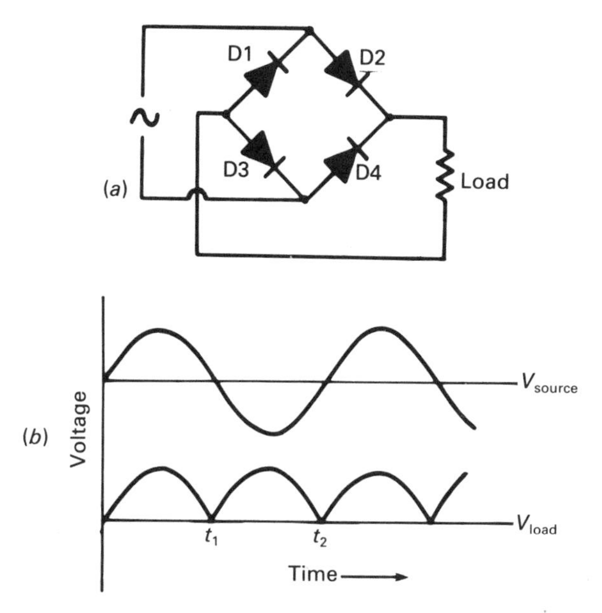

Figure 2.11. (*a*) The full-wave rectifier and (*b*) its input and output voltages.

Now we consider the application of two specialized diodes.

Zener diodes. Consider a real reverse-biased diode (refer to Figure 2.7). Let us look in detail at the reverse-bias region (see Figure 2.12). There is a relatively small resistance for the reverse-bias condition up to about 0.1 V or so. The corresponding current for larger reverse-bias voltages is referred to as the *saturation current.* In an actual diode there is a small slope in the I–V curve as a result of leakage, so R is finite but very large. This does not continue forever. For large reverse-bias voltages something else happens—*breakdown.* This is shown in Figure 2.13 and occurs at the reverse-bias voltage at which the resistance suddenly becomes very small. There are two factors that contribute to the breakdown:

1. Zener current
2. Avalanche effect

Table 2.2. Bias conditions for the full-wave rectifier (F = forward, R = reverse)

	Diode			
Time	D1	D2	D3	D4
$0<t<t_1$	R	F	F	R
$t_1<t<t_2$	F	R	R	F

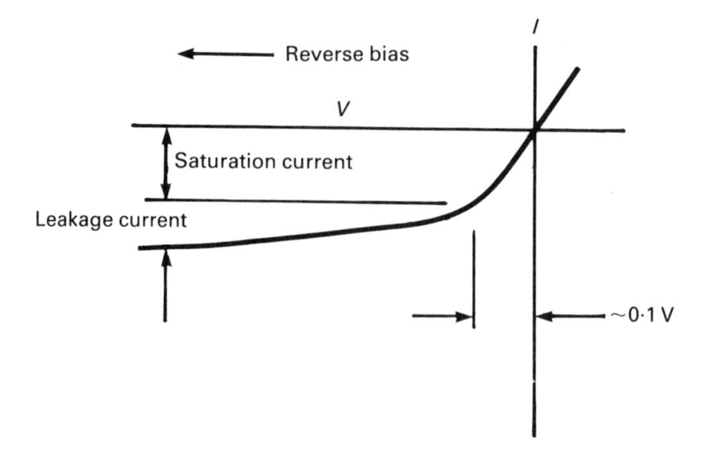

Figure 2.12. Current–voltage characteristics for the reverse-biased diode.

Beginning with the Zener current, the reverse-bias condition shifts the energy levels of the n side relative to the p side. Still assuming that $\mathscr{E}_a = eV_a < (\mathscr{E}_a - \mathscr{E}_0)$ (see Figure 2.1), the top of the valance band on the p side is still at a lower energy than the bottom of the conduction band on the n side. Now look at the electron flow from the p side to the n side (refer to Figure 2.14). The normal electron flow is from the conduction band of the p side to the conduction band of the n side. Here the electrons are minority carriers and are not very common, hence the small reverse current. If the bias is increased then $\Delta\mathscr{E}$ (see Figure 2.14) is decreased and at some point electrons from the valence band of the p side will begin jumping up into the conduction band of the n side. This occurs because of their thermal energy. The electrons on the p side are plentiful in the valence band but normally

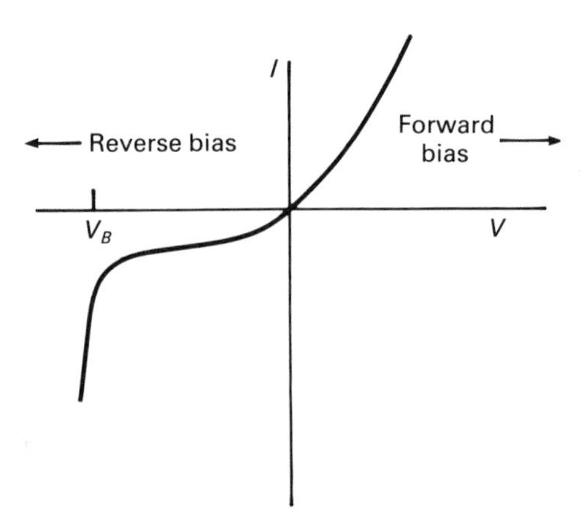

Figure 2.13. Breakdown in a reverse-biased diode.

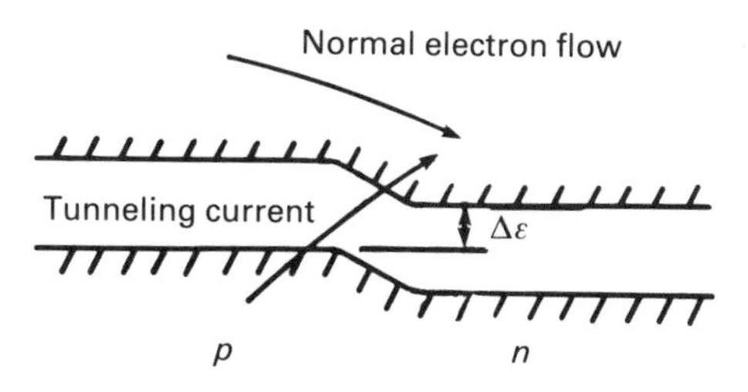

Figure 2.14. Energy level diagram for the reverse-biased diode, showing the tunneling current.

cannot contribute to the conduction. Under these circumstances they can do so and the current that results is referred to as the Zener current.

As we increase the reverse bias, something else happens. The electric field across the depletion layer becomes larger. When a charge carrier gets into the depletion region (by diffusion), it is swept to the appropriate side of the junction. The larger the electric field across the depletion layer, the faster the charges are swept out. When the sweeping becomes fast enough, a charge that is being carried across the depletion layer gains sufficient energy to cause an ionization event—a simultaneous creation of an electron and a hole. It does this by knocking a valence-band electron up into the conduction band. Now, instead of one charge moving across the depletion layer, we have *three*. These can in turn cause additional ionizations before being swept entirely out of the depletion region. So, for large reverse-bias voltages, there is a multiplication of current. This is called the *avalanche effect* (for obvious reasons).

The net result of these effects is a reverse-bias curve as shown in Figure 2.13. The breakdown voltage V_B is a function of many things, but above all it is a function of the doping levels in the materials. Zener diodes have very high doping levels and as a result have low V_B values. These can be in the range of a few volts.

Now let us see how we can use the Zener diode. Consider the circuit shown in Figure 2.15. Note that the diode symbol is somewhat different, to indicate that this is a Zener diode. Usually the value of the breakdown voltage is written next to the diode on the circuit diagram. This circuit will keep the voltage drop across the load equal to a constant value, namely V_B,

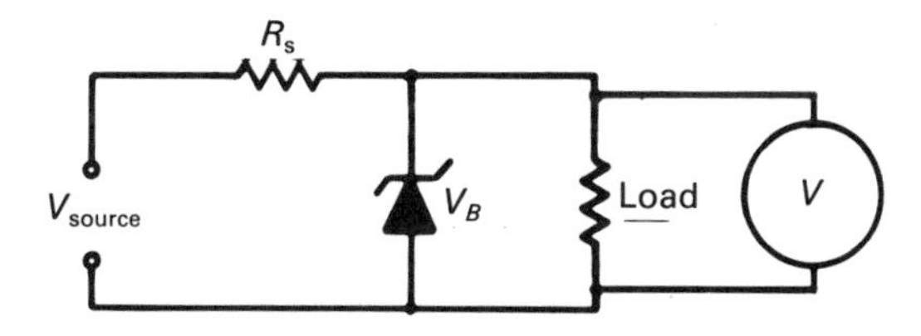

Figure 2.15. The Zener diode used as a voltage regulator.

as long as the source voltage $V_{\text{source}} > V_B$ (although V_{source} cannot be too large). It works as follows. Suppose $V_{\text{load}} > V_B$: the diode will be in the breakdown region and will conduct quite well. The more it conducts, the lower V_{load} will be. But if $V_{\text{load}} > V_B$, the diode will not conduct and V_{load} will increase. The equilibrium value of V_{load} will be V_B—right on the curved part of the I–V curve.

So even if V_s changes significantly, V_{load} remains constant. How well it regulates depends on the proper choice of R_s. The device is a voltage divider and the choice of R_s depends on the characteristics of the Zener diode as well as on the resistance of the load.

Tunnel diodes. Tunnel diodes are very heavily doped, $\sim 10^{19}$ impurity atoms per cubic centimeter, compared with $\sim 10^{15}$ impurities/cm^3 for a normal diode. The resulting depletion region is very small, $\sim 10^{-6}$ cm thick, compared with perhaps 10^{-4} cm for a normal diode. These devices are normally used in the forward-bias mode. The I–V curve is shown in Figure 2.16.

It is interesting to note that the region from about 50 to 300 mV (the exact values vary between diodes) has a resistance that is *negative*! The explanation of Figure 2.16 is as follows. Because of the high level of doping, the energy bands actually overlap, as shown in Figure 2.17*a*. It is easiest to look at the zero-temperature case. The Fermi energy lies halfway between the top of the valence band of the p side and the bottom of the conduction band of the n side. At $T = 0$ the states are filled up to $\mathscr{E}_F$. Figure 2.17 shows what happens when we apply a forward bias. For a small bias, electrons from the conduction band of the n side can tunnel across the gap to empty

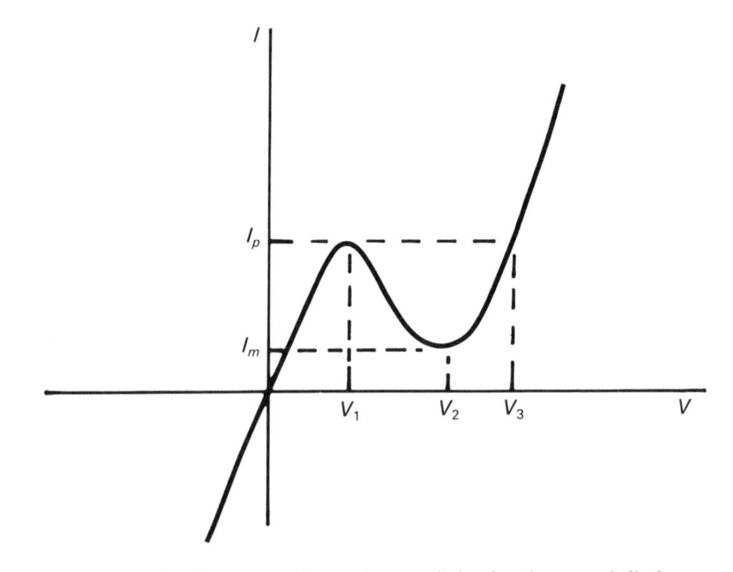

Figure 2.16. Current–voltage characteristics for the tunnel diode.

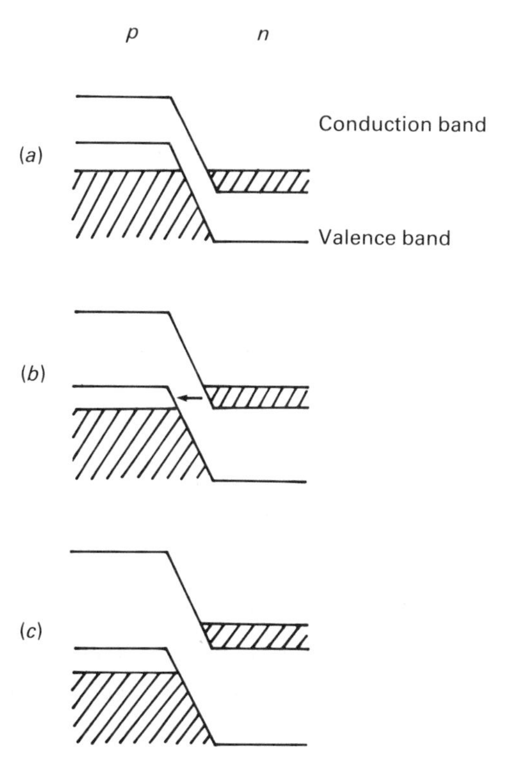

Figure 2.17. Tunnel diode energy level diagrams for different forward-bias voltages: (*a*) zero bias; (*b*) biased to the point of maximum tunneling current; (*c*) biased to the point where the tunneling current is again zero.

states in the valence band of the *p* side. They can do this easily because there are unoccupied states at the same energy there. Even a small electric field can cause them to tunnel. Figure 2.17 shows that as the forward bias increases, the overlap of states increases, and the tunnel current increases. This is the case in Figures 2.17*b* and *c*. However, when the bias is too large,

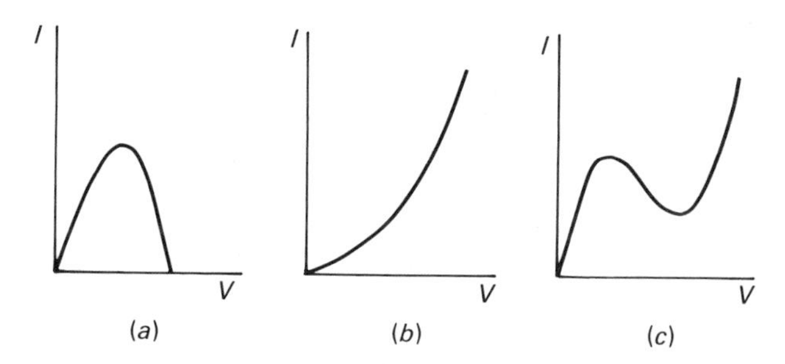

Figure 2.18. Forward-bias current components in the tunnel diode: (*a*) tunneling current; (*b*) diffusion current; (*c*) total current.

Table 2.3. Typical characteristics of tunnel diode

Material	I_p/I_m	V_1(mV)	V_2(mV)	V_3(mV)
Germanium	8	55	350	500
Gallium arsenide	15	150	500	1100
Silicon	3.5	65	420	700

some of the *n* conduction-band states have no corresponding (in energy) *p* valence states to tunnel to. So the tunneling current decreases until, as in Figure 2.17*c*, it goes to zero. Figure 2.18*a* shows the tunneling current as a function of forward-bias voltage. Recall that this was at $T = 0$. The argument for the tunneling current does not change for room temperature, but we have to add the usual forward-bias diffusion current, as shown in Figure 2.18*b*. The total current is shown in Figure 2.16.

Values of the parameters shown in Figure 2.16 are different for different semiconducting materials, but some typical values are shown in Table 2.3.

To complete our discussion of tunnel diode characteristics, let us consider the reverse-bias case. Figure 2.19 shows that as the conduction-band energy on the *n* side is reduced by a reverse bias, tunneling can readily take place. The larger the bias, the greater the tunneling current. This accounts for the reverse-bias curve in Figure 2.16.

There are several applications for tunnel diodes; one of the most common is as a switch. If we apply a bias voltage of V_1 the switch is ON. If we apply a bias voltage V_2 the switch is OFF. That is, in one case it carries a "large" current and in the other case a "small" current. The advantage of this is that the current being switched is the tunneling current, which is very fast compared to the slow diffusion current. Thus the tunnel diode can be used as a very fast switch. Switching times in the order of 50×10^{-12} s (50 ps) can be obtained. This is useful in many types of digital signal processing.

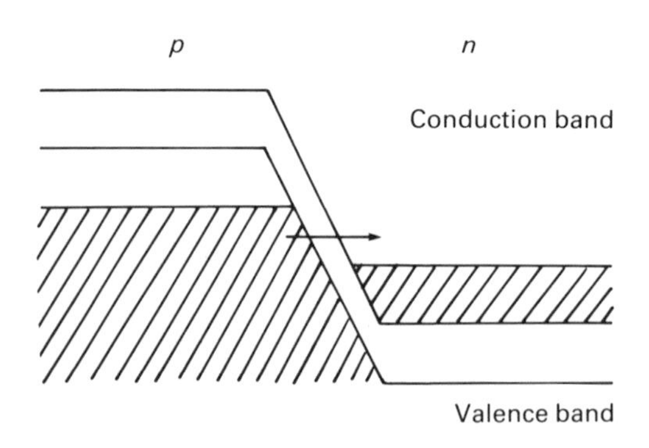

Figure 2.19. Energy level diagram for the reverse-biased tunnel diode.

2.2. Field-effect transistors

a. JFETs

There are two main classes of transistors: the conventional bipolar transistors and unipolar or field-effect transistors (FETs). Each of these types is in turn available in a number of varieties. Figure 2.20 shows the "transistor family tree." Because the FETs are the easiest to understand, we begin with the description of their operation.

Field-effect transistors (FETs) are of two types: junction field-effect transistors (JFET) and metal–oxide–semiconductor field-effect transistors (MOSFET). MOSFETs are sometimes also called insulated gate field-effect transistors (IGFETs). We begin with the JFET.

JFETs are of two types: n-channel and p-channel. Our discussion here begins with a description of an n-channel JFET, although an analogous description of a p-channel device may be given. We begin with a slab of n-doped semiconducting material. A layer of heavily doped p^+ material is formed on either side of the n material. Electrical contacts are made to the two p^+ materials and to the two ends of the n material. This is illustrated in Figure 2.21. The two p^+ materials are connected electrically as illustrated. The connections indicated are referred to as the source (S), the drain (D), and the gate (G). Figure 2.22*a* shows the operation of the JFET with no gate connection. Since the source–drain region is of an n-type material, the majority carriers are electrons. The majority carriers enter the device

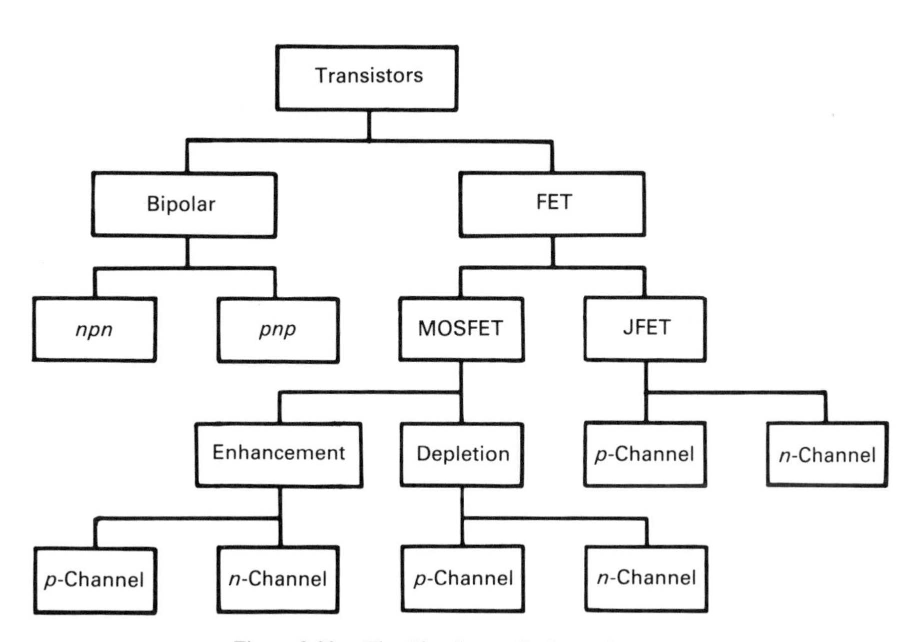

Figure 2.20. The "family tree" of transistors.

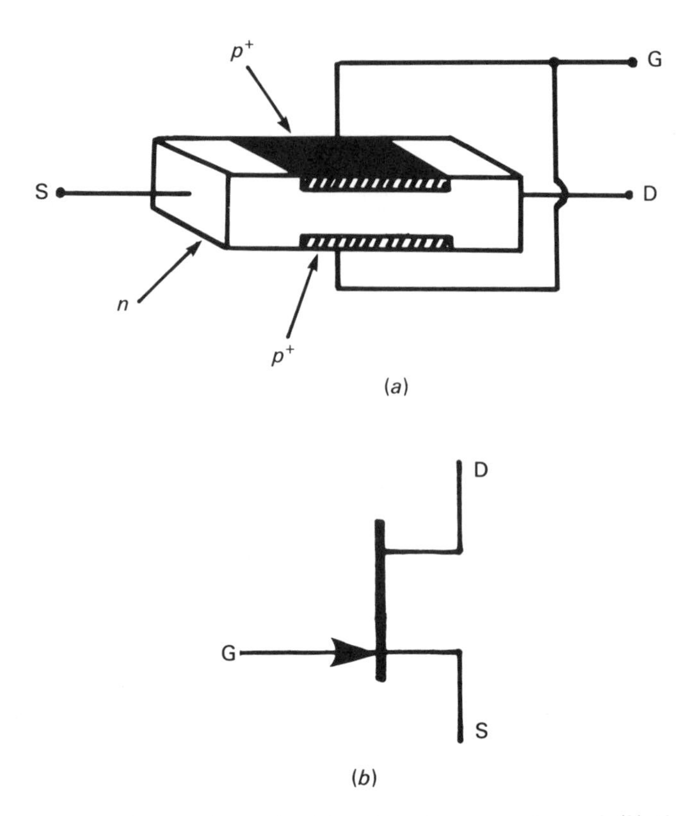

Figure 2.21. The n-channel JFET. (a) The physical construction and (b) the schematic symbol. For a p-channel the direction of the arrow in the symbol is reversed.

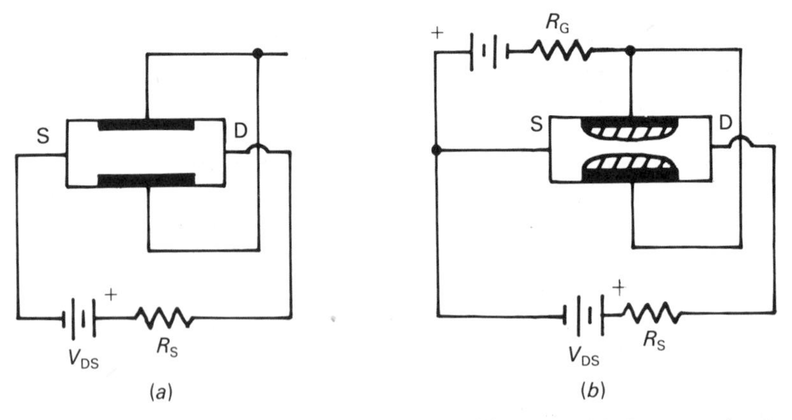

Figure 2.22. (a) Source–drain bias for the JFET and (b) circuit with the connection for gate bias as well.

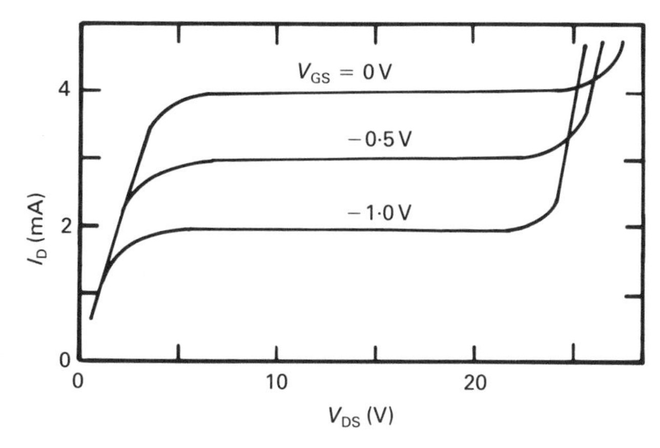

Figure 2.23. FET output characteristics.

though the source and exit from the drain, as illustrated by the bias voltage shown. In the normal mode of operation, the gate is reversed-biased relative to the source–drain. This is illustrated in Figure 2.22*b*. The normally small depletion region at the junction of the p^+ and n materials is increased in size when this junction is reverse-biased. Since the n material is less heavily doped than the p^+ material, the depletion region extends primarily into the n material (indicated by the shaded region in Figure 2.22*b*). The electrical conductivity of the n-type region is fairly large, as this is a doped material. However, the conductivity of the portion of the n region that is in the depletion region is very low—recall that the depletion region has "no" charge carriers present. The larger the reverse gate–source bias V_{GS}, the larger is the depletion region. As the size of the depletion region increases, the effective cross-sectional area of the n type "conductor" decreases. Thus, for a given V_{DS}, increasing V_{GS} decreases the drain current I_D. These current characteristics are illustrated in Figure 2.23. Let us first look at the curve for a fixed V_{GS}. We see that this is an ohmic device at low V_{DS}. This is indicated by the linear increase in I_D with V_{DS}. However, as V_{DS} is increased, I_D "flattens out."

The reason for this can be seen in Figure 2.24. For a given V_{GS}, the

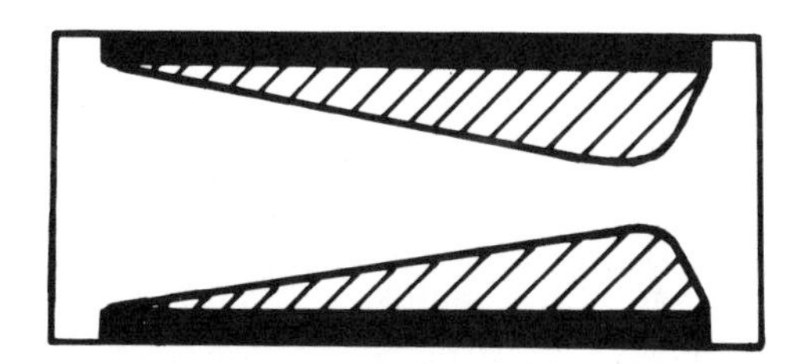

Figure 2.24. FET device illustrating the pinch-off effect.

volume occupied by the depletion region is roughly independent of V_{DS}. However, the shape of the region is not. As the voltage drop across the source–drain increases, the electric field along the x axis increases, and this causes a distortion of the depletion region. This results in a "pinch-off" of the conduction channel and an increase in the dynamic resistance. This corresponds to the flattened region in Figure 2.23. At higher bias voltages, an avalanche breakdown effect occurs and we see a sudden drop in the dynamic resistance. Increasing V_{GS} in general decreases the drain current, and for large enough V_{GS} a "pinch-off" effect occurs here as well. If we plot I_D as a function of V_{GS} for a given V_{DS} in the flat region of Figure 2.23, we find

$$I_D = I_{D0}\left(1 - \frac{V_{GS}}{V_{PO}}\right)^2 \tag{2.11}$$

where I_{D0} is the drain current for $V_{GS} = 0$ and V_{PO} is the pinch-off voltage. Figure 2.25 shows this behavior.

Now that we know basically how the JFET works, we can look at three simple applications.

Constant-current source. A simple and useful JFET circuit is illustrated in Figure 2.26. This device acts as an adjustable constant-current source. From Figure 2.23, we see that for V_{DS} between about 5 and 25 V, the current flowing through the JFET source–drain is essentially independent of V_{DS}. Since the current flowing through the load is the same as the current flowing through the JFET, this is also constant, regardless of changes (within reason) of V_{supply} or R_L. As you can see from Figure 2.23, this "constant" current value is a function of V_{GS}. In the circuit of Figure 2.26, the gate bias voltage is adjusted by changing R_c and this in turn varies the current through the load. While this circuit is a very simple and easy way of

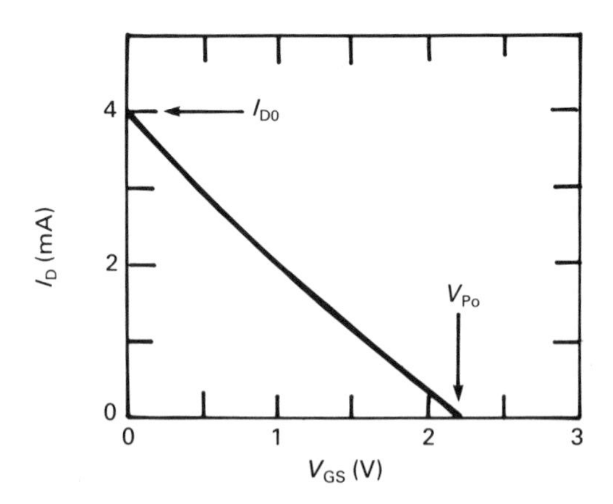

Figure 2.25. I_D as a function of V_{GS} for a typical JFET.

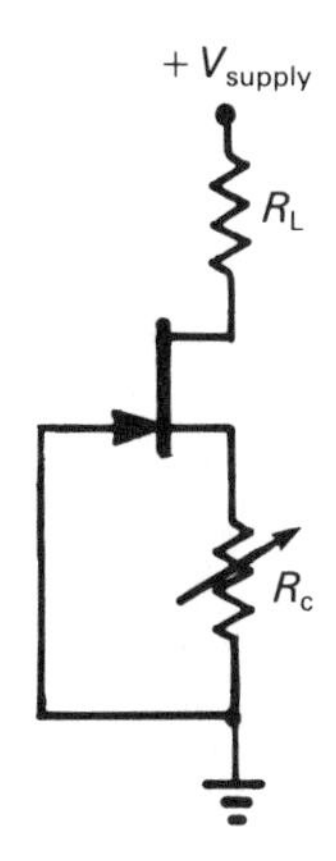

Figure 2.26. A simple JFET constant-current source.

providing a constant-current source, it is not an especially good one. Part of the problem comes from the fact that the curve shown in Figure 2.23 is not really flat. Another problem is that JFET current characteristics are particularly sensitive to the ambient temperature. As a result, circuits such as that shown in Figure 2.26 keep the current constant only to about 5% or so over a typical range of operating conditions. If this is sufficient for your needs, you do not have to build anything any more complicated.

Amplifier. Let us now consider an amplifier built from a JFET (see Figure 2.27). This configuration is called the common-source mode. We offer a rather phenomenological explanation of how this circuit works.

A supply voltage is connected between the drain and ground. As this is an *n*-channel JFET, the current flows from the drain to the source and

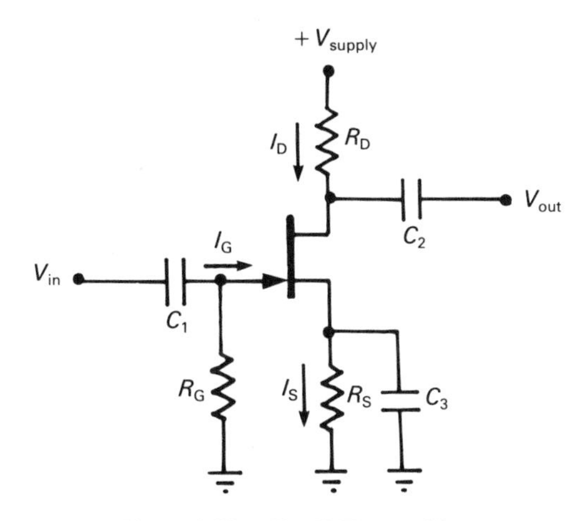

Figure 2.27. The JFET amplifier.

comprises majority electron carriers. We could build an analogous device from a p-channel JFET. The input signal forms the gate bias voltage. We operate the device with a fairly large V_{DS}, so we are in the flat region of the curve in Figure 2.23. As we change V_{GS} (the input), I_D changes. The voltage drop across the source–drain is related to I_D and to the resistors R_D and R_S, as well as the supply voltage.

As I_D changes, V_{out} will change. Let us look at the problem somewhat more quantitatively. Consider the case in which $R_D = 2\,\text{k}\Omega$ and calculate the change in the voltage at V_{out}. As the inputs and outputs of this device are capacitively coupled, we will not worry about absolute voltage levels but only about the changing part. We could go through a great deal of circuit analysis, but we can learn the same things much more easily if we approach the problem graphically. Consider Figure 2.23. If V_{DS} is in the range of 15 V then we are clearly in the flat region of the I–V curves. If we change V_{GS} (the input) by about 1 V, we see that I_D changes by about 2 mA. That is, when V_{GS} changes from 0 to −1 V, I_D changes from 4 to 2 mA.

We calculate the voltage V_1 at the output of the JFET, prior to the output capacitors. The gate current is small so we can, without difficulty, ignore it in this treatment of the problem. The situation is illustrated in Figure 2.28.

From the currents, we can easily calculate the voltage V_1. This is merely

$$V_1 = 15 - R_D I_D \tag{2.12}$$

Table 2.4 gives calculated values for this problem. This means that

$$\Delta V_{out} = -4\,\Delta V_{GS} \tag{2.13}$$

We have constructed an amplifier with a gain of −4; that is, it is inverting. A detailed circuit analysis of Figure 2.27 would yield similar results.

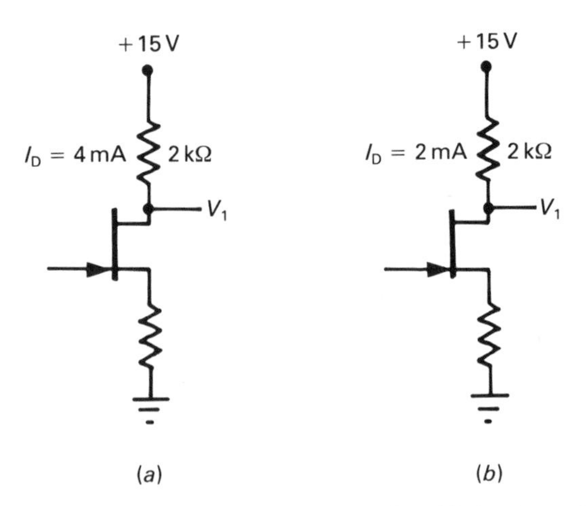

Figure 2.28. Current relationships in the JFET amplifier (*a*) for $V_{GS} = 0$ V and (*b*) for $V_{GS} = -1$ V.

Table 2.4. Quantities for the FET amplifier problem

V_{GS}(V)	I_D(mA)	V_1(V)
0	4.0	7.0
−1	2.0	11.0

Amplifiers made from JFETs, and MOSFETs too, are particularly useful in applications where we require very high input impedance. Such a situation arises in the amplification of the output from photodetectors and radiation detectors.

Voltage variable resistors. The resistance between the source and the drain of a JFET is a function of the bias voltage between the source and the gate. If V_{DS} is kept low enough (see Figure 2.23), the device is ohmic; the larger the V_{GS}, the larger the resistance. This is a useful and commonly employed device. With it we can control by means of a voltage almost anything that could be controlled by the use of a resistance. An example is the gain of an operational amplifier. Consider the circuit shown in Figure 2.29*a*. This is a standard noninverting op-amp configuration, with the gain given by the expression (see Section 3.1 for details on the use of operational amplifiers)

$$\frac{V_o}{V_i} = \frac{R_i + R_f}{R_i} \tag{2.14}$$

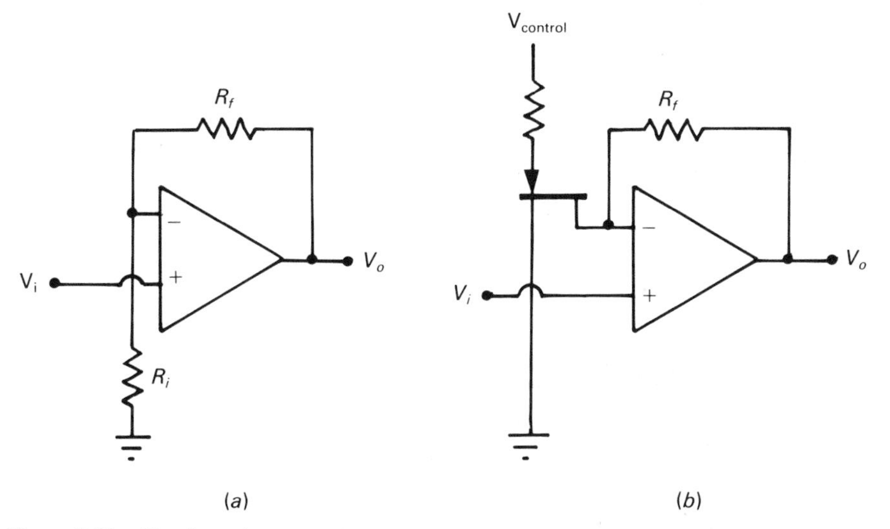

Figure 2.29. Non-inverting operational amplifier circuits with (*a*) fixed gain and (*b*) voltage variable gain.

For the circuit in Figure 2.29*b*, the gain is given by

$$\frac{V_o}{V_i} = \frac{R_{\mathrm{JFET}} + R_f}{R_{\mathrm{JFET}}} \tag{2.15}$$

Of course, we want to choose R_f so it is not too much smaller or too much larger than the value of R_{JFET} with control voltages between 0 and V_{PO}.

b. MOSFETs

MOSFETs come in two varieties: depletion MOSFETs and enhancement MOSFETs. The enhancement MOSFETs are of either *n*-channel or *p*-channel type, while the depletion MOSFETs are generally only of the *n*-channel type. We begin with the enhancement type. In keeping with our previous discussion, we consider the construction and operation of an *n*-type device. We begin with a slab of lightly doped *p* material. This is called the *substrate* or sometimes the *base* (not to be confused with the base

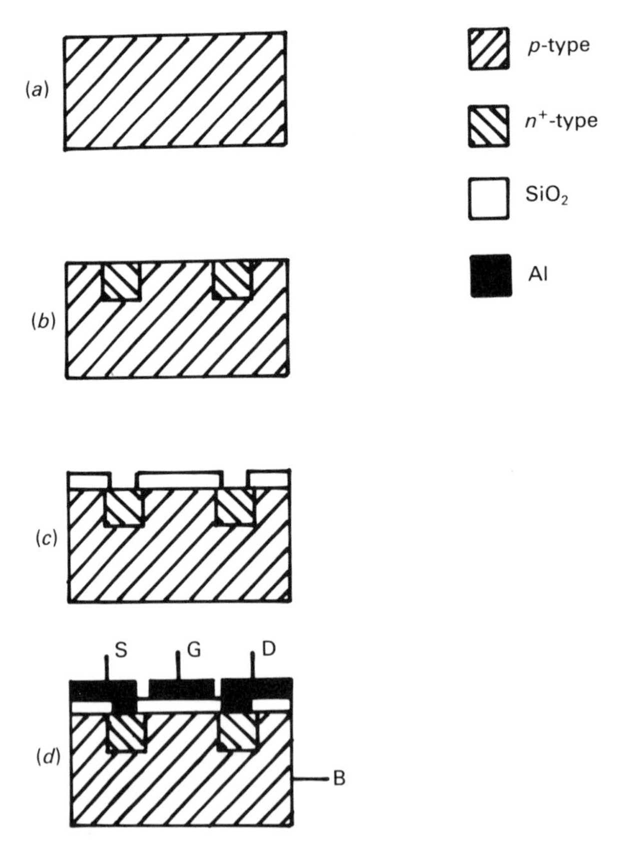

Figure 2.30. The construction of an *n*-type enhancement MOSFET: (*a*) *p*-type substrate; (*b*) n^+ islands diffused in; (*c*) insulating layer applied; (*d*) conducting layer applied.

of a bipolar transistor, which will be discussed later). Into this material we diffuse two islands of heavily doped n^+ material. The process is shown in Figure 2.30. The doping of selected areas in MOSFETs is generally accomplished by *masking*. Those areas we do not want to dope are protected by putting down a layer of aluminum. Doping is done by subjecting the surface of the semiconductor to a beam of appropriate ions or to a gas at elevated temperature that contains those ions. Next, a layer of insulating material, usually silicon dioxide (SiO_2), is deposited on the surface of the semiconductor. Openings are left over the two n^+ regions. Finally a conducting material (aluminum) is deposited on the top of the device as shown in Figure 2.30. Electrical contacts are made to the two n^+ regions through the holes in the insulating material and to a layer of conduction material placed over the region between the n^+ regions, but electrically isolated from the p material by the SiO_2.

It works as follows. The gate is biased (+) relative to the substrate [(−) for a p-channel enhancement MOSFET]. The central layer of aluminum, the SiO_2, and the p material in the gate region form a parallel-plate capacitor. Electrons, which are minority carriers in the p-region, are attracted towards the edge of the substrate by the electric field set up in the SiO_2 by the applied gate bias (see Figure 2.31). Note that the substrate material is not heavily doped, so there is not an excessive number of majority carriers present. Now this build-up of (−) minority carriers between the two n^+-doped regions forms a conduction channel in which the carriers are electrons. The larger the gate bias, the larger is this region, and the greater the conductivity between the source and the drain. This region is referred to as an *inversion layer,* because the populations of the majority and minority carriers are inverted. The reason for its formation is quite different from the reason a depletion layer forms at a semiconducting junction.

The symbol for an enhancement MOSFET is shown in Figure 2.32. In some cases, a connection for the substrate is not shown and it is understood that this is connected internally to the source.

Consider now the depletion MOSFET (this is more like the JFET), for example, the n-type. We begin with a piece of lightly doped p material

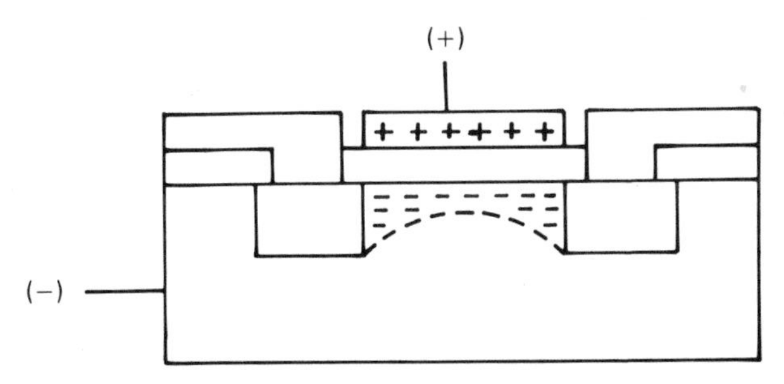

Figure 2.31. The operation of an n-channel enhancement MOSFET.

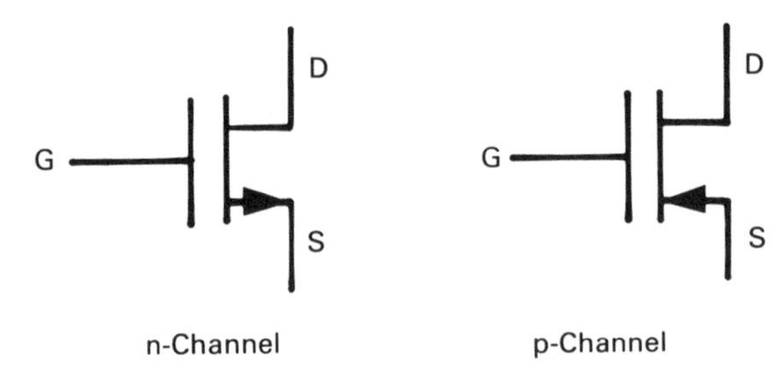

Figure 2.32. Schematic symbols for MOSFETs.

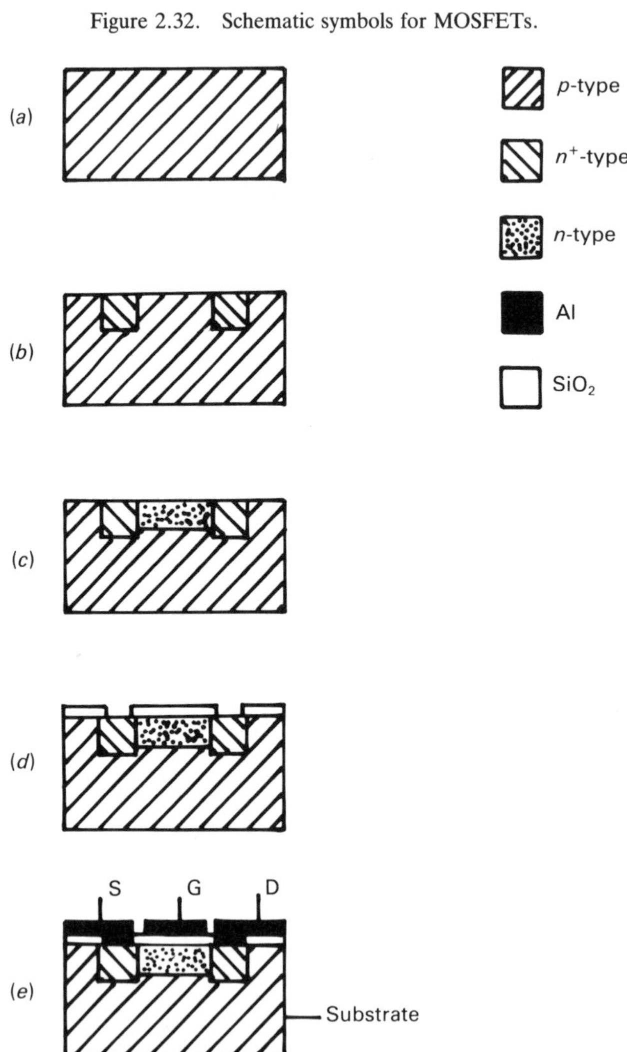

Figure 2.33. Construction of an n-type depletion MOSFET; (a) p-type substrate; (b) n^+ regions diffused in; (c) n region diffused in; (d) insulating layer applied; (e) conducting layer applied.

(again the substrate). Into this we diffuse two n^+ (heavily doped) regions as illustrated in Figure 2.33. Next, a region of lightly doped n material is created between the two n^+ regions. This is the "channel." Finally, layers of SiO_2 and aluminum are deposited as for the construction of the enhancement MOSFET (see Figure 2.33).

In the unbiased condition, the n material forms a conduction channel between the two n^+ regions (source–drain) and current is carried by majority electron carriers. Under most conditions, the depletion MOSFET is operated with the gate at (−) potential relative to the substrate [(+) for a p channel depletion MOSFET]. This is just the opposite of that used with the enhancement type. Figure 2.34 shows how this works. Again the gate–substrate acts as a parallel-plate capacitor. The induced electric fields cause a "depletion" of the majority electron carriers in the region between the n^+ areas. Thus, the size of the conduction channel is reduced and the conductivity is proportionately decreased.

A common use of FETs, particularly MOSFETs, is as switches. In this context they are often used in digital integrated circuits, referred to as CMOS. We will return to this topic later. You can see that JFETs and depletion MOSFETs are quite good conductors (source–drain) without any gate bias and that adding sufficient gate bias will cause them to be good insulators (enhancement MOSFETs are precisely the opposite). Thus we can say that the FET is "ON" or "OFF" depending on whether it is a good or bad conductor, respectively. A gate bias of a few volts (+) or (−) relative to the substrate is usually sufficient to change from one state to another. Table 2.5 gives gate biases for the various conditions. Later we will see how these devices can be used in a variety of circuits.

2.3 Bipolar transistors

The bipolar (or bipolar junction) transistor contains charge carriers of both signs and both of these are imporant in the operation of the transistor. In

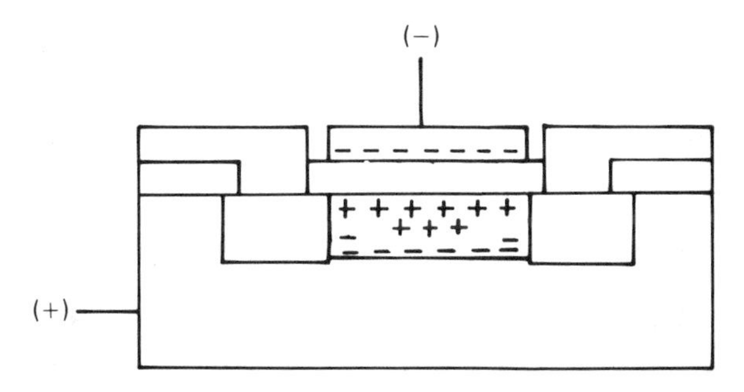

Figure 2.34. Operation of the n-channel depletion MOSFET.

Table 2.5. FET switching characteristics, gate bias voltages

Type	Channel	ON	OFF
JFET	n	0	$-V$
	p	0	$+V$
Enhancement MOSFET	n	$+V$	0
	p	$-V$	0
Depletion MOSFET	n	0	$-V$

this sense, they are more complex than FETs, although they are what we normally mean when we talk about "ordinary" transistors.

Bipolar transistors consist of a piece of doped semiconducting material sandwiched between two pieces of doped semiconducting materials with the opposite type of majority carrier. There are two possibilities, as illustrated in Figure 2.35. The schematic representation of the two types is also shown. The semiconductor in the middle is called the *base*; the two pieces on the outside are called the *collector* and the *emitter*. We will see shortly why they are different. As an example, in this section we consider the operation and use of a *pnp* transistor. The operation of an *npn* transistor follows along similar lines except that the roles of the electron and hole carriers are interchanged.

The normal biasing condition for the bipolar transistor is shown in Figure 2.36. This is referred to as the common-base or CB configuration. In this configuration, the base–emitter junction is forward-biased by a fraction of a volt or so. The B–C junction is reverse-biased, usually by a few volts, but it can be as much as some tens of volts. To see how this works, let us look at the energy-level diagram of the transistor. Figure 2.37 shows the situation for both a biased and an unbiased transistor. We can see how the bias has changed the energy levels. Looking at the current relationships in the transistor, we call the positive (majority) current at the emitter I_E, so that

$$I_E = I_B + I_C \tag{2.16}$$

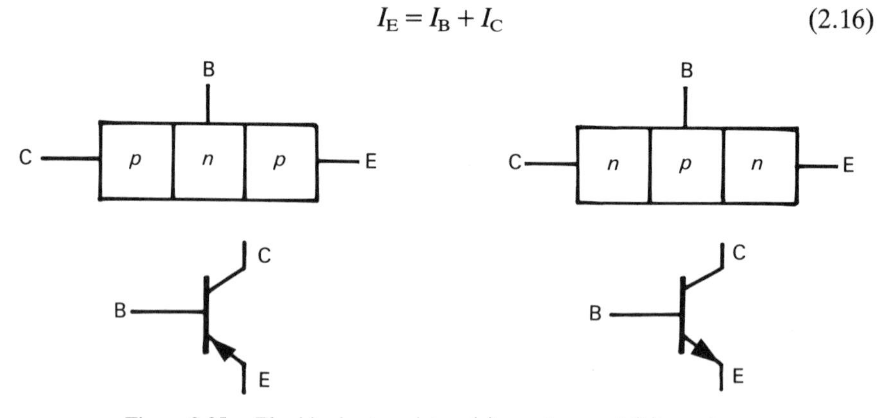

Figure 2.35. The bipolar transistor: (*a*) *pnp* type and (*b*) *npn* type.

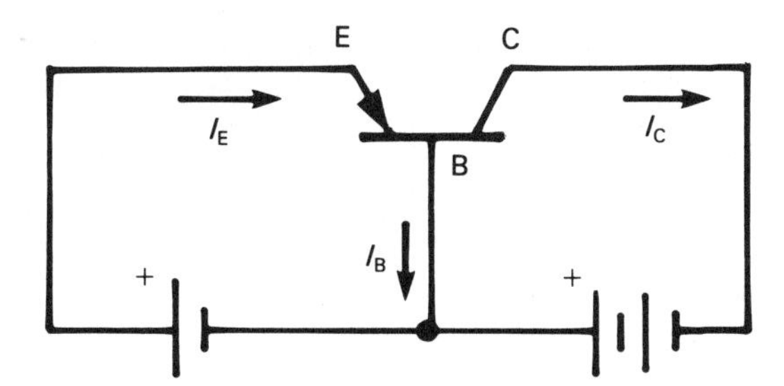

Figure 2.36. The common-base configuration for the *pnp* transistor.

When these majority carriers get into the base, they become minority (positive) carriers. Some will combine with majority (negative) carriers in the base and will contribute to the base current. Some (generally most) will reach the collector. The collector current will then be given by

$$I_C = \alpha I_E + I_{CO} \tag{2.17}$$

This requires some explanation. The quantity α is defined as the fraction of the emitter current that does not recombine with the base current. I_{CO} is the reverse saturation current for the B–C junction. Recall that the reverse B–C bias is large and examine Figure 2.7. We are in the *flat* saturation region on this diagram. Now we can look at equations (2.16) and (2.17) and, using Kirchhoff's junction law $[\sum I_i = 0]$, we see from Figure 2.36 that

$$I_E = I_B + I_C \tag{2.18}$$

and we find I_B as

$$I_B = (1 - \alpha)I_E - I_{CO} \tag{2.19}$$

In principle, we can establish whatever convention we wish for the current direction in this analysis, as long as we are consistent throughout. The

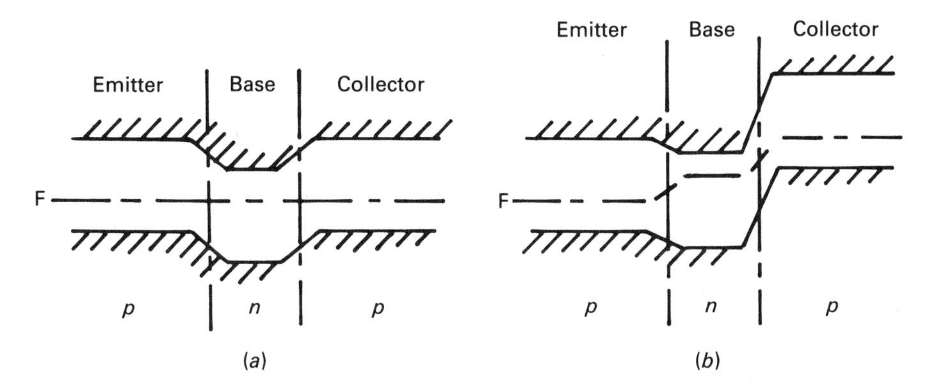

Figure 2.37. The energy level diagram for the *pnp* transistor: (*a*) unbiased; (*b*) biased.

convention we are using is that the direction of the current flow is equivalent to the direction of a flow of positive charge carriers. We see from equation (2.17) that a change in the collector current is given by a change in emitter current as

$$\Delta I_C = \alpha \, \Delta I_E \tag{2.20}$$

From equations (2.17) and (2.19) the derivatives give

$$\Delta I_C = \beta \, \Delta I_B \tag{2.21}$$

where the value of the parameter β is

$$\beta = \frac{\alpha}{1 - \alpha} \tag{2.22}$$

For a typical transistor, the value of α would be in the range 0.9–0.99 or so. This means that the value of β is in the range 10–100. This is generally called the *gain* (or the current gain) of the transistor.

Typically the I_{C0} of a transistor is quite small, so that we can make the approximation from equation (2.17) that

$$I_C \approx \mathrm{I_E} \tag{2.23}$$

(recall that $\alpha \approx 1$). So we see from equation (2.19) that

$$I_B \ll I_E \tag{2.24}$$

That is, the base current is very small. This is an essential prerequisite for the proper design and operation of bipolar transistors. The design of a transistor requires that certain criteria be fulfilled:

1. A physically thin base
2. Emitter more heavily doped than base
3. Collector physically larger than emitter
4. Emitter more heavily doped than collector

The reasons for each of these design criteria are as follows.

1. A small base current is important in obtaining a high gain. If the base is thin, there is less probability of the (majority) carriers from the emitter recombining with base majority carriers and contributing to the base current. Rather, they are likely to arrive at the collector and contribute to the collector current.

2. Having the emitter more heavily doped than the base also decreases the base current by lowering the ratio of base majority carriers to emitter majority carriers.

3. A collector physically larger than the emitter is important for proper heat dissipation. We know $I_E \approx I_C$ but from Figure 2.37 we see that $V_{BC} \gg V_{BE}$. This means that the power dissipated at the collector–base junction is much greater than that at the emitter–base junction, and a greater bulk is necessary to dissipate the excess heat produced there.

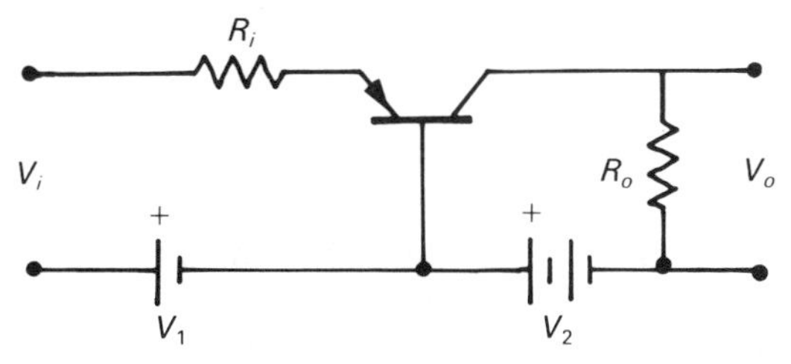

Figure 2.38. The common-base configuration amplifier. The base–emitter junction is forward biased and the base–collector junction is reversed biased.

4. It is important to have a lightly doped collector because the voltage drop across the B–C junction is large (see Figure 2.37) and we must avoid breakdown from the reverse bias.

We look now at a transistor circuit that will actually work; this is not a particularly practical design, but it will demonstrate the principles. This is a common-base configuration amplifier. There are other modes of operation (e.g., common-emitter). The circuit is shown in Figure 2.38. Figure 2.39 shows the current–voltage relationships for a typical transistor. Assume an operating current I_E of (say) 1 mA. We can see from Figure 2.39*b* that the voltage V_{BE} is 0.11 V. If we use a 0.2 V bias in the B–E circuit, we have to choose the resistor R_i to give us this I_E. Here we have considered the impedance of the input to be approximately 0 and for the present $V_i = 0$. This means that Kirchhoff's circuital law for the B–E circuit gives

$$0.2 = 0.11 + (0.001)(R_i) \quad \text{volts} \tag{2.25}$$

yielding $R_i = 90\ \Omega$.

We know that $I_C \approx I_E$, but since the B–C junction is reverse-biased, this current represents a much larger voltage drop than for the B–E junction. We note from Figure 2.39*b* that the B–C bias has essentially no effect on

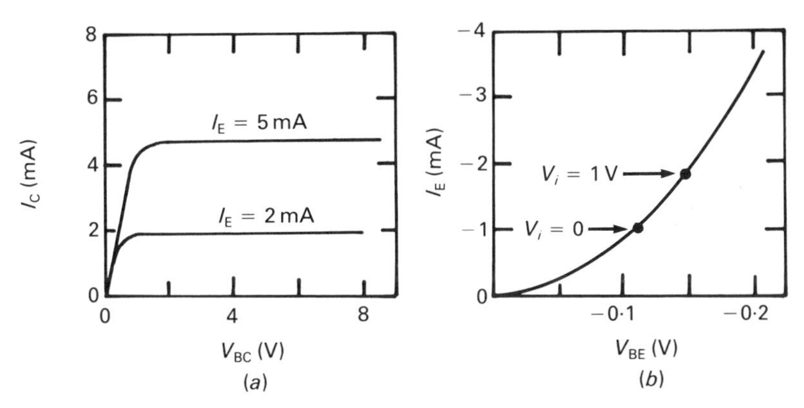

Figure 2.39. Output current relationships and input characteristics for a typical *pnp* transistor.

what happens at the B–E junction. In the B–C circuit, Kirchhoff's law gives

$$V_2 = V_{BC} + (0.001)R_0 \tag{2.26}$$

We can now choose V_2 and R_o to suit our needs. What we have to look at is the I_C–V_{BC} curve. This looks like the reverse-biased part of the diode curve (cf. Figure 2.7) and is shown in Figure 2.39*a*. We want to ensure that V_{BC} is in the flat region of the curve. Choosing, somewhat arbitrarily, $V_{BC} = 4$ V and $V_2 = 12$ V, equation (2.26) gives

$$R_o = \frac{V_2 - V_{BC}}{0.001} = 8\ \text{k}\Omega \tag{2.27}$$

Thus, we see from the schematic that for zero input ($V_i = 0$), the output is $V_o = 8$ V. Since we will, in the end, capacitively couple the input and output and use this as an AC amplifier we do not worry about the DC level of the output. Let us see what happens when we apply a voltage at the input (V_i).

Suppose we put in a pulse of 0.1 V. Now write Kirchhoff's law for the B–E loop as:

$$V_1 + 0.1 = V_{BE} + I_E(90) \tag{2.28}$$

In general we cannot solve equation (2.28) for both V_{BE} and I_E without knowing more about the transistor. When the voltage at V_i is increased, part of this increase is represented by an increase in V_{BE} and part is represented by an increase in $I_E R_i$ (the voltage drop across the input resistor). A detailed analysis of the curve in Figure 2.39*b* can tell us how much a change in V_i changes V_{BE} and how much it changes $I_E R_i$. In this particular case, $V_{BE} \approx 0.13$ V, so from Figure 2.39*b* we find that $I_E \approx 1.9$ mA. Since $I_C \approx I_E$, this change in current is reflected in the B–C part of the circuit as well. Application of Kirchhoff's law for the B–C loop gives

$$V_2 = V_{BC} + (8\ \text{k}\Omega)(1.9\ \text{mA}) \tag{2.29}$$

From this, $V_o = 15.2$ V and we find that

$$\Delta V_o = 72\,\Delta V_i \tag{2.30}$$

We have built an amplifier. This is not a particularly useful amplifier circuit (you can find better ones in any electronics text), but it does demonstrate the operation of a transistor device. Of course, you can make other kinds of transistor amplifiers, such as the common-emitter amplifier shown in Figure 2.40. Similar devices can also be made from *npn* transistors.

Another important use of bipolar transistors is in switching circuits. In this case, the transistor is driven to saturation and operates as a MOSFET does; that is, it is in an ON state or it is in an OFF state. As an example, we consider a particularly useful circuit—the *Schmitt trigger.* (In reality if you require a Schmitt trigger, you will probably find one in the form of an integrated circuit.) Consider the circuit in Figure 2.41. An important feature of this circuit is that the collector resistor for Q_1 has higher resistance than

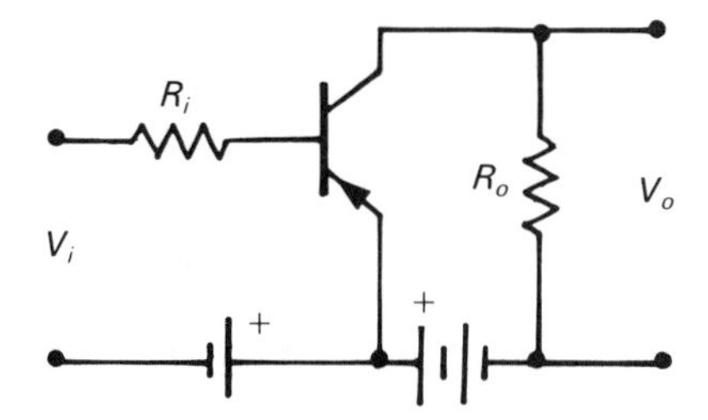

Figure 2.40. The *pnp* common-emitter amplifier.

the collector resistor for Q_2. We will see why this is important. Consider the case when the input is zero ($V_i = 0$). The base current of Q_1 will be zero and the CE current will be small, so the transistor will be OFF. This means that the base of Q_2 will be connected to the (+) supply voltage through R_1 and R_4. Effectively, Q_2 is now ON and the C–E current will be large. Because $R_5 \ll R_2$, the output is essentially connected to ground and we say that the output is low. We should note here that the transistors are not ideal and the OFF state actually represents about ~0.2 V.

If we now increase V_i, we will reach a point at which Q_1 will begin to conduct well and will switch to the ON state. When this occurs, we note that the base of Q_2 will be connected through R_4, R_5, and Q_1 to ground. Hence Q_2 is turned OFF. The output is now connected through R_2 to the $+V$ supply voltage and the output is high. As V_i is reduced again, Q_2 will switch on and Q_1 will switch OFF. Because $R_1 > R_2$, the switching from low to high output does not occur at the same voltage for increasing V_i as the

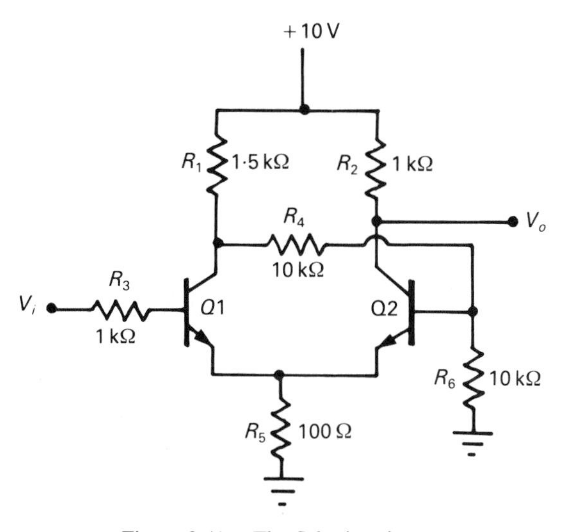

Figure 2.41. The Schmitt trigger.

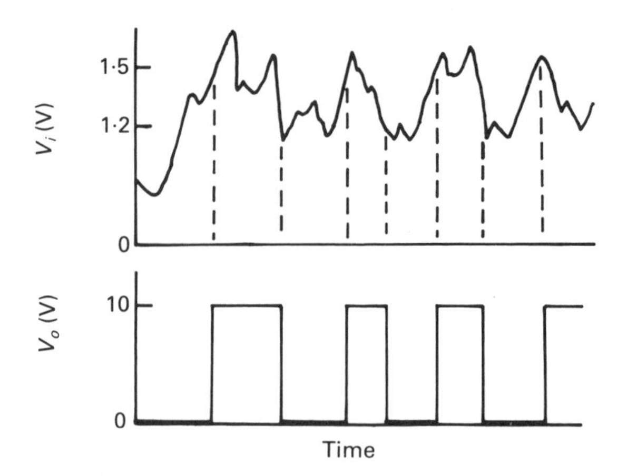

Figure 2.42. The input and output voltage for the Schmitt trigger.

switching from high to low for decreasing V_i. This is the phenomenon of *hysteresis*.

You can see from Figure 2.39*b* that the I–V curve of a transistor becomes steep for V_{BE} greater than a few tenths of a volt. Typically, this curve would saturate (i.e., the transistor would be ON) for $V_{BE} > 0.6$ V.

Some simple circuit analysis on the circuit in Figure 2.41 shows the following.

1. Switching from the low to high output state with increasing V_i will occur for

$$V_i \simeq 0.6 + (10 - V_T)\frac{R_5}{R_5 + R_2} \tag{2.31}$$

where V_T is the C–E voltage drop in the ON condition (~0.2 V). Using values for the circuit, $V_i \sim 1.49$ V.

2. Switching from the high- to the low-output state with decreasing V_i will occur for

$$V_i \simeq 0.6 + (10 - V_T)\frac{R_5}{R_5 + R_1} \tag{2.32}$$

or, for our circuit, for $V_i \sim 1.21$ V.

By choosing the various resistors in the circuit, we can make the switching voltages whatever we want. The hysteresis occurs because $R_1 \neq R_2$.

The purpose of this circuit is to provide an output that is either high or low from a variable input signal. An example is shown in Figure 2.42. Note that if you input a sine wave this circuit acts as a "squaring circuit," since it produces a square wave of the same frequency.

3
Integrated electronics

Integrated circuits are devices that contain a variety of components—transistors, diodes, resistors, capacitors—arranged in a circuit to perform some particular task. Among a large variety of different integrated circuits are devices that function as amplifiers, switches, oscillators, and counters. They are constructed using epitaxial techniques (the laying down of layers of semiconductors, conductors, and insulators) in the same way that MOSFETs are made. The use of integrated circuits not only simplifies the construction of electronic circuits but also makes the resulting circuit more compact. The construction of a practical home computer would certainly not be possible without the use of integrated circuits. There are two basic types of integrated circuits. The first, referred to as analog or linear devices, are those that deal with analog signals. An analog signal is one for which the voltage level provides the relevant information. The second type of integrated circuit is the digital device. These deal with digital signals, which represent either an ON state (if the voltage is within some particular range) or an OFF state (if the voltage is within some other particular range). While many integrated circuits are designed for specific applications, there are others that are more generally applicable to the construction of circuits in the physics laboratory. We deal here with the more general types of circuits and discuss separately analog and digital devices.

3.1 Analog devices—Operational amplifiers

We begin with a discussion of feedback. Feedback is the procedure of mixing a portion of the output signal of an amplifier with the input. Figure 3.1 shows schematically what this means. The portion of the output signal (αV_{out}) that is returned to the input may be either positive or negative; that is, it either adds to or subtracts from the input signal. In the first case, the feedback is called positive and increases the gain of the amplifier; in the second case, the feedback is negative and decreases the gain of the amplifier. Although it may seem preferable to have a higher gain, it turns out that the advantages of negative feedback generally outweigh those of positive feedback. The reasons are that negative feedback, when used properly, will improve the frequency response of the amplifier and also that

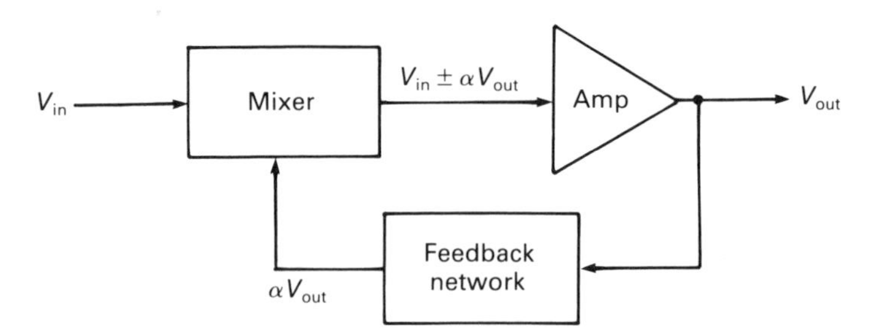

Figure 3.1. The concept of feedback in an electronic circuit.

it will improve the linear relationship between the output and the input of the amplifier.

Operational amplifiers, "op-amps" for short, are integrated-circuit amplifiers that make use of this negative-feedback principle. In general, they contain all of the necessary components for the amplifier except for a few crucial input and feedback components. These are added externally and allow us, by the addition of a few selected components, to build a high-quality amplifier to the required specifications.

Numerous op-amps are on the market today and choosing the correct one for the intended application is often a confusing matter. The 741 is perhaps the most popular general-purpose op-amp and can be used in many applications. It costs about 30 cents. You might find one or several letters appended to either the front or the back of the number; for example, it might be designated μA741C. The letters preceding the number 741 usually have something to do with the particular manufacturer and those following the number refer to some minor differences in specifications. For the most part, you should not have to worry about either of them. The 741 comes in several popular package configurations and these are shown in Figure 3.2.

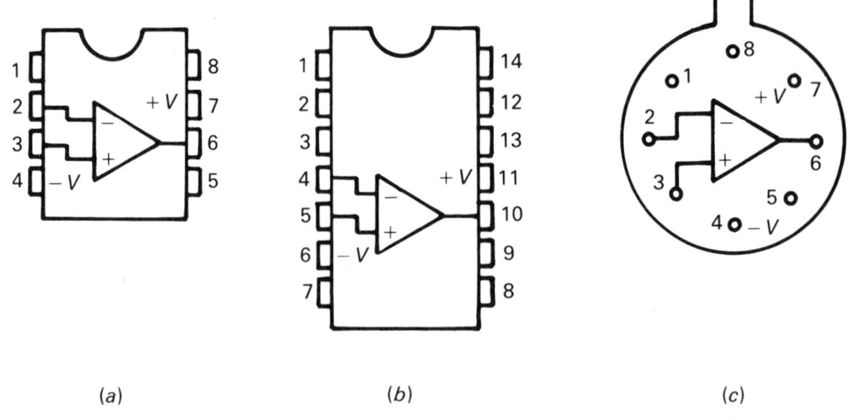

Figure 3.2. 741 op-amp packages (top view): (*a*) 8-pin minidip; (*b*) 14-pin dip; (*c*) 8-pin metal can. $V_{\pm} = \pm 15$ V.

In this section we examine some useful op-amp circuits. For this we need some basic theory of their operation. All we really need to remember are the following two rules.

1. The inputs draw no current. This is in the ideal case; the 741 actually draws about 0.08 μA. We ignore this in future discussions.
2. The output does whatever is necessary so that the feedback keeps the (+) and (−) inputs at the same potential.

As we shall see, these two rules will allow us to analyze the majority of op-amp circuits.

We do not attempt here to describe all of the uses of op-amps, but rather to give you an idea of some of the things you can do with them and to give you an appreciation of their versatility. Throughout the remainder of the book you will find other more specific applications of the simple op-amp.

a. The inverting op-amp

Consider the circuit shown in Figure 3.3. This is the inverting op-amp circuit. Note that drawings of integrated circuits generally do not show the connections for the supply voltages, but it is understood that these have to be connected.

Let us apply the two fundamental rules to this circuit. Rule 2 tells us that, since point 2 is at ground potential, point 1 must also be at ground potential. This means that the potential drop across R_i is equal to V_{in} and the voltage drop across R_f is equal to V_{out}. If we now apply rule 1 along with Kirchhoff's junction law, we find that to prevent the op-amp from drawing current it is necessary that

$$\frac{V_{out}}{R_f} = -\frac{V_{in}}{R_i} \tag{3.1}$$

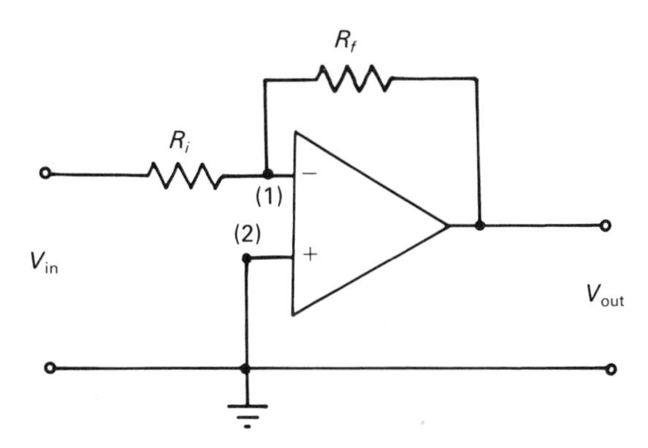

Figure 3.3. The inverting op-amp circuit.

From this we obtain the gain as

$$\text{gain} = \frac{V_{out}}{V_{in}} = -\frac{R_f}{R_i} \tag{3.2}$$

We see that the ratio of V_{out} to V_{in} is negative, so the amplifier is inverting. By adjusting the resistors in the circuit we can control the gain of the amplifier. Figure 3.4 shows some examples of the input and output signals for an inverting op-amp.

An important aspect of the operation of an op-amp is its frequency response. If we input a sinusoidal signal into the circuit shown in Figure 3.3 and measure the gain, we would find that equation (3.2) would be valid at low frequencies. At sufficiently high frequencies, however, we would find that the gain decreased as the frequency was increased. This phenomenon is known as *rolloff.* Figure 3.5 shows the results of such an experiment. The results are generally plotted on a log–log scale. This is referred to as *a Bode plot.* The figure shows that the 741 has a rolloff that begins at around 100 kHz. Some other op-amps have better high-frequency response and the rolloff occurs at a correspondingly higher frequency.

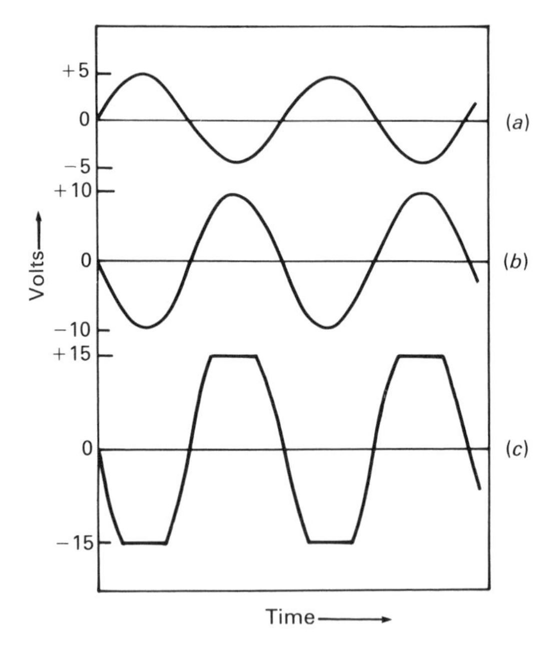

Figure 3.4. Characteristics of the inverting op-amp circuit: (*a*) input signal, (*b*) output for gain 2; (*c*) output for gain 5. Note that the amplifier will not output more than about 80% of the supply voltage.

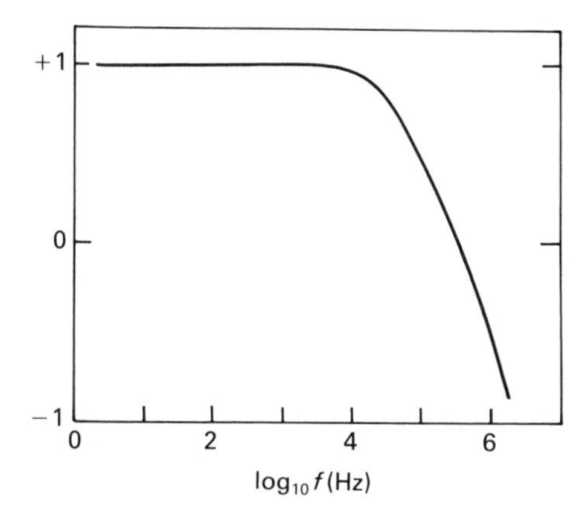

Figure 3.5. Bode plot for the 741 op-amp in the inverting configuration. V_{in} is 1 V and the DC gain is 10.

b. The noninverting op-amp

This configuration is shown in Figure 3.6. The circuit analysis is as follows. Point 1 must be at the same potential as point 2, i.e. V_{in}. This means that the voltage drop across R_f must be $(V_{out} - V_{in})$. Application of Kirchhoff's junction law at point 1 tells us that

$$\frac{V_{out} - V_{in}}{R_f} = \frac{V_i}{R_i} \tag{3.3}$$

This yields

$$\frac{V_{out}}{V_{in}} = \frac{R_f + R_i}{R_i} \tag{3.4}$$

Hence the gain is positive (noninverting) and can again be controlled by the choice of the external resistors.

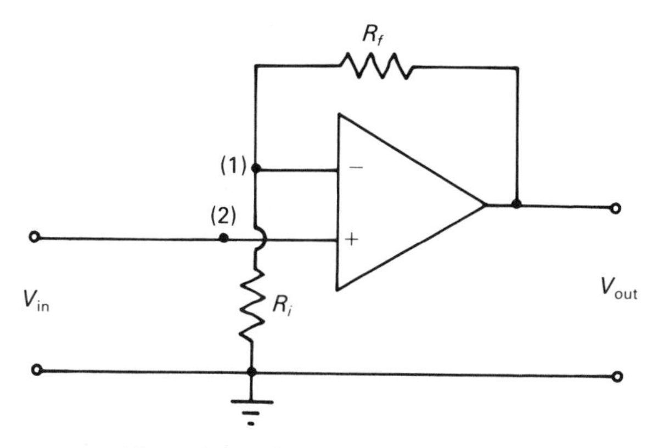

Figure 3.6. The noninverting op-amp.

c. *Addition and subtraction circuits*

Op-amps are convenient for adding or subtracting voltages from two (or more) sources. Remember, however, that the total output voltage can never be more than the supply voltage. Examples of circuits are shown in Figure 3.7. For the addition circuit, we can easily show that the output will be

$$-V_{\text{out}} = \frac{R_f V_1}{R_{i1}} + \frac{R_f V_2}{R_{i2}} \tag{3.5}$$

If we merely want to add the voltages, we make the gains of the two portions of the circuit the same by setting $R_{i1} = R_{i2}$.

For the subtraction circuit, the output voltage is given by

$$V_{\text{out}} = (V_2 - V_1)\frac{R_f}{R_i} \tag{3.6}$$

d. *Integration circuit*

Figure 3.8 shows an op-amp configured as a integrator. Since point 1 is at ground potential, the current flowing through R is easily found to be V_{in}/R. The current flowing through the feedback capacitor is given in terms of the voltage V_0 across the capacitor as $-C(\partial V_{\text{out}}/\partial t)$. The $(-)$ sign comes from the definition of the direction of the current in R. We can equate

$$\frac{V_{\text{in}}}{R} = -C\frac{\partial V_{\text{out}}}{\partial t} \tag{3.7}$$

and on integration obtain

$$V_{\text{out}} = -\frac{1}{RC}\int V_i \, dt \tag{3.8}$$

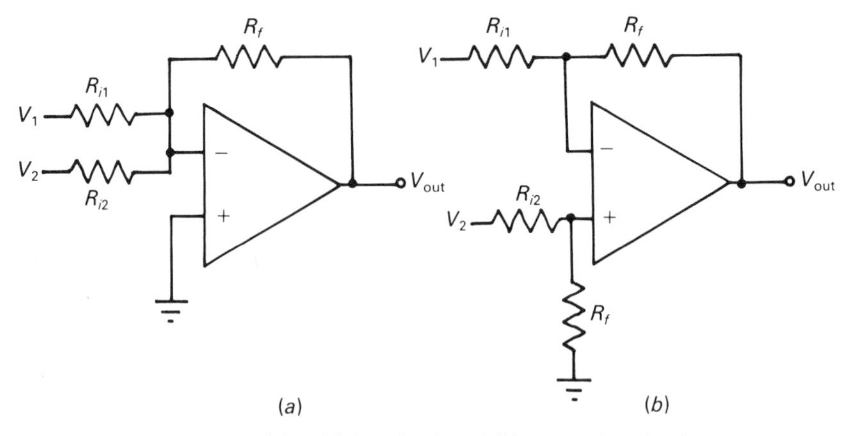

Figure 3.7. (*a*) Addition circuit and (*b*) subtraction circuit.

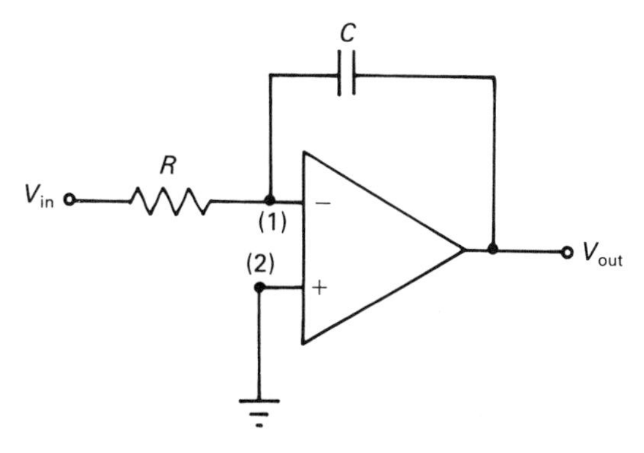

Figure 3.8. An op-amp integration circuit.

We must be careful about using DC signals as inputs for the integrating configuration: this can quickly saturate the output.

e. Differentiation circuit

Figure 3.9 shows the op-amp differentiation circuit. The analysis of this circuit follows along lines similar to that for the integrator. The current in the feedback resistor is $-V_{out}/R$ and the current in the input capacitor is $-C(\partial V_{in}/\partial t)$. Equating these quantities gives

$$V_{out} = -RC\frac{\partial V_{in}}{\partial t} \tag{3.9}$$

A DC signal input into this device will, of course, not pass the input capacitor and will, as we expect, yield an output voltage of zero.

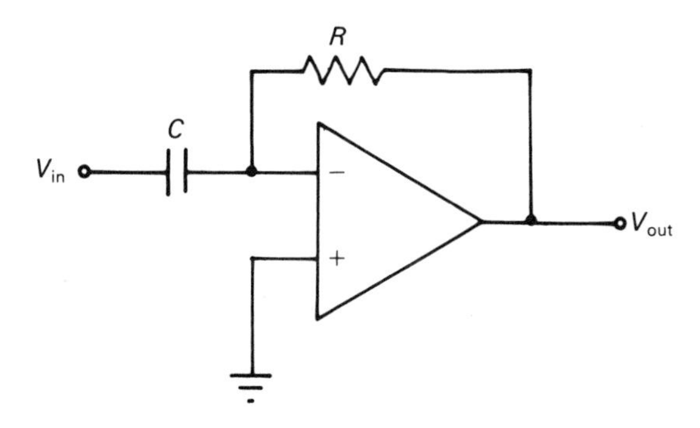

Figure 3.9. An op-amp differentiation circuit.

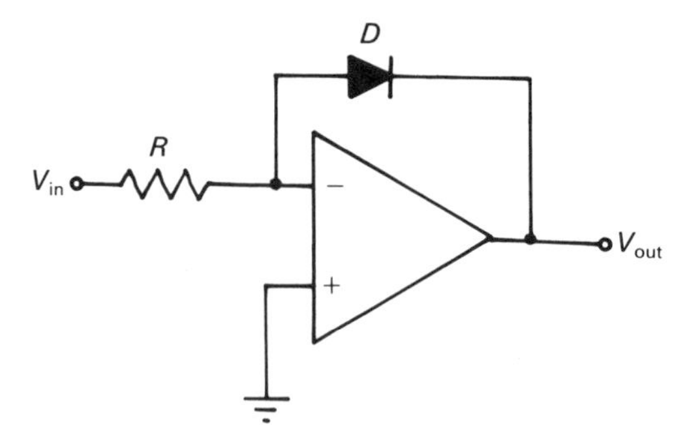

Figure 3.10. The logarithmic amplifier.

f. Logarithmic amplifier

Consider the circuit shown in Figure 3.10. Note that the circuit requires that the input voltage be positive. This is determined by the direction of the feedback diode but also follows from our knowledge that the logarithm of a negative number is not defined. For sufficiently large currents (refer to Section 2.1), the forward-bias diode current is

$$I_d = -I_0 \exp\left(\frac{V_{out}}{V'}\right) \tag{3.10}$$

where I_0 and V' are constants characteristic of the particular diode we are using. The current through R is V_{in}/R and this must equal I_d. Solving (3.10) for V_{out} yields

$$V_{out} = -V'\left[\ln\left(\frac{V_i}{R}\right) - \ln I_0\right] \tag{3.11}$$

Hence, the output is proportional to the logarithm of the input voltage. The gain of this particular circuit is determined by R and by the properties of the particular diode.

3.2 Digital devices

Integrated logic circuits come in a variety of types. Those most commonly used are given in Table 3.1 along with a description of what the name means. While each type of circuit has its own particular uses, CMOS and TTL are those that are most commonly encountered, and we confine our discussion here to these two. We begin with a discussion of the CMOS and TTL gates. This will be followed by a general discussion of the various types of multivibrators.

Table 3.1. Types of logic circuits

Name	Meaning
CMOS	complementary metal–oxide–semiconductor
TTL	transistor–transistor logic
ECL	emitter-coupled logic
DTL	diode–transistor logic
CML	current-mode logic
DCTL	direct-coupled transistor logic
HTL	high-threshold logic
RTL	resistor–transistor logic

a. CMOS gates

The C in CMOS refers to complementary and has to do with the way in which the MOSs are constructed. We will consider the case of enhancement MOSFETs. An example of a typical device is illustrated in Figure 3.11. Note that it is understood that the source is connected internally to the substrate. We see that this device is a combination of a p-channel MOSFET and an n-channel MOSFET with certain elements in common. Let us consider its operation. It is connected to a supply voltage as shown in Figure

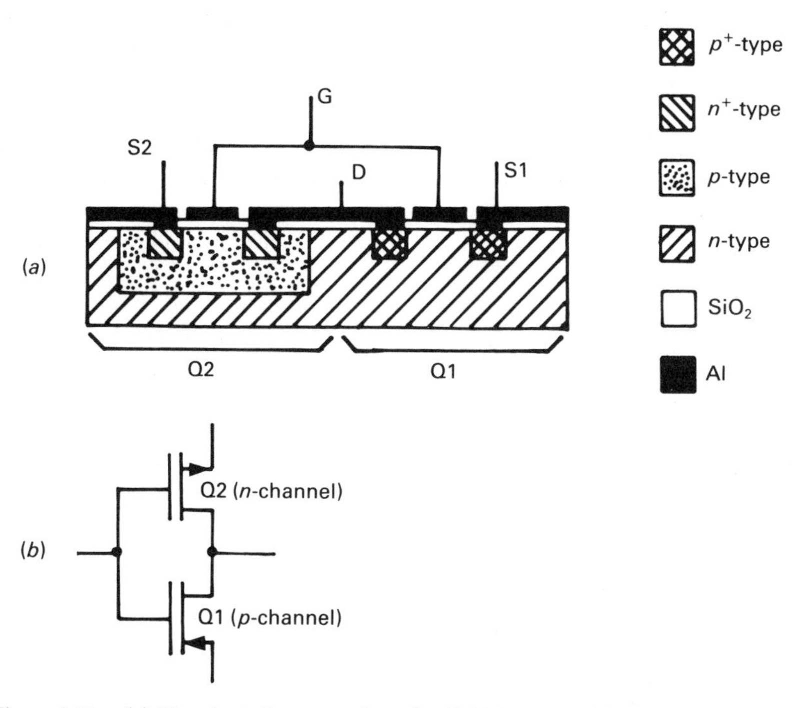

Figure 3.11. (*a*) The physical construction of a CMOS device; (*b*) the equivalent circuit.

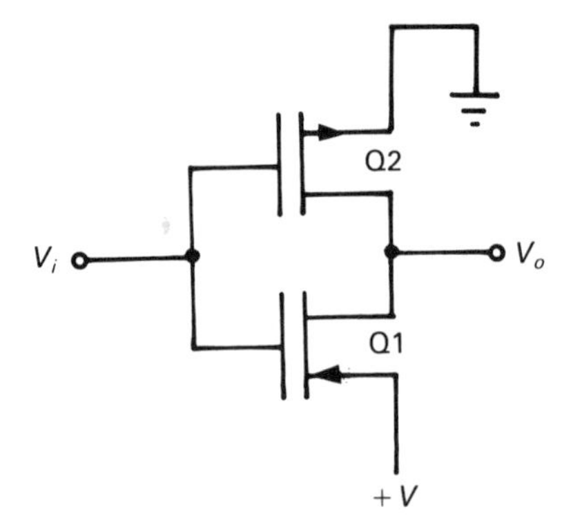

Figure 3.12. The construction of a CMOS inverter (NOT) from the device shown in Figure 3.11.

3.12. A voltage V_i is applied to the gates (the input). Typically this voltage will be either 0 or $+V$. This will refer to either a logical 0 or a logical 1, respectively. For the case where $V_i = 0$, we have $V_{GS} = 0$ for Q2 and from Table 2.5 we see that Q2 is OFF. For Q1, $V_{GS} = -V$, since the gate is at a potential of V below the source. From Table 3.2, we see that Q1 is ON. This means that the output, V_o, *is connected to* $+V$. This condition is shown in Table 3.2. Now when V_i is at potential $+V$ then a similar analysis shows that Q2 is ON and Q1 is OFF. This means that the output is connected to ground and is at zero potential. Table 3.2 again shows the properties of this circuit. We can readily see that this device acts as a inverter and is generally referred to as a NOT gate. Since we have defined $+V$ as logical 1 and ground as logical 0, we refer to this device as a positive logic circuit. We could construct a similar negative logic circuit with the logic levels defined as in Table 3.3. The construction of a negative logic inverter is illustrated in Figure 3.13. From Table 3.2, the analysis of the circuit as shown in Table 3.4 follows.

This is a simple method of constructing a CMOS inverter, although it is not the only way. It is rare that we would actually build such a circuit. Rather, integrated circuits are available that perform as CMOS inverters. We will consider this in more detail later. An interesting thing about the circuits shown in Figure 3.12 and 3.13 is that they consist only of MOSFETs and no other components. This is true of all CMOS gates, although

Table 3.2. Voltages for Figure 3.12

V_i	Q2	Q1	V_o
0	OFF	ON	$+V$
$+V$	ON	OFF	0

Table 3.3. Logical states for positive and negative CMOS logic

Logical	Positive logic	Negative logic
0	0	0
1	$+V$	$-V$

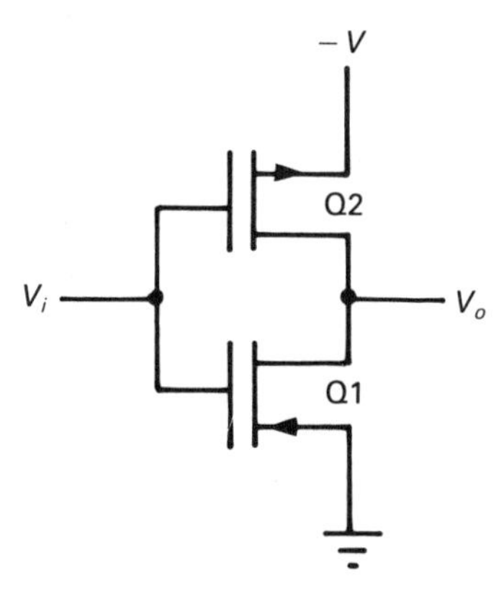

Figure 3.13. A negative logic CMOS inverter.

Table 3.4. Voltages for Figure 3.13

V_i	Q2	Q1	V_o
0	ON	OFF	$-V$
$-V$	OFF	ON	0

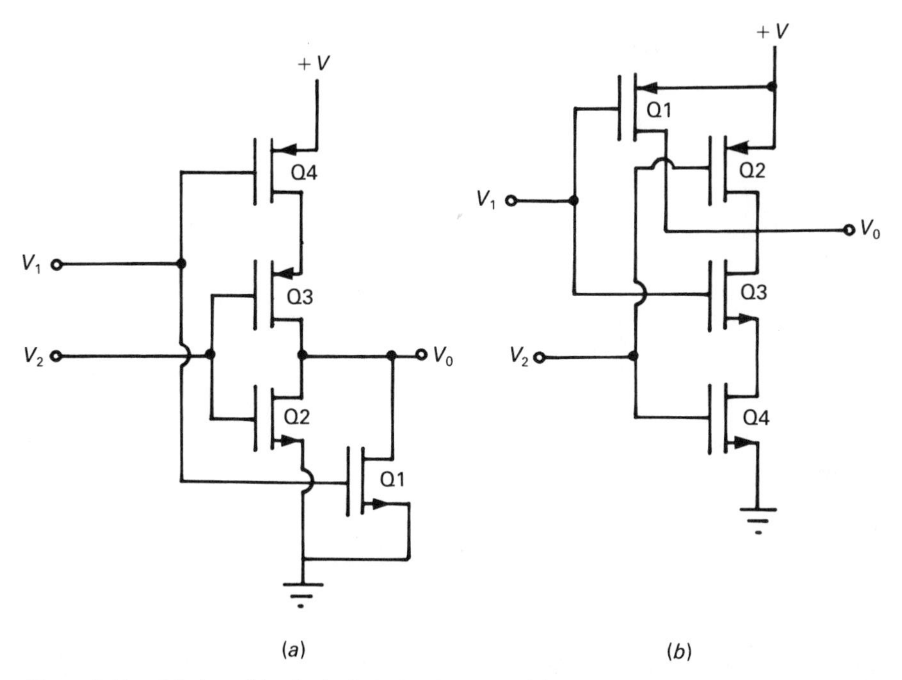

Figure 3.14. (*a*) A positive logic CMOS NOR gate; (*b*) a positive logic CMOS NAND gate.

Table 3.5. Operating conditions for circuits in Figure 3.14

Type	V_1	V_2	Q1	Q2	Q3	Q4	V_o
NOR	0	0	OFF	OFF	ON	ON	$+V$
	0	$+V$	OFF	ON	OFF	ON	0
	$+V$	0	ON	OFF	ON	OFF	0
	$+V$	$+V$	ON	ON	ON	OFF	0
NAND	0	0	ON	ON	OFF	OFF	$+V$
	0	$+V$	ON	OFF	OFF	ON	$+V$
	$+V$	0	OFF	ON	ON	OFF	$+V$
	$+V$	$+V$	OFF	OFF	ON	ON	0

commercially available devices may have other components that serve to protect the circuit.

Other basic gates are the NAND and the NOR gates. These are shown in Figure 3.14. An analysis of these circuits follows along the lines of that for the inverter. The status of the various transistors, as well as the output for the different input conditions, for these two circuits is shown in Table 3.5.

Other CMOS gates, that is AND and OR gates, are constructed from combinations of the gates already described. Thus, an AND gate is formed from a combination of a NAND gate and a NOT gate. In CMOS circuitry we can write

$$\text{AND} = \text{NAND} + \text{NOT} \qquad \text{OR} = \text{NOR} + \text{NOT} \tag{3.12}$$

The CMOS AND gate is illustrated in Figure 3.15. Of course, all of the

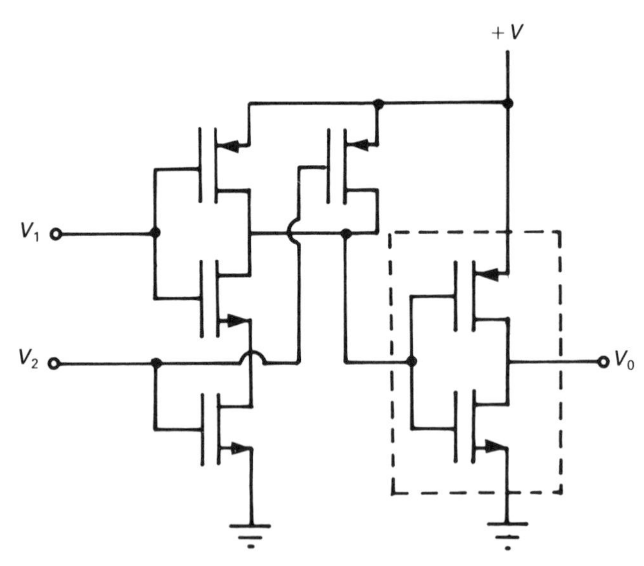

Figure 3.15. The construction of a positive logic CMOS AND gate from a NAND gate and a NOT gate. The portion of the circuit in the dashed rectangle is the NOT gate.

gates mentioned here can be constructed to operate on negative logic signals as well.

The various gates are summarized in Table 3.6. The standard numbering scheme for CMOS gates uses a four-digit number beginning with the digit 4. The commonly used CMOS gates are listed in Table 3.7. Note that there are NOR, NAND, OR, and AND gates that have more than two inputs. These are described as follows. A two-input NAND or NOR gate will produce an output given by

$$Q = \overline{AB} \qquad \text{or} \qquad Q = \bar{A} + \bar{B} \tag{3.13}$$

respectively. For three-input devices, the outputs would be defined as

$$Q = \overline{ABC} \qquad \text{or} \qquad Q = \bar{A} + \bar{B} + \bar{C} \tag{3.14}$$

respectively. Similarly, we can define the operation of gates with a larger number of inputs. Many more CMOS devices are available and can be found in manufacturers' catalogs.

One final important aspect of CMOS devices is that, they have very high input impedance and draw very little current, they are susceptible to damage by static electricity. Static electricity can develop very large voltages with negligible current and this can permenantly damage the semiconducting materials used in CMOS devices. For this reason, CMOS circuits generally come packaged on pieces of conductive plastic foam, usually black. You should be very careful about touching the pins of the integrated circuit when they are not connected to anything. Further, in connecting a CMOS device you should not leave any pins unconnected; rather, any unused pins should be tied either to $+V$ or to ground. Some CMOS devices are now manufactured with internal protection against damage from static electricity, but if you are uncertain, it is best to be cautious.

b. TTL logic gates

TTL logic circuits consist not only of bipolar transistors but also of resistors and sometimes diodes. For this reason, the operation of TTL gates is more complex than that of CMOS gates and we deal with this type of device in somewhat more general terms.

As an example, we illustrate a TTL NOR gate in Figure 3.16. It operates as follows. If A or B is a logical 1 then a collector–emitter current will flow through either T1 or T2, respectively. This means that a large base current will be applied to either T3 or T4. Thus, either T3 or T4 will have a large collector–emitter current and the corresponding base currents for T5 and T6 will be determined by the choice of the resistors in the circuit. In the case under discussion, T5 will carry a large collector–emitter current and is said to be ON while T6 will carry a small collector–emitter current and is said to be OFF. Consequently the output V_o will be connected to ground. Table 3.8 explains the status of the transistors in this circuit. The same arguments

Table 3.6. Symbols and logic levels for the basic logic gates

Gate	Symbol	A	B	Q
NOT	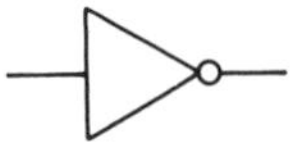	0	—	1
		1	—	0
NOR	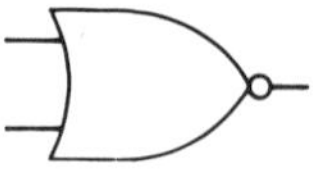	0	0	1
		0	1	0
		1	0	0
		1	1	0
NAND	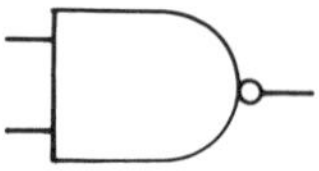	0	0	1
		0	1	1
		1	0	1
		1	1	0
OR	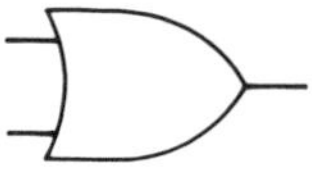	0	0	0
		0	1	1
		1	0	1
		1	1	1
AND	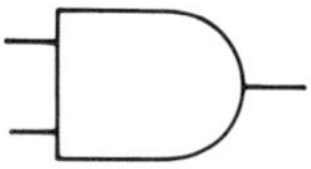	0	0	0
		0	1	0
		1	0	0
		1	1	1

Table 3.7. Common CMOS gates

Type	Number of inputs	CMOS No.
NOT	2	4069
NOR	2	4001
	3	4025
	4	4002
	8	4078
NAND	2	4011
	3	4023
	4	4012
	8	4068
OR	2	4071
	3	4075
	4	4072
AND	2	4081
	3	4073
	4	4082

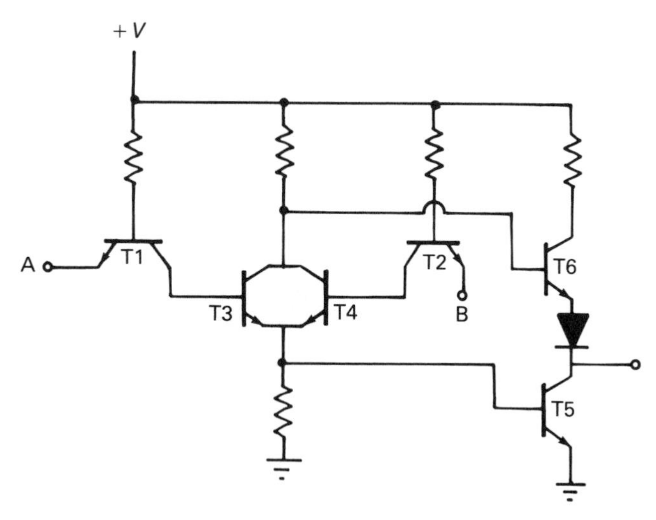

Figure 3.16. A TTL NOR gate.

apply for the case when A and B are both logical 1. If, on the other hand, A and B are both logical 0, then neither T3 nor T4 will be on and the corresponding base current of T5 and T6 will determine T5 to be OFF and T6 to be ON. V_o will therefore be equal to $+V$.

The operation of other TTL logic gates follows along similar lines and you can find appropriate schematics in many modern electronics texts.

For TTL devices, the logical 1 level is defined as greater than 1.4 V and the logical 0 level as less than 1.4 V. These are defined (for most devices) as having the nominal values of +5 and 0 V, respectively.

The industry standard for numbering TTL gates uses a four- or five-digit number beginning with 74. These devices are referred to as the 7400 series. The numbers will often have letters appended to them, as in SN7408. SN is a manufacturers' code. Sometimes you will see letters in the middle of the device number as in 74LS08. The LS refers to a specific type of TTL circuit: the code LS refers to the low-power Schottky variety. The four or five digits alone, however, will tell you what kind of device you have. Table 3.9 gives the common TTL integrated circuits you are likely to use.

Table 3.8. Logical conditions for the TTL NOR gate

A	B	T1	T2	T3	T4	T5	T6	V_o
0	0	OFF	OFF	OFF	OFF	OFF	ON	1
1	0	ON	OFF	ON	OFF	ON	OFF	0
0	1	OFF	ON	OFF	ON	ON	OFF	0
1	1	ON	ON	ON	ON	ON	OFF	0

Table 3.9. TTL logic circuits

Type	Number of inputs	TTL No.
NOT	1	7404
NOR	2	7402
	3	7427
	4	7425
	8	74260
NAND	2	7400
	3	7410
	4	7420
	8	7430
OR	2	7432
	3	—
	4	—
AND	2	7408
	3	7411
	4	7421

In building a circuit you will have to choose between CMOS and TTL (or in some cases another less common series of logic devices). Whereas CMOS has the advantage that it requires less power, TTL is much faster. It is possible to interface between CMOS and TTL, but it is not a simple matter. You not only have to worry about voltage levels but about currents, too. For example, does the output of one device have enough current to trigger the input of the next one? Horowitz and Hill's book, *The Art of Electronics,* has many good tips on this sort of thing, but as a beginner you would do best to stick to building circuits with only one type of logic.

c. *Multivibrators*

There are three basic types of multivibrators, which are distinguished by the behavior of their two output states as follows.

1. *Astable multivibrators.* Both output states of these devices are quasistable. The device will remain in one output state for a period of time and then switch to the other output state. After a while it will switch back to the first state, and so on. Hence the astable multivibrator is a square-wave or pulse generator.
2. *Monostable multivibrators.* These have one stable and one quasistable output state. Normally the device will be in the stable state. A trigger

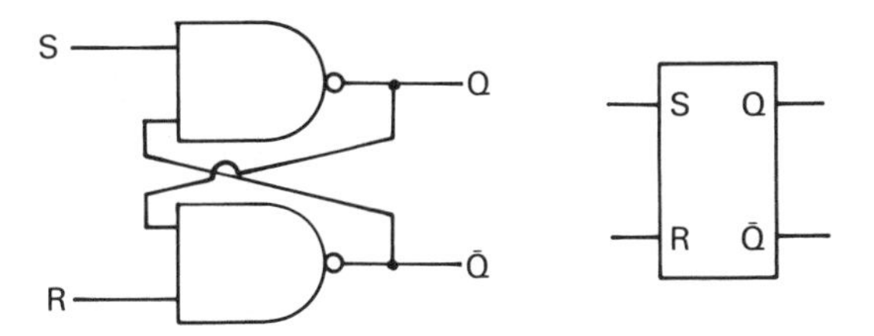

Figure 3.17. The RS flip flop and its schematic symbol.

pulse will cause the device to switch to the quasistable state for a period of time after which it will switch back to the stable state. The output will remain in the stable state until another trigger pulse is input. The device is generally referred to as a one-shot.

3. *Bistable multivibrator.* This device has two stable output states. It will remain in a given state until an appropriate signal is input into the device in order to change it. These devices are called flip-flops and will be discussed first below. There are several kinds of flip-flops and we begin with the simplest.

The RS flip-flop. The RS flip-flop consists of two NAND gates connected as shown in Figure 3.17. Consider the situation in which both inputs R and S are logical 1. There are two equally stable output configurations; $Q = 0$ and $\bar{Q} = 1$ or $Q = 1$ and $\bar{Q} = 0$. Consider the case $Q = 0$ and $\bar{Q} = 1$. Now we allow one of the inputs to go momentarily to logical 0 ($R = 0$). Q will be forced to 1 and this will make $\bar{Q}$ go to 0. Thus the output has changed to the other stable state. The only way to get the flip-flop to return to its original output state is to allow the other input, S, to go to logical 0.

The clocked RS flip-flop. It is often convenient to have some additional control over the switching of the flip-flop. This can be done as shown in Figure 3.18. Changes in the input states R and S can only affect the output state when a clock pulse is present at the clock input. Thus, the clock pulse can be used to sample the status of R and S periodically to see if they have changed.

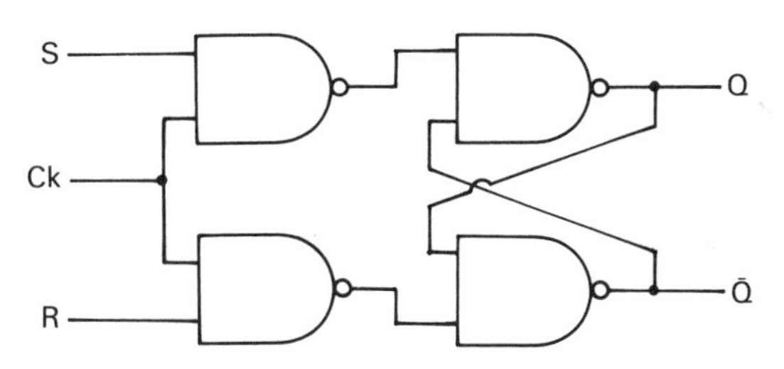

Figure 3.18. A clocked RS flip flop.

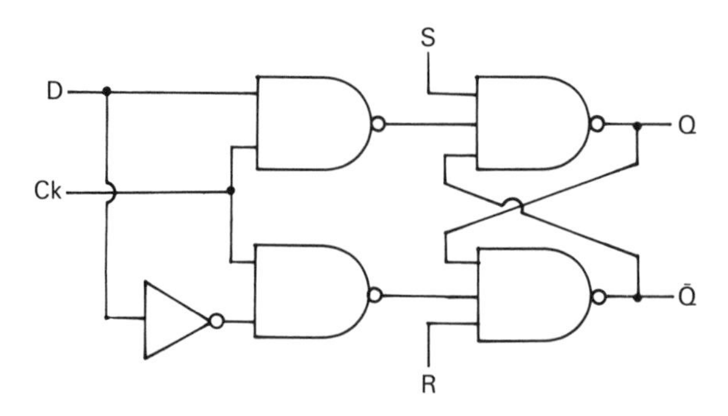

Figure 3.19. The D flip flop.

The D flip-flop. The D flip-flop is illustrated in Figure 3.19. The R and S inputs are used to preset the Q and $\bar{Q}$ output state. The input is at D and is split into two parts; one of these is inverted. When a clock pulse is input into the device, the output is determined by the status of D. If D is high; then Q and $\bar{Q}$ change state. If D is low, there is no change of Q or $\bar{Q}$. At any time we can change or reset the output state by appropriate inputs into R and S.

The JK flip-flop. The JK flip-flop is the most complex of the flip-flops described here and is shown in Figure 3.20. It consists of two clocked RS flip-flops, one of which drives the other. The information from the JK inputs is transferred to the output only when a clock (Ck) pulse is present. Table 3.10 gives the state of the output in terms of the input and the previous output state. The "clear" input (Cl) is normally held high. A low pulse into

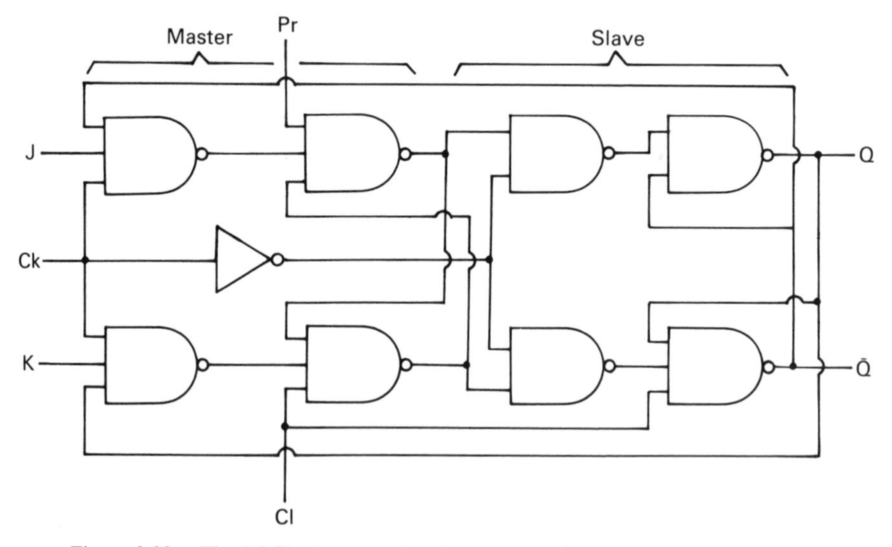

Figure 3.20. The JK flip flop showing the master and slave portions of the circuit.

Table 3.10. Truth table for the JK flip-flop

J	K	Q_f	$\bar{Q}_f$
0	0	Q_i	$\bar{Q}_i$
0	1	0	1
1	0	1	0
1	1	$\bar{Q}_i$	Q_i

Cl will reset the output state to $Q = 0$ and $\bar{Q} = 1$. A "preset" input (Pr) can be used to set the output state to whatever is desired. When not in use or when not specified, Cl and Pr should be held high.

Flip-flop applications. One of the common applications of flip-flops is discussed here. You will find more uses of these devices discussed in the section on computer interfacing. Figure 3.21 shows the circuit for a binary counter. As R and S are held permanently high, the state of the output Q of the leftmost flip-flop in the figure will change every time a pulse is input into the device. This means that it will require two complete pulses at the input to result in one complete pulse at Q of the leftmost flip-flop. Similarly it will require two complete pulses at the output of the leftmost flip-flop to yield one complete pulse at the output (Q) of the next flip-flop, and so on for the other flip-flops. Thus the Q states will represent the binary equivalent of the number of pulses that have been detected. The simple four-bit counter illustrated here will count up to binary 1111. The subsequent pulse will reset all of the Q states to 0. Of course, we could build a binary counter with a larger number of bits merely by adding more flip-flops to the circuit.

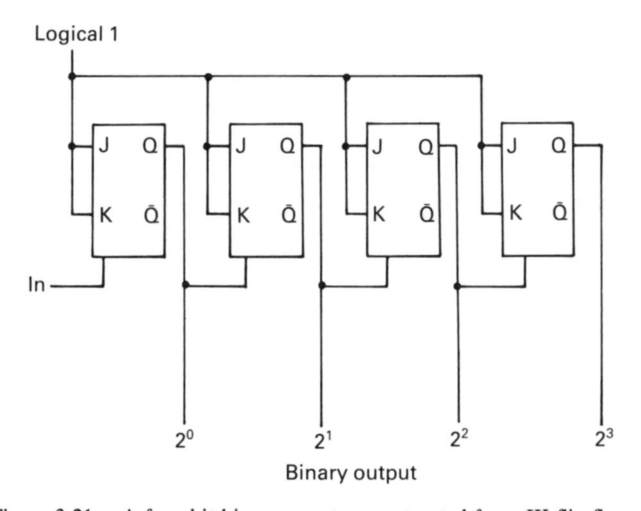

Figure 3.21. A four-bit binary counter constructed from JK flip flops.

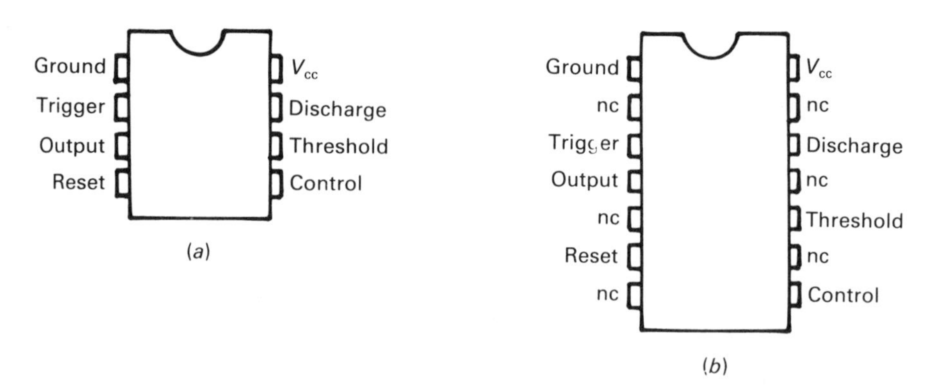

Figure 3.22. 555 timer packages: (*a*) 8-pin minidip; (*b*) 14-pin dip.

Monostable and astable multivibrators. The 555 timer is a common device that can be used as either a monostable or astable multivibrator. This device can be used in either mode of operation by choosing appropriate external circuitry. Standard package configurations for the 555 timer are shown in Figure 3.22.

555 monostable operation. A block diagram of the 555 in this configuration is shown in Figure 3.23. In the idle state the trigger input is held high. At this time, C2, the lower comparator holds the S input of the flip-flop low. Similarly C1 holds the R input of the flip-flop high, and the Q output of the flip-flop will be high. This signal keeps the transistor Q1 in the ON condition and connects the output to ground. Now let us assume that the trigger input momentarily goes low. This means that the comparator C2 will cause the flip-flop S input to go high and the flip-flop output changes state—that is, Q = 0. This turns OFF the transistor Q1 and the output will be connected through Q2 to $+V_{cc}$. As the external capacitor is no longer grounded, it begins to charge. This is because it is connected through R to $+V_{cc}$. The charging time is given by RC. When the voltage at the (+) input of the comparator C1 becomes large enough, the R input of the flip-flop will go high, changing the state of Q. Thus the output is connected again to ground through Q1. This means that a short trigger pulse at the output will result in a output pulse of a fixed length. For the 555 the details of the internal circuitry results in a output pulse length of 1.099 RC. Additional trigger pulses during this time interval have no effect on the output. The output pulse, however, can be cut short if the reset is disconnected from $+V_{cc}$. That is, the reset can be held at $+V_{cc}$ independently of the supply voltage and it can be disconnected in order prematurely to terminate the output pulse. Examples of this operation are shown in Figure 3.24. We are able, through our choice of the external components R and C, to control the length of the output pulse.

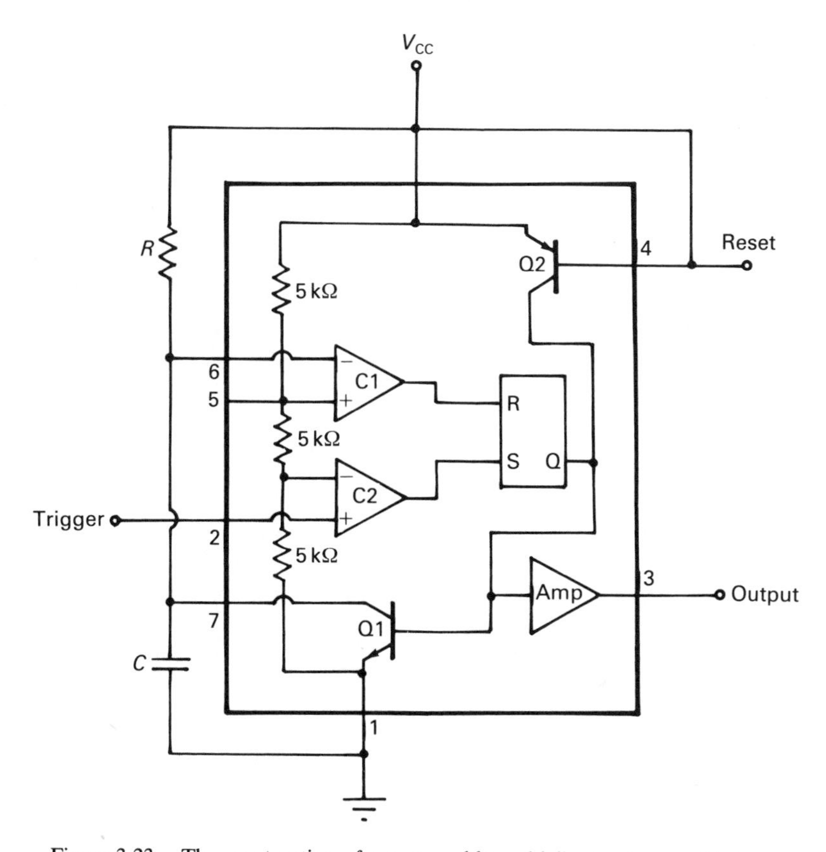

Figure 3.23. The construction of a monostable multivibrator from a 555 timer.

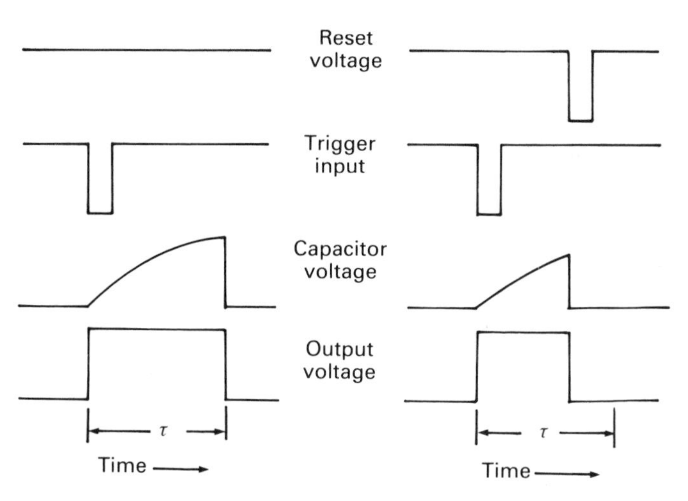

Figure 3.24. Operation of the monostable 555 (*a*) without reset and (*b*) with reset.

555 Astable operation. The astable mode of operation of the 555 timer is illustrated in Figure 3.25. C' is merely a filter capacitor and is not important to the discussion of the operation of the device. When the supply voltage is initially turned on, the output is high. At this time C begins to charge through R_1, the time constant for charging being determined by $C(R_1 + R_2)$. When the voltage drop across C becomes sufficient, the comparator C2 changes the state of the flip-flop and thus drives the output low by turning Q1 on. The comparator Cl begins to discharge the capacitor C through R_2 with a time constant of CR_2. When the voltage drop across C becomes small enough, comparator C1 causes the flip-flop state to change again, driving the output high. This process continues indefinitely. The voltage relationships are illustrated in Figure 3.26.

An analysis of the 555 circuit shows that the total time for one oscillation period is given by

$$\tau = 0.6931(R_1 + 2R_2)C \tag{3.15}$$

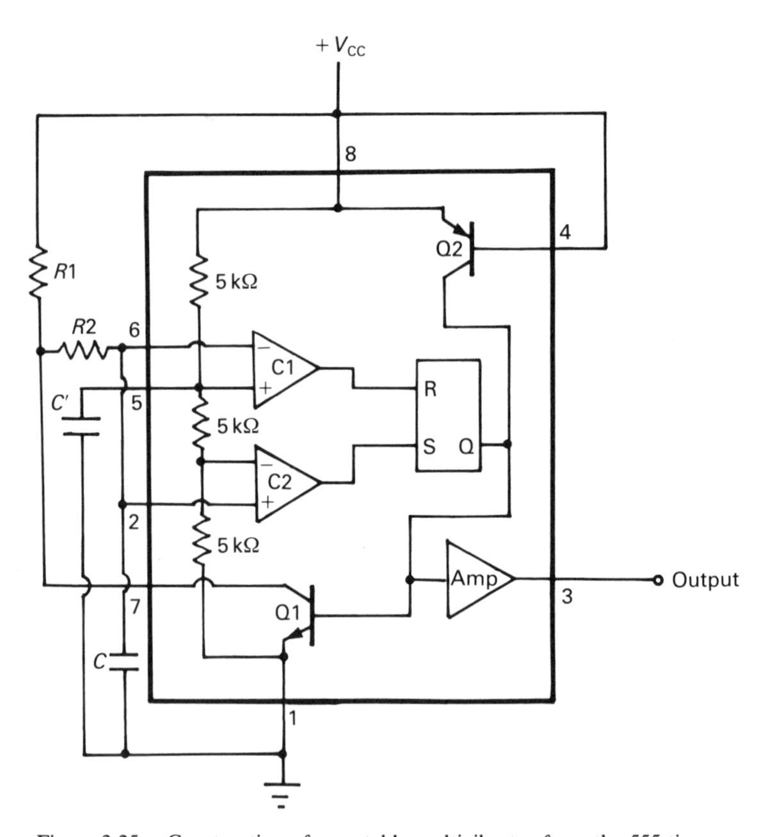

Figure 3.25. Construction of an astable multivibrator from the 555 timer.

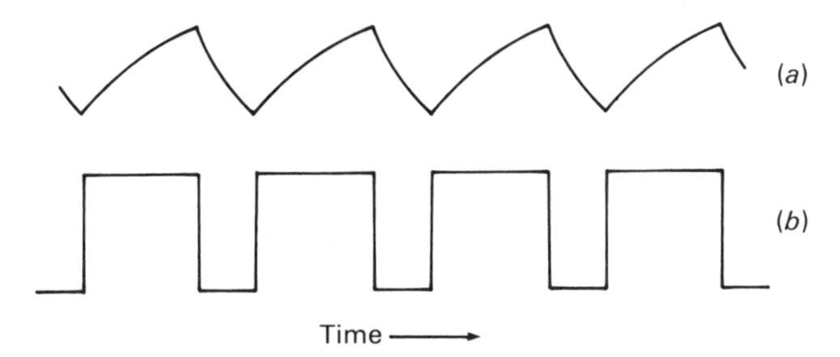

Figure 3.26. Voltages in the astable 555: (*a*) voltage drop across the capacitor; (*b*) the output voltage.

The duty cycle is determined to be

$$D = \frac{\tau'}{\tau} = \frac{R_1 + R_2}{R_1 + 2R_2} \tag{3.16}$$

The output voltages are normally 0 for the low state and $\sim +V$ for the high state.

Frequency-to-voltage converter. A useful frequency-to-voltage converter can be constructed from a 555 timer as illustrated in Figure 3.27. A periodic waveform is input into the device. The Schmidt trigger is used to square up the waveform if it is not already a square wave. This signal is then

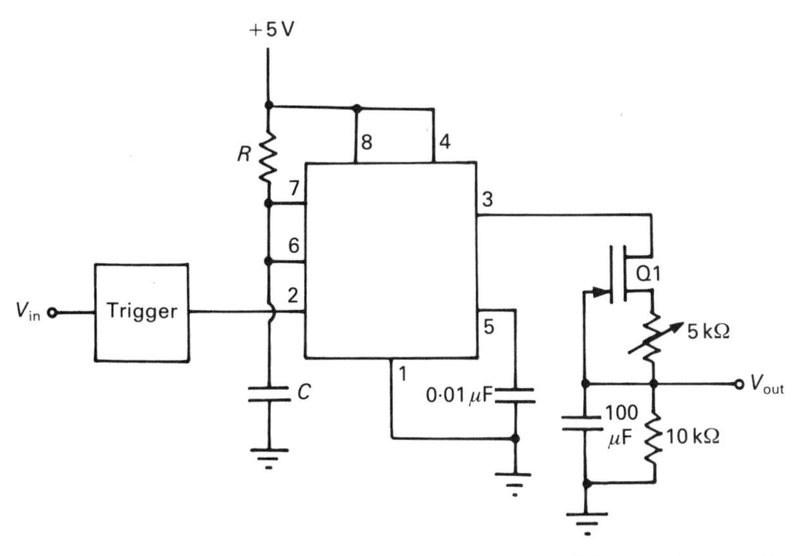

Figure 3.27. A frequency-to-voltage converter based on the 555 timer. (Adapted from H. Berlin *The 555 timer, applications sourcebook with experiments,* reprinted by courtesy of E & L Instruments, an Interplex Electronics Co.)

input into the trigger input of the 555. The trigger input will produce pulses at the output of the 555 with the same frequency as the input signal. The pulses, however, will have a duration determined by the time constant RC. Thus, as the input frequency changes, the duty cycle of the 555 output will change. The FET acts as a constant-current source, with the 10 kΩ resistor providing the load. The voltage across the 10 kΩ resistor will be averaged by the 100 μF capacitor and the voltage at the output of the circuit will be determined by the duty cycle of the wave produced by the 555. Thus, the output is a voltage proportional to the input frequency. The values of R and C must be chosen to be compatible with the range of frequencies of interest. Several values of either R or C, or both, can be chosen using a switch.

4
Signal processing

4.1 Noise

Noise in the general sense refers to anything that obscures the quantity we are trying to measure. It can refer to interference, often from electromagnetic fields, or to random noise that is often thermal in origin. The various types of noise are discussed here.

a. Interference

In dealing with noise it is important to know the origin of the noise and its frequency spectrum. It is difficult to generalize in the case of interference. The nature of the interference depends on what we are trying to measure. Any outside parameter that couples to the parameter of interest is a concern. For example, if we were making sensitive Hall-effect measurements we might be concerned about the direction of the Earth's magnetic field, but we probably would not worry about a small amount of mechanical vibration from automobiles outside the building. On the other hand, if our aim were to perform optical interferometry we would not care about small magnetic fields, but too many cars passing by could be disastrous. The method of dealing with interference depends not only on the origin of the offending signal but on the nature of the experiment as well.

Sometimes the experiment can be isolated from the interference, as is done by placing optics experiments on massive stone tables or by placing r.f.-sensitive experiments inside a shielded (wire-mesh) room. If isolation is insufficient or inappropriate, then filtering sometimes works. This is particularly effective against 60 Hz interference from power cables. Since the frequency spectrum consists of a sharp peak at 60 Hz (and often additional sharp peaks at the harmonics of 60 Hz), a slot filter (band rejection) is often ideal.

Normal (passive) filters consist of combinations of resistors, capacitors and inductors arranged in such a way that the impedance does not permit signals of the undesired frequency to pass. Improved filtering characteristics can be obtained with active filters. These additionally use an operational amplifier. We will discuss in general terms the design of filters in the next section. Before going to the trouble of filtering out interference, it is often effective to see whether we can eliminate its source. Sometimes, improperly placed power cables are responsible and we can easily separate signal-carrying cables from those that which are causing the interference. Ground loops are also often a problem. These are caused by connecting grounds in a circuit to different points. It is important in a circuit to have only *one* ground and to connect all ground leads to that one point.

Many of the techniques described later for noise reduction are suitable for eliminating interference.

b. Johnson noise

Johnson noise is the thermally induced noise that is present in any resistive element. It is the electronic analogue of the random motion of gas molecules in a box, which is known as Brownian motion. In noise terminology, Johnson noise is *white noise.* This means that the noise power per unit frequency is independent of frequency; that is, it has a *flat* frequency spectrum. To put it another way: if we were to measure the noise power between frequencies of, say, 100 Hz and 101 Hz, it would be the same as between, say, 1000 and 1001 Hz. The r.m.s. (root-mean-square) noise voltage is proportional to the square root of the resistance of the resistor, to the temperature and to the range of frequencies. We write

$$V_J = [4k_B TRB]^{1/2} \tag{4.1}$$

where T is in K, R in Ω, and B is the bandwidth in Hz. For an example, let us take a 10 KΩ resistor at room temperature. Over a band width of, say, 1 kHz the r.m.s. Johnson noise voltage is

$$V_J = 0.41\ \mu\text{V} \tag{4.2}$$

The 1 kHz band width can be in any part in the frequency spectrum; so we would measure 0.41 μV between 1 kHz and 2 kHz and also between 100 kHz and 101 kHz. This noise is present in all resistive elements and represents noise we could never eliminate. The Johnson noise is independent of the design of the resistor; so a cheap carbon resistor would have the same Johnson noise as an expensive precision wire-wound resistor of the same resistance. This is not necessarily true of other types of noise.

Although the r.m.s. voltage of the noise is given by equation (4.1), the distribution of noise voltages is a Gaussian centered at zero volts. That is, at any given time the probability that the noise will be in the voltage range V to $V + dV$ is given by

$$P(V, V + dV) = \frac{dV}{V_J\sqrt{2\pi}} \exp\left(\frac{-\frac{1}{2}V^2}{V_J^2}\right) \tag{4.3}$$

when V_J is given by (4.1)

c. Shot noise

Shot noise is present in any circuit that carries a current. Its relative importance varies inversely with current. The origin of shot noise is as follows: A current I consists of a flow of electrons of the form

$$I = e\langle N \rangle \tag{4.4}$$

where $\langle N \rangle$ is the *average* number of electrons passing through a plane normal to the electron flow per unit time. For time in seconds and e in coulombs, I is in amperes. However, during any given time interval not necessarily exactly $\langle N \rangle$ electrons will pass the reference point. The distribution about $\langle N \rangle$ is given by $\langle N \rangle^{1/2}$, as for any statistical process; see Figure 4.1. This noise is a direct result of the quantization of the electric change and is given by (r.m.s. value)

$$I_s = (2eIB)^{1/2} \tag{4.5}$$

Like Johnson noise, shot noise is white noise. The shot noise as a percentage of the circuit current measured over, say, 10 kHz as a function of I is shown in Figure 4.2. We see that shot noise can be important for currents of less than one nanoampere or so and is a serious problem in the picoampere to femtoampere range.

d. 1/*f* noise

$1/f$ noise is generally the most significant contribution to the noise (excepting interference). It depends a good deal on the construction of the components and the quality of the contacts. It can, unlike shot and Johnson noise, be reduced by careful choice of components. Its power spectrum is proportional to $1/f$ (hence the name). Given this dependence, the noise energy per frequency decade (rather than per unit frequency) is constant over the spectrum (over which the noise extends—it can be cut off below

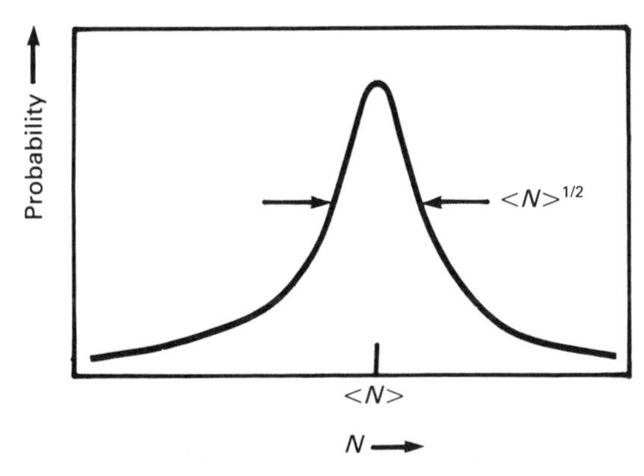

Figure 4.1. Number N of electrons flowing per unit time in an electric current.

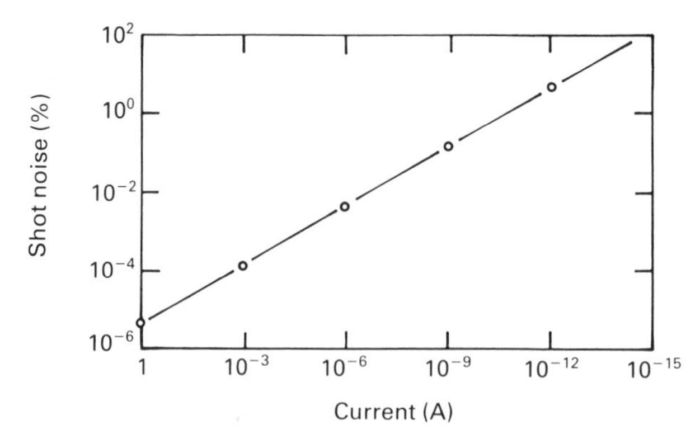

Figure 4.2. Relative importance of shot noise for different currents over a band width of 10 kHz.

and/or above certain frequencies). This type of noise spectrum is called *pink* noise.

For a resistor, the $1/f$ noise is proportional to the DC current flowing through it. The typical noise per frequency decade ranges from about 10^{-6} percent for high-quality wire-wound resistors to about 5×10^{-4} percent for inexpensive carbon composition resistors.

e. Signal-to-noise ratio

It is convenient to measure the quality of the signal in terms of a signal-to-noise ratio S/N defined in decibels (dB) to be

$$\mathrm{S/N} = 10 \log_{10} \frac{V_s^2}{V_n^2} \tag{4.6}$$

where V_s and V_n are the signal and noise voltages, respectively, measured over the same bandwidth. If the signal has a frequency spectrum that extends over a limited bandwidth, then increasing the bandwidth beyond this point will only decrease the S/N. This is because, in general, V_n will increase without increasing V_s.

In the remainder of this chapter we discuss various methods for improving the S/N of an experiment.

4.2 Filters

There are all kinds of filters for all kinds of applications and there are elaborate and complex methods for designing the best filter for your needs. In view of this, it is not surprising that entire books (a lot of them) have been written on filters. Some examples are given in the references. We will not get carried away with too many details; rather, we will review the basics and then give some practical general-purpose filter circuits and discuss their characteristics.

We divide filters into two general categories: (1) passive filters and (2) active filters. Passive filters consist only of resistors, capacitors, and inductors; active filters consist of, among other things, operational amplifiers. We begin with passive filters.

a. Passive filters

An analysis of the characteristics of a filter circuit requires some knowledge of the response of circuit elements to AC signals. The general rules for the analysis of filter circuits are given in this section. The reader who is interested in a more thorough introduction to AC circuit theory can find this information in many introductory physics texts. The references for this chapter give three texts that are particularly recommended—Halliday and Resnick's *Physics,* Reimann's *Physics: Electricity, Magnetism and Optics,* and Purcell's *Electricity and Magnetism.* At a more advanced level, a good text is Kerchner and Corcoran's *Alternating-current Circuits.*

In a DC circuit the total resistance is the sum of the individual resistances in the circuit, summed in an appropriate way that depends on whether they are connected in series or parallel. In an AC circuit the impedance is analogous to the resistance and can be summed in a similar manner. It is important to note that the AC impedance has not only a real component but an imaginary component as well. It is this imaginary component that is responsible for the phase shifts in an AC circuit. The various components we are interested in are resistors, inductors, and capacitors. The contributions to the total impedance for those components are as follows.

1. The resistance

$$Z_R = R \tag{4.7}$$

2. The inductive reactance

$$Z_L = i\omega L \tag{4.8}$$

3. The capacitive reactance

$$Z_c = (i\omega c)^{-1} \tag{4.9}$$

It should be noted that for real components these three contributions cannot necessarily be separated. that is, a real inductor will have not only inductance but resistance and probably capacitance as well. Here we deal with ideal components.

Consider the simple circuit shown in Figure 4.3. This is a voltage divider circuit consisting of a resistor and a capacitor in series, the output taken across the capacitor. We can write the total impedance seen by V_{in} as

$$Z_T = R + (i\omega c)^{-1} \tag{4.10}$$

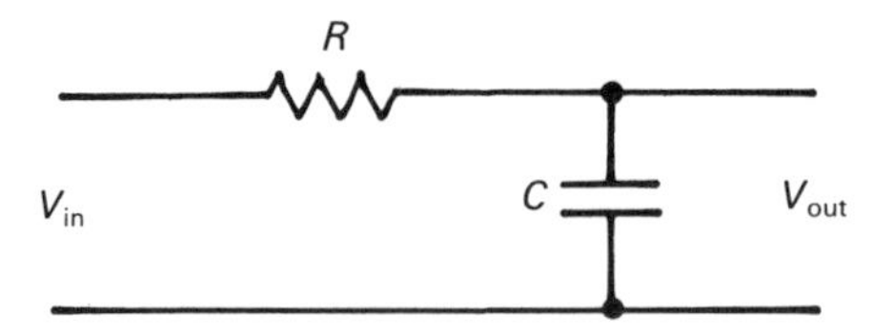

Figure 4.3. A simple first-order low-pass filter.

The ratio of V_{out} to V_{in} is given by

$$\frac{V_{out}}{V_{in}} = \frac{1}{i\omega C}\left(R + \frac{1}{i\omega C}\right)^{-1} \tag{4.11}$$

We can now solve for V_{out} in terms of V_{in}. In the present case we are primarily concerned with the amplitude of V_{out} and not its phase, so we write

$$|V_{out}| = (V_{out}V^*_{out})^{1/2} \tag{4.12}$$

and from (4.11) we find

$$|V_{out}| = |V_{in}|\left[\frac{1}{\omega^2C^2R^2 + 1}\right]^{1/2} \tag{4.13}$$

We can now plot this function in terms of $\omega = 2\pi f$; we show this in Figure 4.4. As is customary, we have plotted this in terms of $\log \omega$. We see that $V_{out} = V_{in}$ for $\omega \ll 1/RC$; that is, all the voltage is passed at low frequencies. For $\omega \gg 1/RC$ all the voltage is dropped across the resistor.

For filters the ratio of input to output voltages is generally quoted in decibels, defined by equation (4.6). For the low-pass filter, equation (4.13) shows that at $\omega = 1/RC$,

$$20 \log(2^{-1/2}) = -3 \text{ dB} \tag{4.14}$$

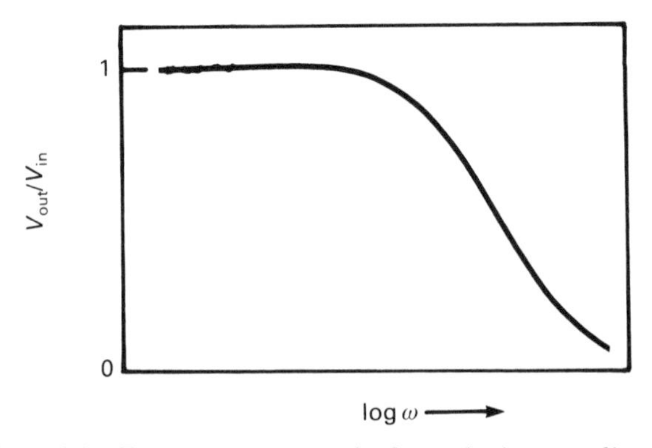

Figure 4.4. Frequency response of a first-order low-pass filter.

That is, the output is down by 3 dB at this frequency. The amount by which the output has decreased is called the *rolloff*. The rolloff is generally measured in dB per octave or dB per decade (octave = a factor of 2 in frequency; decade = a factor of 10 in frequency). For $\omega > 1/RC$ we find that for this filter the rolloff is 6 dB/octave or 20 dB/decade. This is typical of simple filters. These two component filters are referred to as first-order filters, and they come in low-pass and high-pass varieties. The various combinations that can be used and the frequency response of each are illustrated in Table 4.1. A band pass or a band reject filter can be constructed by using three (or more) components. Two simple examples are shown in Figures 4.5 and 4.6 along with their frequency responses.

Several first-order filters of the same type can be combined to produce a filter of a higher order: n first-order filters make an nth order filter. In general, the higher the order the sharper the rolloff, usually of 6 dB/octave per order. How sharply a particular filter needs to rolloff depends on how close the bandwidth of the signal you are interested in observing is to the noise you are interested in eliminating. For example, you do not need a

Table 4.1. First-order filter circuits

Circuit	Filter type	V_{out}/V_{in}
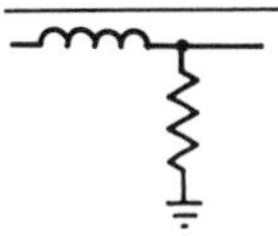	Low pass	$\left(1+\frac{\omega^2L^2}{R^2}\right)^{-1/2}$
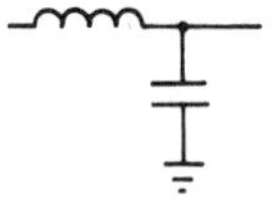	Low pass	$\lvert(1-\omega^2LC)^{-1}\rvert$
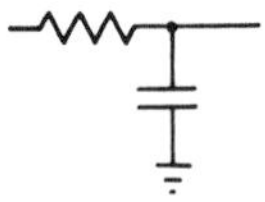	Low pass	$(1+\omega^2R^2C^2)^{-1/2}$
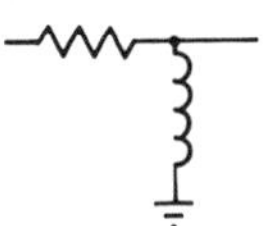	High pass	$\left(1+\frac{R^2}{\omega^2L^2}\right)^{-1/2}$
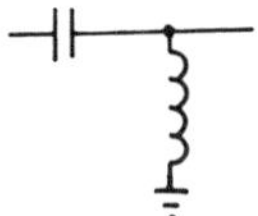	High pass	$\left\lvert\left(1-\frac{1}{\omega^2LC}\right)^{-1}\right\rvert$
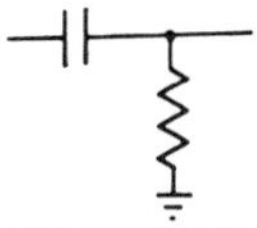	High pass	$\left(1+\frac{1}{\omega^2R^2C^2}\right)^{-1/2}$

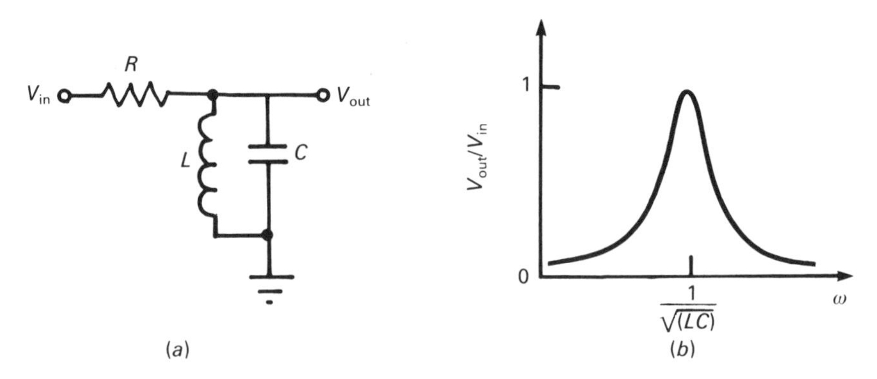

Figure 4.5. (*a*) A simple band pass filter; (*b*) its frequency response.

very good filter if you want to remove 60 Hz interference from a 10 kHz signal. On the other hand, if you want to extract a 100 Hz signal from 60 Hz interference then you need something much better.

There are numerous methods of putting together different combinations of first-order filters to form higher-order filters that have names such as *Butterworth filters, Chebychev filters,* or *Bessel filters*. Although active filters are generally better at fairly low frequencies, the rolloff of most op-amps at 100 kHz or so makes active filters useless above this. Thus, higher-order passive filters are best for higher frequencies. Similar higher-order band pass and band reject filters can be constructed as well.

Rather than going into extensive design criteria for filters in general we discuss the practical applications of one particular filter, the 60 Hz band reject filter. Figure 4.7 shows a circuit that is a good example of this kind of filter. Filter performance depends on the source and load impedances. We will consider a typical situation: a source impedance of 50 Ω and a load impedance (of an oscilloscope, for example) of 1 MΩ. Figure 4.8 shows the

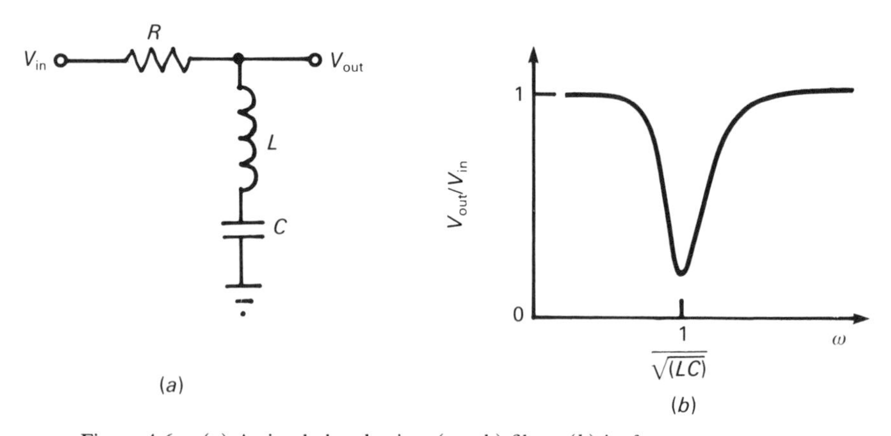

Figure 4.6. (*a*) A simple band reject (notch) filter; (*b*) its frequency response.

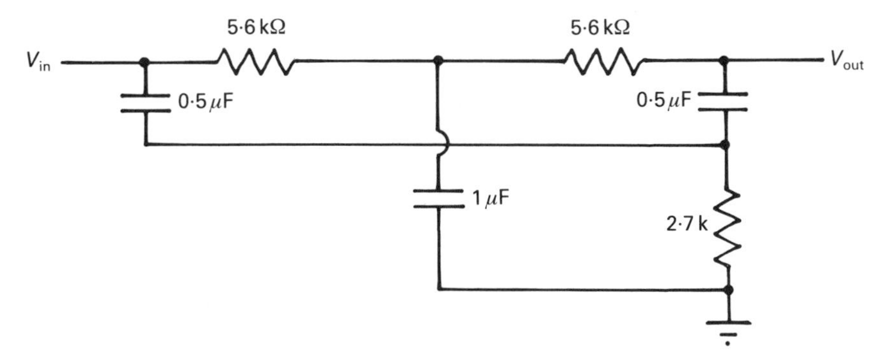

Figure 4.7. A 60 Hz band reject filter.

performance of this filter and demonstrates the usefulness of this type of circuit in rejecting noise at a particular frequency while allowing components of the signal at other frequencies to be passed.

b. Active Filters

There are a variety of ways of using op-amps to make better filters. Two problems that arise in the use of filters can be solved easily with op-amps: (1) the dependence of the filter characteristics on the input and output impedance; and (2) the attenuation of the signal due to the impedance of the filter. Let us see how op-amps help solve these problems.

In Figure 4.9*a* we show a conventional *RC* high-pass filter. In Figures 4.9*b–d* we show some analogous active filters. In the circuit in Figure 4.9*b* the op-amp is used to buffer the filter from the load. In this way the load impedance has no effect on the characteristics of the filter circuit. In this particular case the op-amp circuit is of unity gain. This op-amp configuration is called a *voltage follower*. Figure 4.9*c* shows a similar circuit in which the effects of source impedance have also been buffered by a voltage

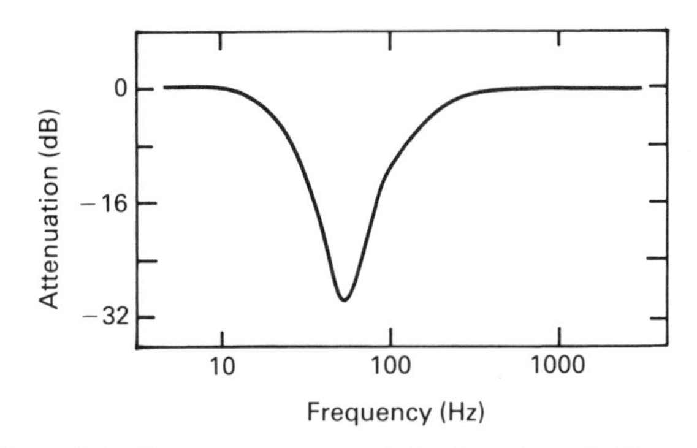

Figure 4.8. Frequency response of the filter shown in Figure 4.7.

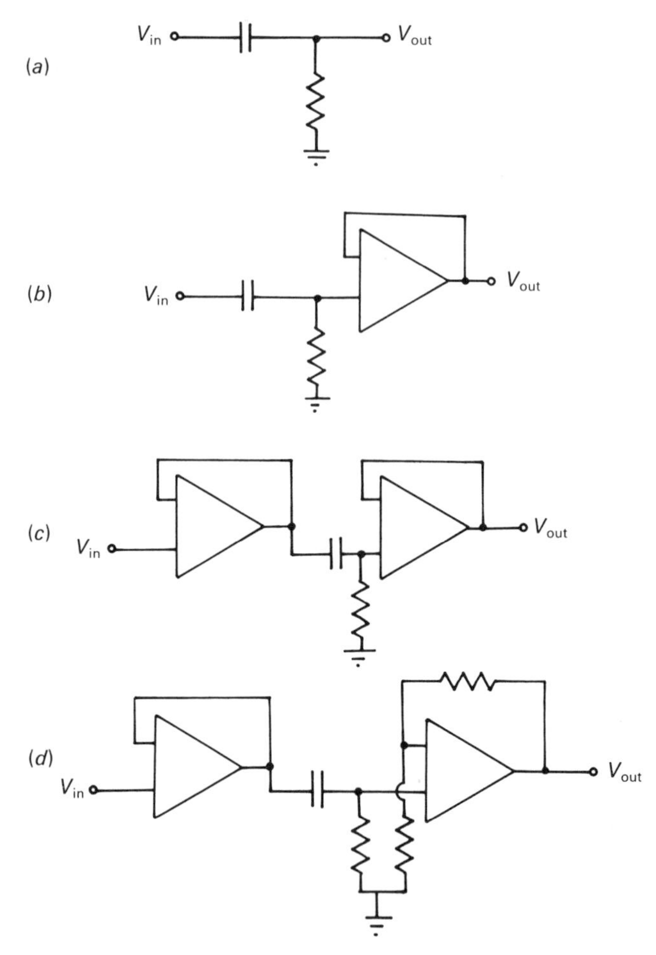

Figure 4.9. High-pass filters: (*a*) passive; (*b*) active with load buffering; (*c*) active with source and load buffering; (*d*) active filter with additional gain.

follower. In Figure 4.9*d* we show a filter in which the output has been supplemented by an adjustable-gain op-amp. Figure 4.10 illustrates how higher-order filters can be constructed using op-amps.

There are a number of other ingenious things we can do with op-amps as far as filters are concerned. One of these is illustrated in Figure 4.11. There are source and load buffers at the input and output, respectively. At low frequencies the signal is blocked from entering the center op-amp by the capacitor C_1; it therefore passes through R_3 to the output buffer. In this sense it is a conventional *RC* low-pass filter. Now comes the tricky part: high-frequency signals pass through C_1 and also through R_3. Through R_3 they go directly to the output buffer. From C_1 they are inverted by the center op-amp before reaching the output buffer via capacitor C_2. As this

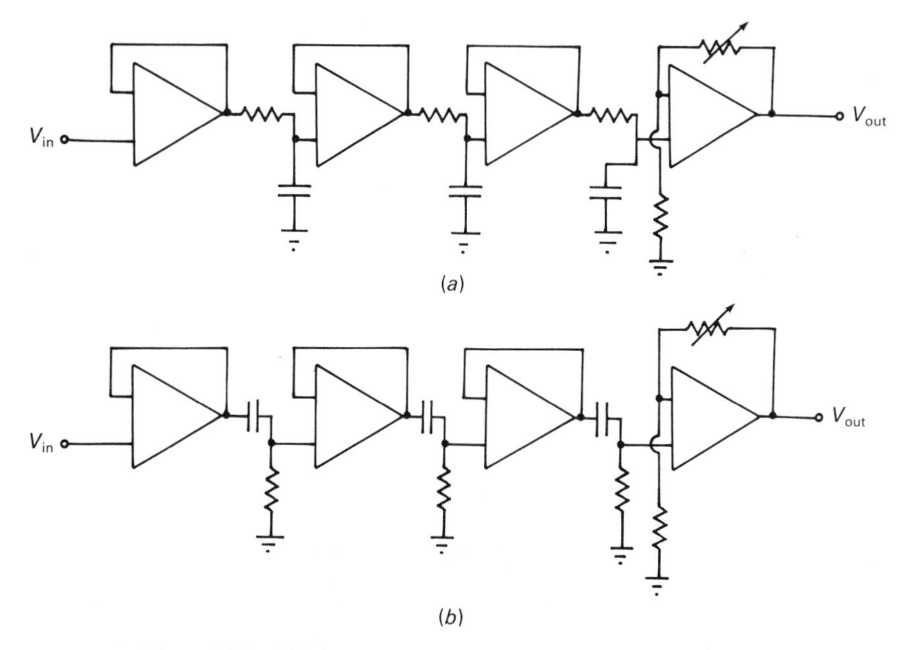

Figure 4.10. Third-order active filters: (*a*) low-pass; (*b*) high pass.

high-frequency signal is inverted it tends to cancel the high-frequency signal that arrived through R_3. This significantly increases the rolloff.

The filter design books listed in the references will serve as a guide for more circuits and to help you choose filter circuit components properly.

4.3 Signal averaging

Signal averaging is applicable to many types of experiments. Let us begin with some general consideration on how we go about making an experimental measurement.

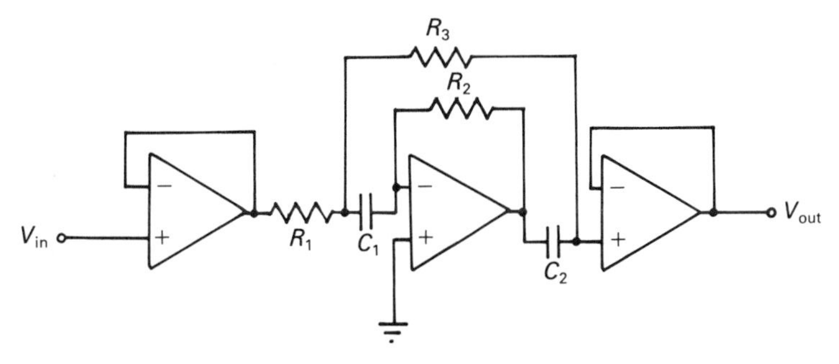

Figure 4.11. An active low-pass filter circuit.

In many cases a physics experiment does not consist of a single measurement of a particular quantity, but rather of a series of measurements of a quantity as a function of another quantity. For example we may want to measure the current through a semiconducting device as a function of the bias voltage.

The quantity we control in the experiment is referred to as the *extrinsic* parameter. The quantity we measure is referred to as the *intrinsic* parameter. In many experiments it is the case that there is more than one extrinsic parameter. It may be that there is more than one intrinsic parameter. Which quantities are intrinsic and which quantities are extrinsic is determined by the nature of the specific experiment. For example, current could be the intrinsic variable (and voltage the extrinsic variable) as in the example of the measurements on the semiconducting device given above. On the other hand, if we wanted to study the magnetic field H produced by a solenoid, the current might be the extrinsic variable and H the intrinsic one.

Conventionally, the extrinsic parameter x is varied over some relevant range of values and the intrinsic parameter y is measured. Signal averaging refers to the situation in which the extrinsic parameter is varied repeatedly over the same range of values and the intrinsic parameter is averaged (or summed) for each respective value of the extrinsic parameter. Mathematically we write

$$y = f(x_0) \tag{4.15}$$

The measured value of y in a single measurement is actually

$$y = f(x_0) + N \tag{4.16}$$

where N is the noise at the time the measurement was made. If we make n measurements of y at x_0 then we find

$$\langle y \rangle = \frac{1}{n} \sum f(x_0) + \frac{1}{n} \sum N \tag{4.17}$$

or

$$\langle y \rangle = f(x_0) + \langle N \rangle \tag{4.18}$$

The quantity $\langle N \rangle$ depends upon the nature of the noise in question, but for Poisson-distributed noise (i.e., random events), we find that

$$\langle N \rangle \propto n^{-1/2} \tag{4.19}$$

Therefore, as n increases, $\langle N \rangle$ decreases relative to $f(x_0)$. In fact from equation (4.6) we see that n measurements will improve the S/N relative to the S/N from a single measurement, $(\mathrm{S/N})_0$, by

$$\mathrm{S/N} = 10 \log_{10} n(\mathrm{S/N})_0^2 \tag{4.20}$$

A common method of performing signal averaging is with a multichannel analyser. This is sometimes referred to as multichannel scaling. The *multichannel analyser* (*MCA*) consists of v channels or memory locations. Most commonly v is a power of 2; typically 256, 512, 1024, 2048 or 4096. Often the memory can be subdivided into 2, 4, 8 or even 16 subgroups, each of which can be used independently to store data. These channels are generally numbered from 0 to 255, 511, 1023, 2047 or 4095, respectively. The MCA begins in channel 0. During some specified time τ, known as the *dwell time*, the MCA will remain in channel 0 and will count pulses. These are added to the "sum" in the specific memory location associated with channel 0. The value of τ can be specified by the user within certain limits typically from ~0.1 ms to a few seconds. After remaining in the 0th channel for time τ, the analyser moves to channel 1. During the next τ all counts are added to the sum in the memory location associated with channel 1. This process repeats until channel $v - 1$ is reached. The entire process takes a time τv. At this this point the process repeats, beginning again in channel 0. Most MCAs have two modes: in one (*recurrent*) a sweep is initiated by the MCA immediately after the proceeding one has been completed; the other mode (*triggered*) requires an external signal to initiate each sweep. In some cases the first channel (channel 0) or even the first two or three channels are reserved for information (such as the number of sweeps completed) and are not used to accumulate data.

The object of the experiment is to produce an extrinsic parameter that is commensurate with the sweep of the MCA. That is, every time the MCA is in some channel i, the extrinsic variable should have the same value. This means that the intrinsic parameter (noise excluded) will have the same value every time the MCA is in channel i. Thus, performing each sweep is like making a series of measurements of y versus x, and while the real part of the signal will continue to add linearly in the number of sweeps the noise will add only according to equation (4.19) and the S/N will increase as equation (4.20).

Three aspects of the MCA need consideration (a) the nature of the input, (b) the method of synchronizing the MCA sweep to the extrinsic parameter and (c) the method of outputing the data from the MCA.

a. Input

Some experiments, such as many nuclear physics experiments, produce pulses directly; in many cases these can be input into the MCA directly. More commonly, in other areas of physics, the signal we are interested in measuring is an analog signal. This must be converted into a series of pulses in order to be recorded by the MCA. If we are interested in measuring a voltage, we want the number of pulses input to the MCA per unit time to be proportional to that voltage. The device most commonly used for this purpose is a V/F (*voltage-to-frequency converter*) or a VCO (*voltage-controlled oscillator*). Two possible input configurations for MCA operation

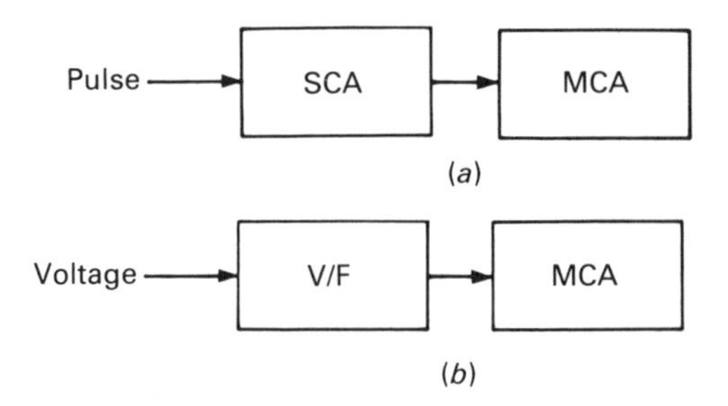

Figure 4.12. MCA input configurations for (*a*) pulse input and (*b*) voltage input.

are shown in Figure 4.12. In the pulse-detecting mode a single-channel analyser (SCA) is typically placed on the input. This enables us to set maximum and/or minimum pulse voltage that will be counted by the MCA. More details on the SCA and its operation will be given in the section on nuclear instrumentation (see Chapter 12).

b. Synchronization

There are various methods of controlling the extrinsic variable in a way that is synchronized with the sweep of the MCA. The best method is determined by the particular quantity we want to control as well as by the nature of the experiment. Most MCAs provide a various signals that can be used for this purpose. Figure 4.13 shows some examples for a hypothetical 16-channel MCA. (MCAs this small do not exist, but the picture would be too messy if we used 512 channels). Except for the ramp, most signals are TTL-compatible† (0–5 V). We give an example of the kind of things that can be done.

MCA example. We wish to vary an externally applied magnetic field from $-B$ to $+B$. We decide to ramp the field up and back down during each MCA sweep. To do this we use the circuit shown in Figure 4.14, which uses the MSB as the trigger. The first op-amp inverts the signal and shifts the voltage so the square wave is symmetric about zero. The second op-amp integrates the square wave to provide a triangular wave, and the third op-amp shifts the triangular wave so that it is symmetric about zero. Figure 4.15 shows the voltages at various points in the circuit. This output drives a programmable power supply (+ and − voltage output) for an electromagnet. The choice of the components in the integration circuit will depend on the dwell time and the characteristics of the programmable power supply. R' depends on R, C and τ. The voltage V_1 in Figure 4.15 is found to be

$$V_1 = 1.25\tau v/(RC) \tag{4.21}$$

† Transistor–transistor logic.

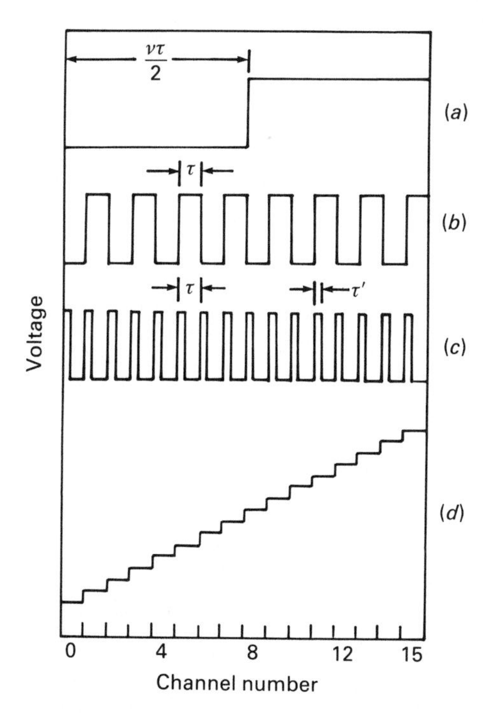

Figure 4.13. Typical MCA output signals for a hypothetical 16-channel analyser (τ = dwell time per channel): (*a*) most-significant bit (MSB); (*b*) least-significant bit (LSB); (*c*) channel advance signal; (*d*) digital ramp.

and from this R' is

$$R' = (5)(20\ \text{k}\Omega)/V_1 \tag{4.22}$$

c. *Output*

Essentially all modern MCAs have a CRT, such as an oscilloscope, to display the data stored in each of the channels. Many MCAs have an on-screen readout of one channel at a time. A cursor allows the channel of

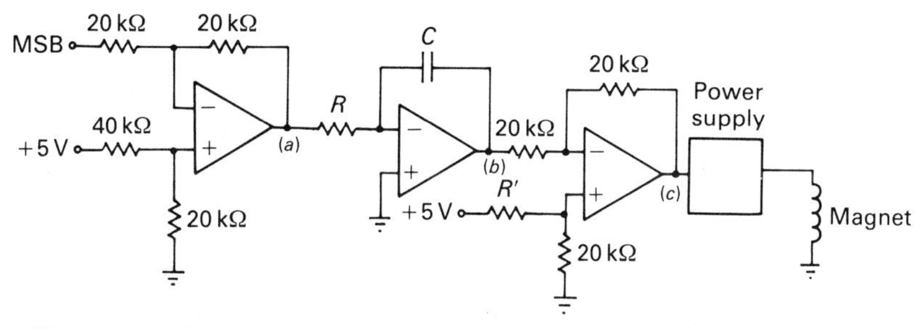

Figure 4.14. Control circuit to ramp magnet power supply from MCA MSB signal.

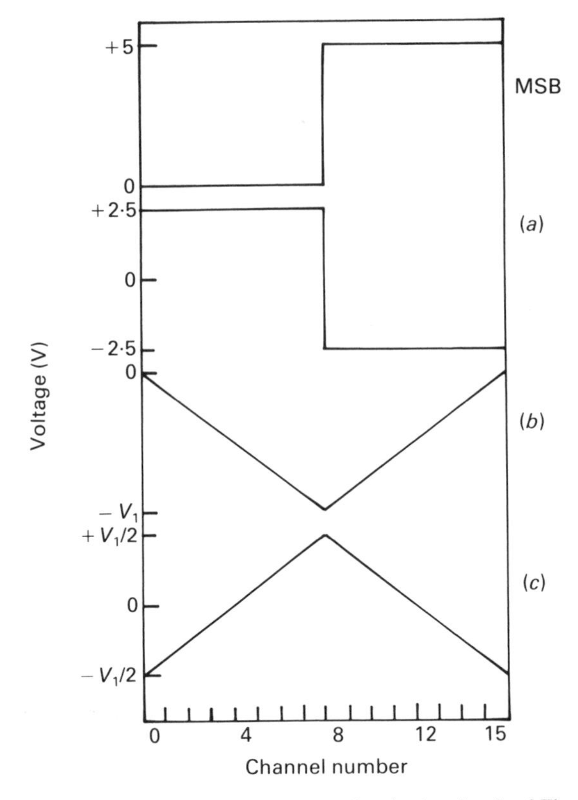

Figure 4.15. Voltages at the indicated points in the circuit of Figure 4.14.

interest to be selected. Most commonly, the data are output via a serial interface (RS-232) to a printer or to microcomputer or mainframe computer. The details of computer interfacing will be dealt with in Section 4.7. Some (more expensive) MCAs have a built-in interface for an x–y recorder. This allows an immediate hard copy of the data, plotted on an appropriate scale, to be obtained.

4.4 Pulse-height analysis

In nearly all cases, pulse-height analysers (PHAs) and multichannel analysers are incorporated in the same piece of equipment. Thus the term MCA is generally used to refer to a device that has these two modes of operation. You will usually see a switch somewhere on the device that enables you to change the mode.

The MCA accepts pulses in the PHA mode but the *voltage* of the pulses is the important factor. The v channels of the MCA represent different voltages, beginning with $V = 0$, which corresponds to channel 0, and ending with $V = V_{max}$, which corresponds to channel v. Each channel i represents a voltage range of V to $V + \Delta V$, where

$$V = \frac{iV_{max}}{v} \tag{4.23}$$

and

$$\Delta V = \frac{V_{max}}{v} \tag{4.24}$$

The example of the hypothetical 16-channel MCA is illustrated in Figure 4.16. For each pulse that enters the MCA with a particular voltage V, one count is added to the sum in the memory location that corresponds to the channel representing a voltage range that includes V. V_{max} is chosen to include the voltage of the largest voltage pulses of interest. Most analysers have $V_{max} = 10$ V. In many cases this can be adjusted by adjustment of the conversion gain. If we cannot adjust V_{max} to a reasonable value with controls available on the MCA, then we can always modify the range of the voltage of the pulses by the insertion of an amplifier.

In the PHA mode the MCA will continue to accept and process pulses until the device is manually turned off or until some preset time limit is reached. It is common to reserve channel 0 for a number equal to the length of time in seconds for which the device has been accumulating data. The result is that we get a histogram of the number of pulses as a function of their voltage. As the number of available channels is large, ΔV is small, and very good voltage resolution in the histogram can be obtained. As in Figure 4.12 an SCA can be inserted before the analyser to block out pulses of too high or too low a voltage.

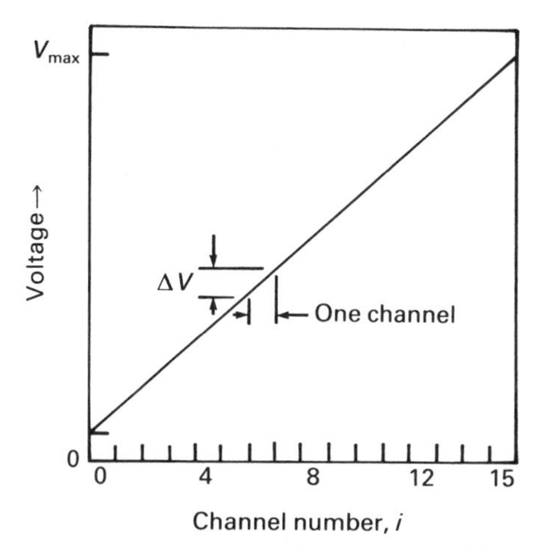

Figure 4.16. Voltage–channel relationship for the PHA mode of the multichannel analyser.

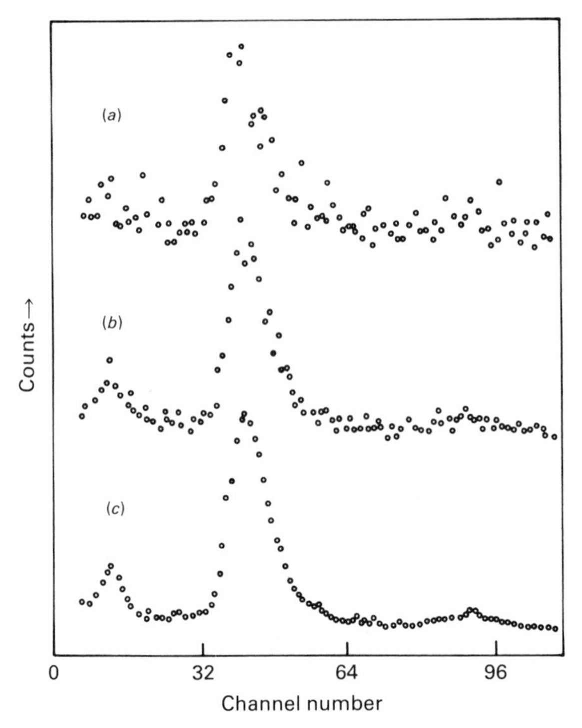

Figure 4.17. Example of noise elimination by signal averaging: spectrum accumulated for (*a*) 10 s; (*b*) 100 s, and (*c*) 1000 s.

Pulse height analysers are invaluable in the field of nuclear physics but is useful in many other fields as well. The energy spectra of radioactive sources, as shown in Chapter 12, are collected with a pulse-height analyser.

As with the MCA, the ability of the PHA mode to deal with noise relies on the statistical nature of the process producing the pulses and the fact that the signal will accumulate faster than the noise. Figure 4.17 illustrates this point. In Figure 4.17*a* the spectrum of the radioactive source was accumulated for 10 s; in Figure 4.17*b* accumulation proceeded for 100 s; in Figure 4.17*c* accumulation proceeded for 1000 s. Observe the clear increase in the S/N of the spectrum.

4.5 Phase-sensitive detection

Phase-sensitive detection is an effective method of eliminating noise from many types of experiments. Although it is particularly effective when used in conjunction with multichannel scaling, it is frequently useful in its own right.

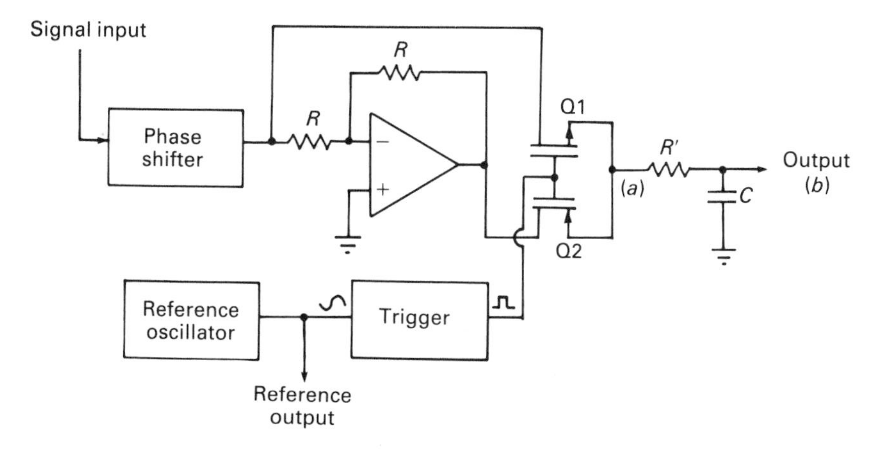

Figure 4.18. Block diagram of the lock-in amplifier.

A very simplified schematic of a phase-sensitive detector is shown in Figure 4.18. This is generally referred to as a lock-in amplifier. The amplifier is of unity gain and is inverting. The trigger provides a voltage that is sufficient to switch the FETs. We see from the FET circuit that the output is connected alternately to the output of the amplifier or to the signal itself at a frequency equal to that of the square-wave output of the trigger. Thus the output is switched from an in-phase to an out-of-phase input signal at the reference oscillator frequency. The phase shifter accounts for phase lags in the experiment; for the time being we will assume that it has no effect.

Let us look at the output for some different situations.

Case 1. *Sine-wave input at a frequency equal to the reference frequency and in phase with the reference frequency*. Figure 4.19 shows the relationship between the input, the reference, output at (a), and the output at (b). The output at (b) has been filtered through the RC circuit, the time constant of which is larger than the period of the reference signal. We see that the device acts as a full-wave rectifier because during the positive half of the input wave cycle the output is connected through Q1 to the input, and for the negative half of the input wave cycle the output is inverted by the op-amp.

Case 2. Sine wave input at frequency equal to the reference frequency but $\pi/2$ *out of phase*. This situation is illustrated in Figure 4.20. The filtered output at (b) is now zero because the average at (a) is zero.

A derivation of the amplitude of the output signal is as follows. The input signal is given by

$$V_i = V_0 \sin(\omega t + \psi) \tag{4.25}$$

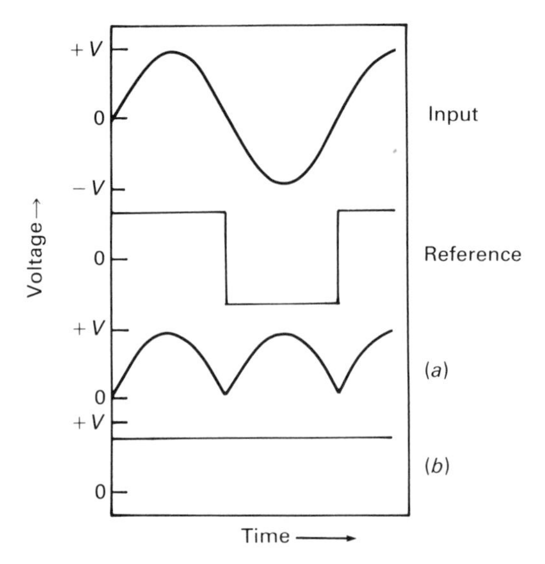

Figure 4.19. Case 1 as described in the text for the lock-in amplifier. The locations of points (*a*) and (*b*) are shown for the circuit in Figure 4.18.

where ψ is the phase difference between the input signal and the reference. We find that the output signal at (*b*) is

$$V_b = (2/\pi)V_0 \cos \psi \tag{4.26}$$

where the $2/\pi$ factor results from the filtering process.

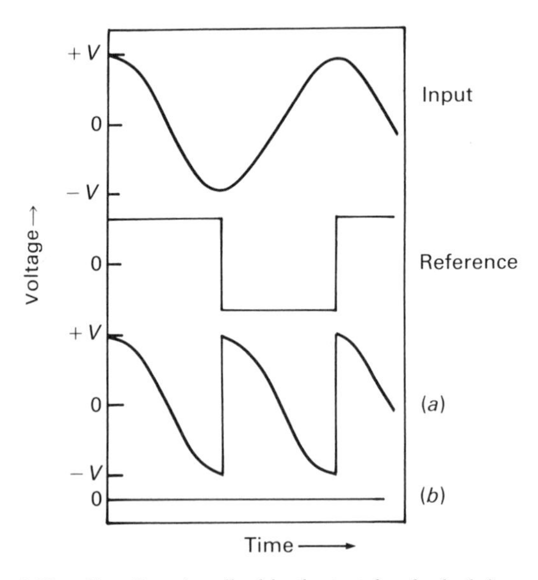

Figure 4.20. Case 2 as described in the text for the lock-in amplifier.

If the frequencies of the input and the reference are different (assumed, to begin with, by a small amount $\Delta\omega$), we write the input as

$$V_i = V_0 \sin(\omega + \Delta\omega)t \quad (4.37)$$

This represents a slowly varying phase shift of the form

$$\psi = \Delta\omega t \quad (4.28)$$

We therefore obtain a slowly varying output of the form

$$V_b = (2/\pi)V_0 \cos(\Delta\omega t) \quad (4.29)$$

When the frequency difference is large, the oscillations in V_a are rapid and they are filtered out by the *RC* network. Hence the voltage at (*b*) is zero.

Thus the lock-in amplifier measures signals only at frequencies very close to the reference frequency, and it measures the in-phase components of those signals. Any noise present is not detected, because any component of the noise with the right frequency will have random phase.

Now all we need to do is to make sure that the signal we are detecting is at the right frequency and with the right phase. Phase problems due to wires, etc., can always be adjusted out with the phase shifter. Getting the signal at the right frequency is easy; we modulate the appropriate extrinsic parameter. Let us look at the example we gave for the MCA system—that of measuring some intrinsic quantity as a function of applied magnetic field. Consider the experimental system shown in Figure 4.21. A large magnetic field (the extrinsic quantity) is applied to the experiment. In addition a small oscillating magnetic field is applied by a small electromagnet (represented by the modulation coils). As a result of this modulation of the extrinsic parameter, the intrinsic parameter that is input into the lock-in amplifier is modulated at this same frequency as well. The reference signal from the lock-in is used to modulate the extrinsic quantity at the proper frequency. Thus the lock-in amplifier accepts only the signal due to the modulated intrinsic parameter and rejects all of the noise (which has either the wrong phase, or the wrong frequency, or both).

A convenient modification of the arrangement shown in Figure 4.21 is to

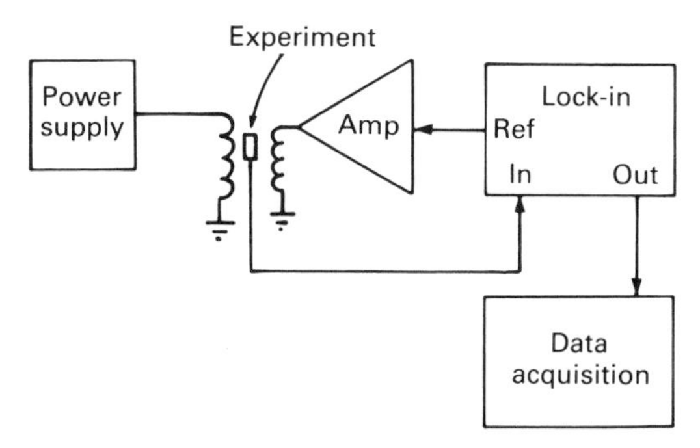

Figure 4.21. Experimental apparatus using a lock-in amplifier to modulate an externally applied magnetic field.

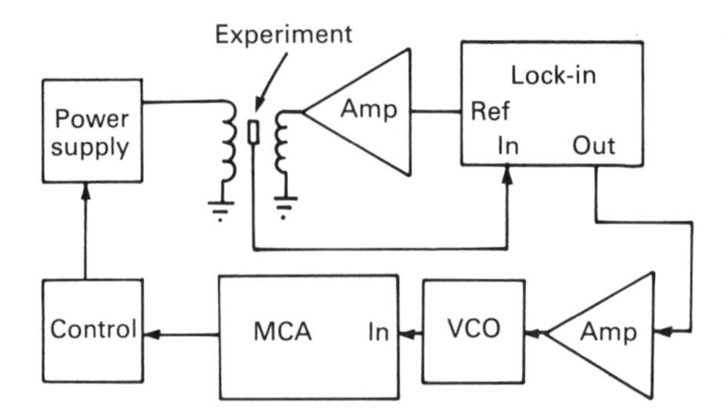

Figure 4.22. The use of signal averaging in conjunction with phase-sensitive detection.

use an MCA both to control the major component of the extrinsic quantity and to signal-average the data. Figure 4.22 shows how this could be done. Certainly the control of the extrinsic parameter may be modified in order to accommodate the purpose of the experiment.

So far we have implied that the modulation is small compared with the total range of extrinsic parameters of interest. This is not always the case. Two situations can arise: (1) the modulation is small (generally sinusoidal); (2) the modulation is a large square wave. We consider these two cases below.

a. Small modulation

Consider a hypothetical relationship between intrinsic and extrinsic parameters as shown in Figure 4.23. The range of x values chosen over which to make measurements is determined by the location of the interesting features in y, in this case the peak. Figure 4.23*a* shows that when the modulation

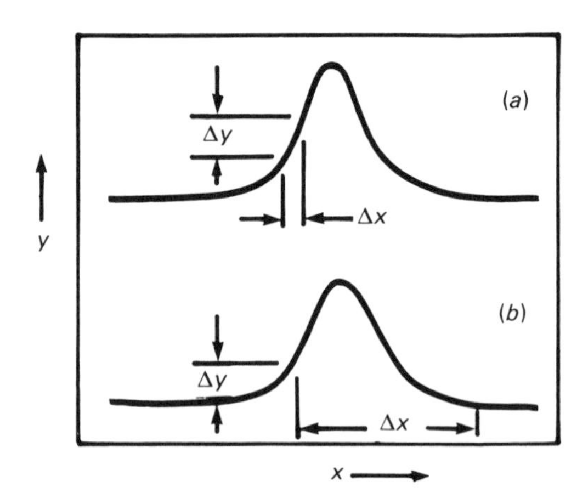

Figure 4.23. The effects of (*a*) small and (*b*) large modulation signals on the quantity measured by the lock-in amplifier.

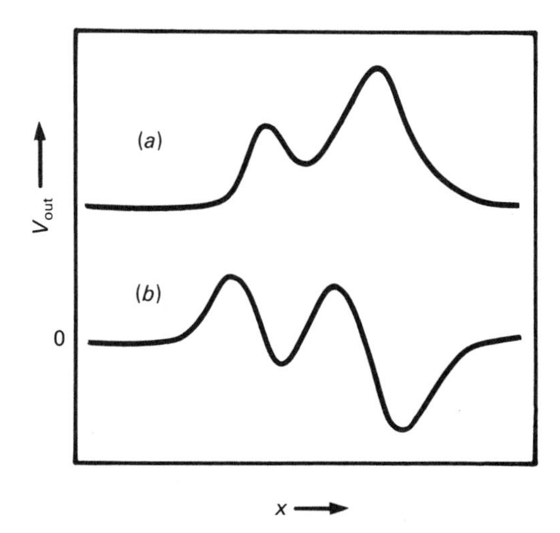

Figure 4.24. (*a*) The functional relationship $y = f(x)$ and (*b*) the signal measured by the lock-in amplifier using a small sinusoidal modulation.

signal is small the size of the change in the intrinsic quantity that is detected is related to the slope of the functional relationship between y and x. Thus we can write

$$\Delta y = \Delta x \frac{\partial y}{\partial x} \tag{4.30}$$

and the in-phase signal measured by the lock-in amplifier is

$$V_{\text{out}} = \frac{2}{\pi} \Delta x \frac{\partial y}{\partial x} \tag{4.31}$$

This means that we measure the first derivative of the signal. The lock-in output for the relationship shown in Figure 4.23*a* is illustrated in Figure 4.24.

b. Large modulation

As illustrated in Figure 4.23*b*, large modulation signals yield a Δy equal to $y(x)$. This is because a *zero* point on the $y = f(x)$ curve is used as a reference. Thus the in phase component of the lock-in amplifier output will look the same as the functional relationship between y and x. In this case we have to be sure that the shape of $y(x)$ is suitable and that Δx is large enough that we are referencing Δy to a zero point on the function.

From a practical standpoint it is important to make a proper choice of the modulation frequency. We cannot make the frequency too high in many cases because of the response time for the extrinsic parameter. If the frequency is too low then we can run into problems with the $1/f$ noise. It is best to avoid frequencies near 60 Hz or any of its low harmonics (i.e., 120 or 180 Hz). A few hundred hertz is generally suitable, but careful consideration should be made of the experimental details.

4.6 Digital-to-analog converters and analog-to-digital converters

Digital-to-analog (DAC) and analog-to-digital (ADC) converters comprise instruments that operate with digital signals to devices which operate with analog ones. While experiments generally deal with analog signals, it is nearly always necessary to convert these to digital signals if we want to process this information using a microcomputer, or for that matter even using an MCA. Most MCAs have ADCs built into their input. Conversely, if we want to use a computer to control an experiment it is possible in many cases to use digital instrumentation that connects directly to the digital output of the computer. Frequently, however, it is necessary to convert the digital output of the computer to an analog signal in order to, say, program a power supply. We begin with a description of the DAC, as this is the easiest to understand and will allow us to introduce some basic concepts of digital signal representation. As we shall see later, digital signals can be serial or parallel. For the time being, however, we will concentrate on parallel digital signals.

a. *The DAC (digital-to-analog converter)*

Any number may be represented in binary form. For the present discussion we consider the simple case of positive integers, but in general it should be understood that we are not necessarily limited to this case. The binary representation of a number is merely that number converted into base 2. For example the binary representation of the decimal number 37 would be 100101, that is,

$$1 \times 32 + 0 \times 16 + 0 \times 8 + 1 \times 4 + 0 \times 2 + 1 \times 1 = 37 \tag{4.32}$$

Although it takes many more digits to represent a number in binary form the digits need only be 0 or 1 rather than $0, 1, 2, \ldots, 9$ as in base 10. This makes things simple, since an electrical signal need only be either "high" (logical 1) or "low" (logical 0) to represent the digit. Each digit in the binary number is referred to as a bit. Hence 100101 is a six-bit binary number. It is easy to see that an n-bit binary number can represent integers from 0 up to $2^n - 1$. We can represent a digital quantity in the electrical sense by a series of n wires each with either a high or low voltage (logical 1 or 0). There is, of course, a common ground connection.

Figure 4.25 shows a simple circuit for converting this kind of digital signal into an analog one: an 8-bit DAC is illustrated. Although 8 is not always the number of bits, it is not uncommon. The larger the number of bits the better the resolution, as we shall see, A simple analysis of the circuit shows

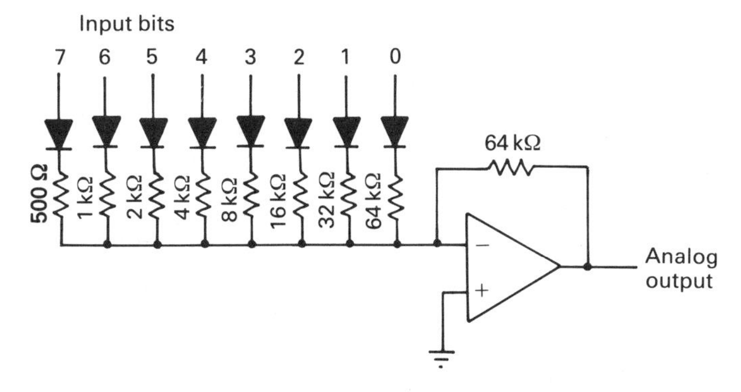

Figure 4.25. The digital-to-analog converter (DAC).

that the analog output signal is given in terms of the signals b_i on the eight input bits ($i = 0$ to 7), where $b_i = 0$ or 1 is represented by 0 V or $+V$, respectively. The output is found to be

$$V_{\text{out}} = -V \sum_{i=0}^{n-1} 2^i b_i \tag{4.33}$$

We see that V_{out} cannot change continuously but is "quantized" in steps of V; the smallest change in V_{out} occurs for a change in b_0 from 0 to 1 (or from 1 to 0).

The maximum output occurs for all $b_i = 1$ and this means that the resolution will be 1 part in $2^n - 1$. The larger the value of n, of course, the better the resolution. A circuit of the type shown in Figure 4.25 provides a convenient method of converting a parallel digital signal (see Section 4.7) from a microcomputer to an analog signal for experimental control. The number of bits of the DAC that you use should be consistent with the number of parallel bits that can be supplied as output by your computer. We shall say more on this in the section on computer interfacing. DACs can be bought for a few dollars, so if you are interested in having something that works well, it is best to buy one. If you are interested in studying how the DAC works, it is useful to build the circuit in Figure 4.25. The F/V (frequency to voltage converter) discussed in Section 3.2 is sometimes called a *time-domain DAC.*

b. The ADC (analog-to-digital converter)

There are a variety of methods of converting analog signals to digital signals; these are in general more complex than the methods for converting digital to analog. We describe one simple method.

Figure 4.26 shows the block diagram of a simple ADC. This is a 4-bit ADC, although in practice ADCs are typically of 8 or more bits. A 4-bit ADC would have a resolution of only $1/2^4$ or about 6%. This is how it

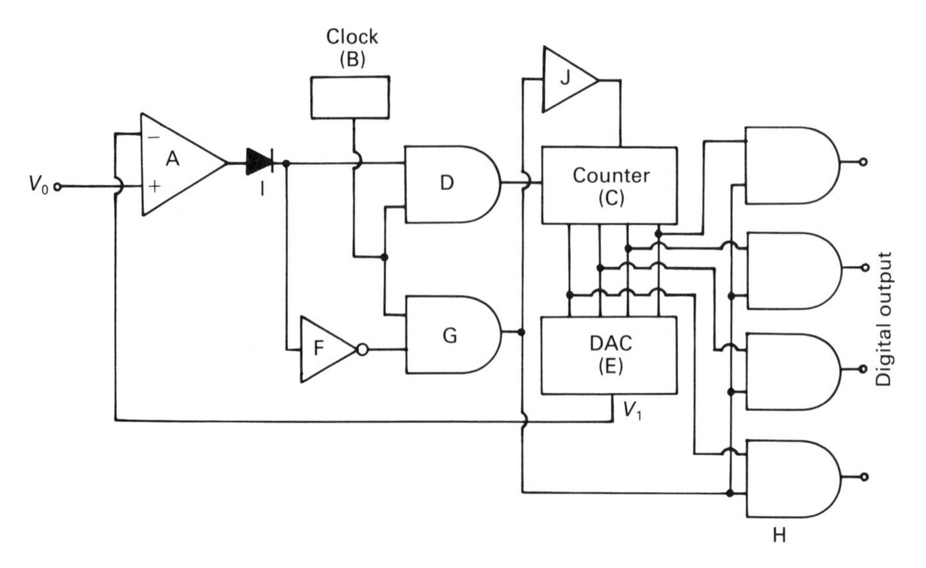

Figure 4.26. A 4-bit analog-to-digital converter (ADC): A = comparator (differential amplifier); B = clock; C = binary (4-bit) counter; D = clock gate; E = digital-to-analog converter (4-bit); F = inverter, G = reset gate; H = output gates; I = diode; J = buffer.

works. A clock B produces pulses at some fixed frequency. These are counted by a 4-bit binary counter C, beginning at 0000, and progressing 0001, 0010, 0011 etc. The digital output of the counter is converted into an analog signal by a DAC. This analog signal V_1 is input into a comparator A (differential amplifier). The other input of the comparator is the analog signal itself (the one we want to digitize)—a voltage V_0. As long as $V_o > V_1$ the comparator outputs a positive voltage and this is allowed to pass the diode I. This signal means that gate D is open whenever a clock pulse is present. Thus as long as $V_0 > V_1$ then the binary counter continues to count; as the counter counts, V_1 increases in steps. The size of the step is determined by the details of the DAC (mostly by the number of bits). This continues until $V_1 \geqslant V_0$ at which time the output of the comparator is 0 or negative, and this is blocked by the diode. Hence gate D is turned off and will no longer pass clock pulses, so that the counter stops counting. This logical "0" from the comparator is inverted by the inverter F to a logical "1". This holds gate G open for one clock pulse. This clock pulse does two things; (1) it opens gates H on which are the logical signals from the counter and outputs the digital (binary) signal; (2) it resets the counter to 0000. Buffer J ensures that the digital signal is output before the counter is reset. Thus the digital signal that is output is the one for which its analog counterpart (as produced by the DAC) is equal to the analog input signal (as determined by the comparator). Hence the output is the digital equivalent of the analog input.

The speed at which the ADC samples the analog input and produces a digital output depends on the speed of the clock and the number of bits in the counter. The number of bits is chosen according to the resolution required. The speed of the clock must be chosen to be compatible with the speed of the other electronics in the circuit.

The successive-approximation ADC (a faster kind of ADC). In this method a similar comparison method is used, but the binary counter is replaced by a programmable chip. This chip first sets the MSB = 1 and all other bits to zero. It then compares an analog signal V_1 to V_0. If $V_0 < V_1$ the MSB is set to zero and the next most significant bit is set to 1; otherwise the MSB is left equal to 1 and the next most significant bit is set to 1 as well. Each bit is "inspected" in this way; it is rather like a binomial search process. This process requires n inspections for n bits, while on average the previous method requires

$$N = 2^{n-1} \tag{4.34}$$

inspections. Thus the successive-approximation ADC is faster by a factor of $2^{n-1}/n$. For $n = 8$ this is a factor of 16, while for $n = 12$ it is a factor of 170, and for $n = 16$ the factor is over 2000.

Typically, we would choose a complete ADC in the form of an integrated circuit. Consideration must be given to what the digital signal will be used for and we must ensure that the output is compatible with the other devices in terms of voltage, speed, and the use of "handshaking" lines. We discuss these aspects further in the following section.

4.7 Standard digital-to-digital interfaces

In Section 4.6 we discussed the conversion of analog signals to digital signals and vice versa. These techniques are useful for converting analog signals from an experiment to digital form for input into a computer and for converting digital signals from a computer to analog form for instrument control. In this section we discuss the connection of equipment with specific digital interfaces to a microcomputer. These digital interfaces are either serial or parallel in nature. Serial signals are those in which the bits are transmitted one after the other along a single electrical connection. Parallel signals are those in which n bits are transmitted simultaneously along n electrical connections. As we shall see, each of these systems has its own advantages and disadvantages. In the past there has been a good deal of confusion over interfacing standards because of the wide variety of techniques used by different manufacturers. It is certainly not possible within the scope of this book to discuss all of the interfacing methods used in recent years. Fortunately, the computer industry now seems to be agreeing upon a fairly small number of interfacing standards. Two of these, RS-232 (serial) and IEEE-488 (parallel), are by far the most commonly used for connecting microcomputers to laboratory equipment. We consider these two in some detail as they are both useful and illustrate nicely the differences between serial and parallel interfacing. However, if you should encounter a piece of equipment or a microcomputer that has a different serial or parallel interface (a situation that is not unlikely), a knowledge of how RS-232 and IEEE-488 work should make it easy for you to decipher the necessary instruction manuals and understand how to use the interface.

a. RS-232 serial interface

Signals sent in the RS-232 standard represent characters (numbers, etc.) by a series of logical bits (1s and 0s). The voltage levels used on an RS-232 line to represent logical 1 and 0 are shown in Figure 4.27. In most cases characters sent on RS232 lines are represented by the ASCII code (*A*merican *S*tandard *C*ode for *I*nformation *I*nterchange). This code represents each character by a *seven*-bit binary number. In transferring data we are primarily concerned with the transmission of numbers but it is obvious that other characters (spaces, decimal points, line feeds, returns, etc.) are necessary as well. Table 4.2 gives the full ASCII code.

We see that the ten digits 0 to 9 correspond to the binary representation 0110000 to 0111001. These bits are transmitted at a particular frequency. This is known as the baud rate and specifies the number of bits transmitted per second. Standard baud rates for data transmission are 75, 110, 134.5, 150, 300, 600, 1200, 2400, 4800, 9600 and 19200. As a character consists of the 7-bit ASCII code and is generally has appended by one or two

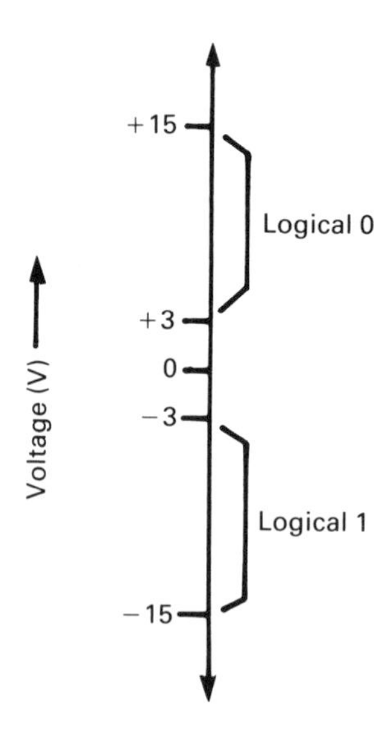

Figure 4.27. Standard logic levels for the RS-232 serial interface.

Table 4.2. ASCII character set, in which the characters are represented by seven bits $b_6b_5b_4b_3b_2b_1b_0$: b_0 is the 2^0 bit, b_1 is the 2^1 bit, etc.

<table>
<tr><th></th><th colspan="8">$b_6b_5b_4$</th></tr>
<tr><th>$b_3b_2b_1b_0$</th><th>000</th><th>001</th><th>010</th><th>011</th><th>100</th><th>101</th><th>110</th><th>111</th></tr>
<tr><td>0000</td><td>NUL</td><td>DLE</td><td>SP</td><td>0</td><td>@</td><td>P</td><td>`</td><td>p</td></tr>
<tr><td>0001</td><td>SOH</td><td>DC1</td><td>!</td><td>1</td><td>A</td><td>Q</td><td>a</td><td>q</td></tr>
<tr><td>0010</td><td>STX</td><td>DC2</td><td>"</td><td>2</td><td>B</td><td>R</td><td>b</td><td>r</td></tr>
<tr><td>0011</td><td>ETX</td><td>DC3</td><td>#</td><td>3</td><td>C</td><td>S</td><td>c</td><td>s</td></tr>
<tr><td>0100</td><td>EOT</td><td>DC4</td><td>$</td><td>4</td><td>D</td><td>T</td><td>d</td><td>t</td></tr>
<tr><td>0101</td><td>ENQ</td><td>NAK</td><td>%</td><td>5</td><td>E</td><td>U</td><td>e</td><td>u</td></tr>
<tr><td>0110</td><td>ACK</td><td>SYN</td><td>&</td><td>6</td><td>F</td><td>V</td><td>f</td><td>v</td></tr>
<tr><td>0111</td><td>BEL</td><td>ETB</td><td>'</td><td>7</td><td>G</td><td>W</td><td>g</td><td>w</td></tr>
<tr><td>1000</td><td>BS</td><td>CAN</td><td>(</td><td>8</td><td>H</td><td>X</td><td>h</td><td>x</td></tr>
<tr><td>1001</td><td>HT</td><td>EM</td><td>)</td><td>9</td><td>I</td><td>Y</td><td>i</td><td>y</td></tr>
<tr><td>1010</td><td>LF</td><td>SUB</td><td>*</td><td>:</td><td>J</td><td>Z</td><td>j</td><td>z</td></tr>
<tr><td>1011</td><td>VT</td><td>ESC</td><td>+</td><td>;</td><td>K</td><td>[</td><td>k</td><td>{</td></tr>
<tr><td>1100</td><td>FF</td><td>FS</td><td>,</td><td><</td><td>L</td><td>\</td><td>l</td><td>|</td></tr>
<tr><td>1101</td><td>CR</td><td>GS</td><td>–</td><td>=</td><td>M</td><td>]</td><td>m</td><td>}</td></tr>
<tr><td>1110</td><td>SO</td><td>RS</td><td>.</td><td>></td><td>N</td><td>^</td><td>n</td><td>~</td></tr>
<tr><td>1111</td><td>SI</td><td>US</td><td>/</td><td>?</td><td>O</td><td>—</td><td>o</td><td>DEL</td></tr>
</table>

Key: Mnemonics and functions of the control codes (ASCII 0000000 to 0011111)

NUL	Null	FF	Form feed	ETB	End transmission block
SOH	Start of heading	CR	Carriage return	CAN	Cancel
STX	Start of text	SO	Shift out	EM	End of medium
ETX	End of text	SI	Shift in	SUB	Substitute
EOT	End of transmission	SP	Space	ESC	Escape
ENQ	Enquiry	DLE	Data link escape	FS	File separator
ACK	Acknowledge	DC1	Device control 1	GS	Group separator
BEL	Bell	DC2	Device control 2	RS	Record separator
BS	Backspace	DC3	Device control 3	US	Unit separator
HT	Horizontal tab	DC4	Device control 4	DEL	Delete
LF	Line feed	NAK	Negative acknowledge		
VT	Vertical tab	SYN	Synchronize		

information bits on either end (we discuss these shortly), the maximum rate of character transmission is about 1/10th of the baud rate.

Whenever data are not being transmitted, the RS-232 line is in the logical 1 state; this is the idle state. When data are transmitted each 7-bit character is proceeded by a start bit. This is *one* logical 0 bit. This is followed by the seven character bits. After this comes an optional parity bit. Finally there are one or two stop bits. The stop bits are logical 1 bits. It is important to note that when a character is transmitted, the lowest-order bit, b_0 (see Table 4.2), is transmitted first.

The parity may be either even or odd and is determined by the number of logical 1 bits in the transmitted character. Table 4.3 shows some examples of the parity bit. Some examples of ASCII data transmission are illustrated in Figure 4.28.

Table 4.3. Determination of the parity bit

Number of logical 1s	Parity	Parity bit "logical"
Even	Even	0
Even	Odd	1
Odd	Even	1
Odd	Odd	0

The standard connector used for making RS-232 connections is the 25-pin subminiature D connector. Table 4.4 gives the signals on the various pins of the RS-232 connector. As you can see, only a small number of connections for serial communication are required. In this sense the 25-pin connector is much larger than is necessary. In fact, in many applications only three connections (pins 2, 3, and 7) are actually required. Some care is required in many cases in connecting two devices together with regard to the use of pins 2 and 3. That is, data transmitted by one device on pin 2 should be received by another device on pin 3, so it is often necessary in RS-232 cables to cross the leads for pins 2 and 3. This is only *sometimes* necessary, so *beware.*

b. IEEE-488 parallel interface

This "standard" method of interfacing has been proposed by the Institute of Electrical and Electronics Engineers (IEEE), hence its name. It is also known as the HPIB (Hewlett-Packard Interface Bus) and the GPIB (General-Purpose Interface Bus).

The IEEE bus consists of three kinds of signals: (1) handshake signals; (2) interface management signals; and (3) data signals. Devices connected to the IEEE bus can be of three types: (1) listener; (2) talker; or (3) controller. A particular device can be one or more of these types. For

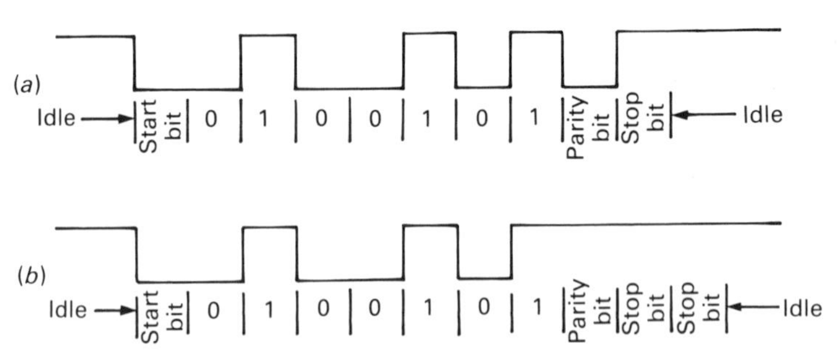

Figure 4.28. Serial transmission of the ASCII letter "R" (1010010): (*a*) one stop bit, odd parity; (*b*) two stop bits, even parity.

Table 4.4. RS-232 interface connections for the 25-pin subminiature D connector

Pin	Signal
1	Protective ground
2	Transmitted data
3	Received data
4	Request to send
5	Clear to send
6	Data set ready
7	Signal ground
8	Received line signal detector
9	+ voltage
10	− voltage
11	nc
12	Secondary received line signal detector
13	Secondary clear to send
14	Secondary transmitted data
15	Transmitter signal element timing
16	Secondary received data
17	Receiver signal element timing
18	nc
19	Secondary request to send
20	Data terminal ready
21	Signal quality detector
22	Ring indicator
23	Data signal rate selector
24	Transmitter signal element timing
25	nc

example, a microcomputer would probably be all of these, while a plotter, for example, would probably be only a listener.

The voltage levels for IEEE-488 are TTL compatible and the logical 0 and logical 1 signals are shown in Figure 4.29. These are active low signals. An exception will be two of the handshake lines (NRFD and NDAC), which are active high signals. We will see later why this is the case.

The IEEE-488 interface uses a 24-pin connector. These have connections on both sides (of different sexes), so the connectors may be stacked, connecting several devices together in parallel as illustrated in Figure 4.30. The status of each device is controlled by the various management lines. Table 4.5 gives the identity of each of the pins on the connector. Data are transmitted in ASCII code over the data lines designated DI01 through DI07 (in order of least- to most-significant bit). DI08 is not used as part of the ASCII code but is used in some devices for transmitting additional information.

It is important to understand the meaning of the various signals, so we explain them briefly.

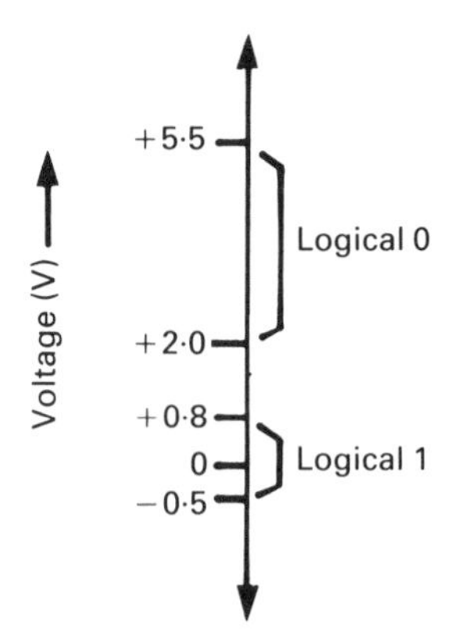

Figure 4.29. Standard logic levels for the IEEE-488 parallel interface.

INTERFACE MANAGEMENT SIGNALS

EOI This has two uses: *identify* (which is a terribly messy problem that we will not delve into any further), and *end.* The end command is used to signify the last character being transferred.

IFC This is a *clear* signal and, when logical 1, will set all other pins to logical 0. Microcomputers or other control devices will frequently set IFC to logical 1 for 100 ms or so when first powered up.

SRQ This is the *service required signal.* It indicates to the controller that a particular device needs service. It is generally not necessary to the operation of the IEEE 488 bus.

ATN This signal tells the various devices on the bus that the characters on the data lines are not to be interpreted according to the ASCII code but rather that they carry special instructions. If you want to use this capability you will need to find out about the specific devices you are using and what instructions you can send them. We will see some examples a little further on.

REN This signal enables remote devices on the bus when it is a logical 1. In many systems this signal is permanently tied low (1).

HANDSHAKING SIGNALS

DAV When a device sends data on the bus, it is the *talker.* At a given time there can be only one talker but there can be many listeners.

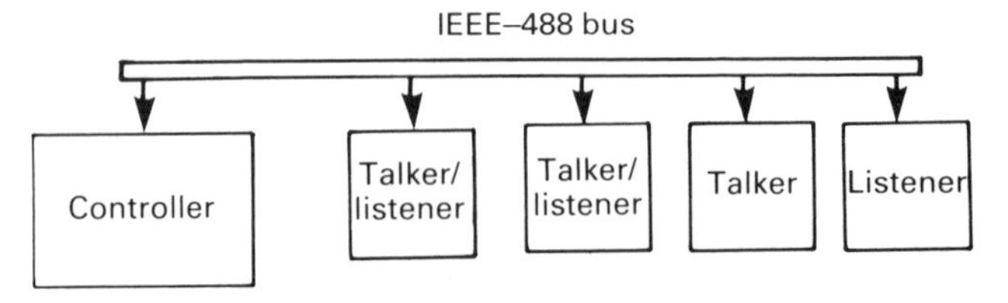

Figure 4.30. The connection of devices on the IEEE-488 bus.

Table 4.5. The IEEE-488 parallel connector pin designations

Pin no.	Designation	Type	Comments
1	DIO1	Data	Least-significant ASCII character.
2	DIO2	Data	
3	DIO3	Data	
4	DIO4	Data	
5	EO1	Management	End or identity
6	DAV	Handshake	Data valid
7	NRFD	Handshake	Not ready for data
8	NDAC	Handshake	Not data accepted
9	IFC	Managemet	Interface clear
10	SRQ	Management	Service request
11	ATN	Management	Attention
12	SHIELD	Ground	Chassis ground
13	DI05	Data	
14	DI06	Data	
15	DI07	Data	Most-significant ASCII bit
16	DI08	Data	Extra data bit
17	REN	Management	Remote enable
18	GND6	Ground	Ground for pin 6 (DAV)
19	GND7	Ground	Ground for pin 7 (NRFD)
20	GND8	Ground	Ground for pin 8 (NDAC)
21	GND9	Ground	Ground for pin 9 (IFC)
22	GND10	Ground	Ground for pin 10 (SRQ)
23	GND11	Ground	Ground for pin 11 (ATN)
24	LOGIC GND	Ground	Ground for pins 5 and 17.

The talker has complete control of the DAV signal and it sets this to a logical 1 when the data it is sending is valid.

NRFD Each *listener* has control of this signal and can use it to signal the talker that it is ready to receive data. Although the signal is called NOT ready for data it indicates to the talker that it IS ready for data when it is a logical 1. The NOT in the name refers to the fact that the logical 1 state for this signal is high rather than low.

NDAC Again the NOT in the name of this signal refers to the logic levels. When this is a logical 1 (high) it signals the talker that each listener has received the data. Each listener has control of this signal.

We see that there are a large number of signals to consider. Fortunately, most controllers (e.g., microcomputers) do the worrying about exactly what signals to send. It is generally sufficient to open the IEEE port for reading or writing and let the computer do the rest.

A typical set of instructions to output the string "IEEE" to the bus might look like this (we have not written the instructions in any particular language for any particular computer so you will have to read the manual

Table 4.6. Commands to send "IEEE" over the IEEE 488 bus

Command no.	IFC	EOI	ATN	Data Lines (DI) 07	06	05	04	03	02	01
1	1	0	0	0	0	0	0	0	0	0
2	0	0	0	0	0	0	0	0	0	0
3	0	0	1	0	1	0	0	1	1	1
4	0	0	0	1	0	0	1	0	0	1
5	0	0	0	1	0	0	0	1	0	1
6	0	0	0	1	0	0	0	1	0	1
7	0	0	0	1	0	0	0	1	0	1
8	0	0	0	0	0	0	1	1	0	1
9	0	1	0	0	0	0	1	0	1	0
10	0	0	1	0	1	1	1	1	1	1
11	0	0	0	0	0	0	0	0	0	0

for your computer to find out specifically how to do this):

1. OPEN IEEE PORT FOR WRITING (PORT #N)
2. WRITE TO PORT #N, "IEEE"
3. CLOSE PORT #N.

The series of instructions sent by the computer might look like those shown in Table 4.6. Note that a "1" in the table represents a logical 1, which is actually a low signal, and a "0" represents a logical 0, or high signal. Let us see what each command does.

Command 1 This clears all the bits on the IEEE bus by setting IFC = 1.
Command 2 Resets IFC = 0—all devices not ready.
Command 3 The ATN = 1 signal indicates that the information or the data lines is a special instruction. This is device specific but the signal shown could be interpreted as follows. The 1 bit in DI06 indicates that a port is to be opened for writing; the last three bits (111) indicate the device number of the port, in this case 7.
Command 4 ASCII "I" data is sent.
Command 5 ASCII "E" data is sent.
Command 6 ASCII "E" data is sent.
Command 7 ASCII "E" data is sent.
Command 8 ASCII carriage return.
Command 9 ASCII Line Feed—the EOI = 1 bit also indicates that this is the last ASCII character to be sent.
Command 10 The ATN = 1 bit indicates a special control signal. The particular bits shown here are only an example. At this point we want to inform all listeners that the controller is relinquishing control.
Command 11 Reset all lines to 0.

The details of the instructions will depend on what specific devices are being used, but the above will give you an idea of what the communications procedure might be like.

RS232 versus IEEE-488. The IEEE bus is clearly superior in terms of data transmission rate. In principle, up to 10^6 bytes per second can be transmitted. In practice, laboratory devices are more likely to transmit at ~5 kbytes per second. This is still much faster than the maximum baud rate over an RS-232 line.

IEEE has some serious disadvantages, however. It is much more susceptible to interference than is the RS-232 line, and as a result it can only be used over fairly short distances (20 meters is an absolute maximum) and in environments where electrical interference is a minimum. RS-232, on the other hand, can be used for much longer distances and is not so susceptible to electrical interference.

c. *Serial-to-parallel conversion and parallel-to-serial conversion*

This is a rather complex business and only the basics will be discussed here.

Serial-to-parallel conversion. Consider the circuit shown in Figure 4.31. This is sometimes known as a *shift register*. This is a 4-bit version. The four output bits are initially zeroed with a clear = 0 pulse. Now each of the four serial bits is input, each accompanied by a clock pulse; the least-significant bit first. The first clock pulse will set b_3 equal to the first bit. The second clock pulse will shift the first bit to b_2 and set b_3 equal to the second bit. This continues until the output $b_3b_2b_1b_0$ is equal to the four input bits in the proper order. Table 4.7 shows the state of $b_3b_2b_1$ and b_0 for the serial input 1011 after each clock pulse.

Parallel-to-serial conversion. The parallel-to-serial converter is a type of shift register. Figure 4.32 shows the circuit for a 4-bit device. The state of

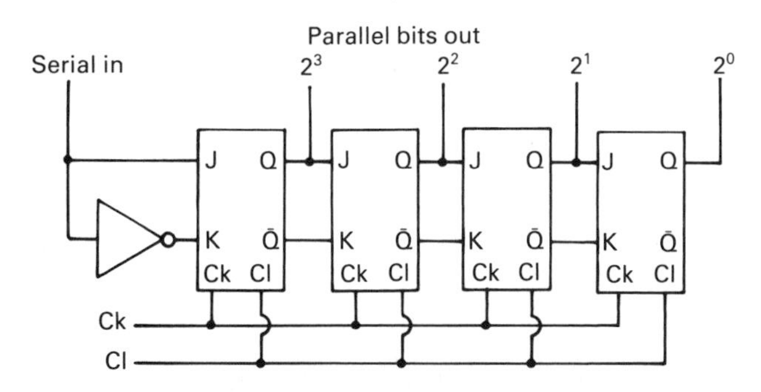

Figure 4.31. A serial-to-parallel converter.

Table 4.7. Operation of a shift register. The series of pulses 1101 was input

Clock pulse	b_3	b_2	b_1	b_0
0	0	0	0	0
1	1	0	0	0
2	1	1	0	0
3	0	1	1	0
4	1	0	1	1

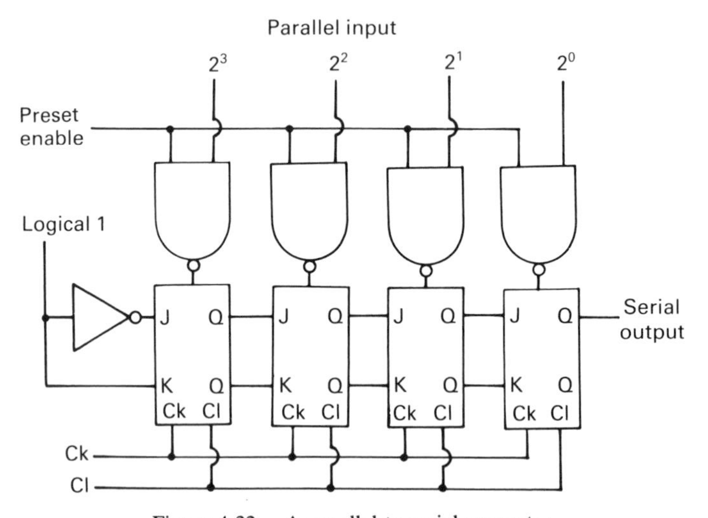

Figure 4.32. A parallel-to-serial converter.

the output of each flip flop (Q) is set by inputting the parallel signal into b_3, b_2, b_1 and b_0 at the same time as a preset enable pulse. A preset "high" pulse coincident with a preset enable pulse will drive the Pr input low and hence Q will go high. Now the serial representation of the number may be read at Q of the lowest-order flip-flop by applying four clock pulses. Each time a clock pulse is input, the register shifts and the next bit appears at the output.

5
Vacuum technology

5.1 Vacuum pumps

The family tree of vacuum pumps shown in Figure 5.1 consists of two major categories—compression pumps and entrapment pumps. The former remove gas molecules from the chamber to be evacuated and convey them to the atmosphere; the latter condense or chemically bind the molecules in a container (this is sometimes even within the chamber to be evacuated). Compression pumps can be subdivided into positive displacement pumps, and kinetic pumps. As Figure 5.1 shows, these may be divided even further. This figure is not intended to be comprehensive, but it does show the relationships of the more common types. Many of these are described in some detail below.

Before discussing specific types of pumps, we begin with some general comments on pumping speed and pressure units.

a. Pumping speed

The change in the pressure within a volume C is given by

$$\frac{dP}{dt} = -S\frac{P - P_s}{C} \tag{5.1}$$

where P_s is the lower limit of the pressure attainable with the pump and S is the speed of the pump. Integration of expression (5.1) yields

$$t = \frac{C}{S}\ln\frac{P_0 - P_s}{P_t - P_s} \tag{5.2}$$

where P_0 is the pressure at $t = 0$ and P_t is the pressure at time t. We can rewrite (5.2) to give

$$P_t = P_s + (P_0 - P_s)e^{-St/C} \tag{5.3}$$

It is clear from this expression that the speed of the pump represents a quantity in units of volume per time. As we shall see later, it is common to connect two vacuum pumps in series so that the exhaust of one is connected to the intake of the other. We can look at this phenomenologically as

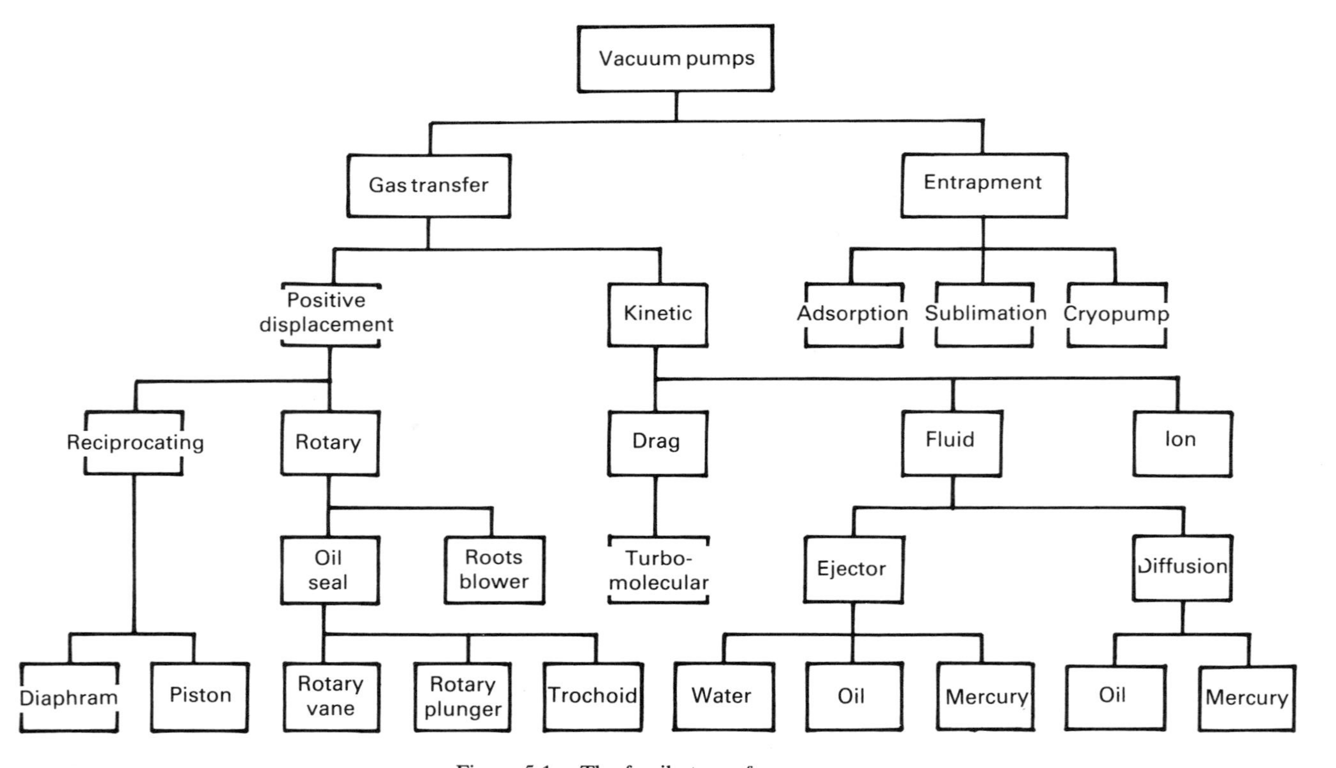

Figure 5.1. The family tree of vacuum pumps.

follows. Gas molecules that leak into the chamber of volume C must go through both pumps to be expelled to the atmosphere. Pump 1 pumps gas at pressure P_0 while pump 2 pumps gas at pressure P_i. In order that the same number of molecules are transported by each pump we must have

$$\frac{S_1}{S_2} = \frac{P_i}{P_0} \tag{5.4}$$

or

$$S_1 = S_2 \frac{P_i}{P_0} \tag{5.5}$$

If we use pumps so that $S_1 > S_2 P_i / P_0$ or $S_2 > S_1 P_0 / P_i$ then the pumping speed is limited by the speed of the slower pump. There are slight modifications to equation (5.5) that result from "resistance" to the gas flow caused by the pipes that connect the pumps together. A typical example may be $P_0 \sim 10^{-3}\, P_i$ (as we shall see later), so if $S_2 = 1\ \mathrm{l\,s^{-1}}$ we would require $S_1 = 1000\ \mathrm{l\,s^{-1}}$ to make full use of all the components of the system. The actual size of the pumps necessary to evacuate a particular chamber depends on the size of the chamber, on how leaky it is, and on how long we are willing to wait for it to pump down.

b. *Notes on vacuum units*

Pressure is a force per unit area. The accepted SI unit of pressure is the Pascal (Pa) defined as $1\ \mathrm{N\,m^{-2}}$. Traditionally, however, vacuum gauges use any of a variety of units; conversions are shown in Table 5.1. The origin of some of these units is as follows.

The bar is a CGS unit equal to 10^6 dyne cm^{-2}.

The atmosphere is, of course, from standard atmospheric pressure.

The Torr is defined as 1 mm of mercury.

The micron is a shortened form of μmHg and is 10^{-3} Torr, the same as mTorr.

PSI (pounds per square inch) is the British unit and is often still seen on inexpensive Bourdon type gauges that cover a large pressure range.

Table 5.1. Pressure unit conversion table

bar	atm	micron	mmH$_2$O	Pa	Torr	mTorr	PSI
bar	1	0.987	7.5×10^5	1.02×10^4	10^5	750	7.5×10^5
atmosphere	1.013	1	7.6×10^5	10332.6	101325	760	7.6×10^5
micron	1.33×10^{-6}	1.32×10^{-6}	1	0.0136	0.1333	10^{-3}	1
mmH$_2$O	9.8×10^{-5}	9.68×10^{-5}	73.55	1	0.102	0.0736	73.55
Pa	10^{-5}	9.87×10^{-6}	7.5	9.804	1	7.5×10^{-3}	7.5
Torr	1.33×10^{-3}	1.316×10^{-3}	10^3	13.596	133.3	1	10^3
mTorr	1.33×10^{-6}	1.32×10^{-6}	1	0.0136	0.1333	10^{-3}	1
PSI	6.89×10^{-2}	6.804×10^{-2}	51706	703.01	6892	51.706	51706

c. *Positive-displacement pumps*

Rotary vane pumps. Rotary vane pumps have a disk-shaped rotor that turns inside a cylindrical chamber. The axis of rotation of the disk is at the center of the disc but not at the center of the chamber. Figure 5.2 shows the principle. Seals are made at the contact between the rotor and the cylinder and between the two vanes and the cylinder. The vanes are "spring loaded" so that as the rotor rotates they remain in contact with the cylinder. The whole thing is sealed by covers on either end. The making of the seals is improved by a quantity of oil. The pump shown in the figure is a two-stage rotary vane pump. The operation of the rotary vane pump is demonstrated in Figure 5.3. As the rotor rotates, the vanes alternately seal off areas in the chamber that are connected to the intake and the exhaust. This enables gas molecules to be transported from the chamber to be evacuated to the atmosphere.

Figure 5.3 illustrates something else too: the operation of the *gas ballast valve*. Frequently, rotary vane and similar types of positive-displacement pumps have a gas ballast valve that works as follows. Normally, the gas molecules that are being transported from the chamber to be evacuated to the atmosphere are at a very low pressure. As a result, it is not easy to force the molecules out into the atmosphere. This is irrelevant when pumping on air, but suppose we were pumping something that contained larger molecules, for example, water vapor. It would be difficult to get rid of these molecules; they would tend to condense inside the pump. The gas ballast helps here. After the vane has moved past the intake port, the ballast valve is opened and air is admitted to the volume of gas being transported. Thus, as the figure shows, when the exhaust is forced out of the pump the vapor goes along with the air. Many pumps have a gas ballast valve that can be opened or closed. It should be opened when pumping on condensable vapors.

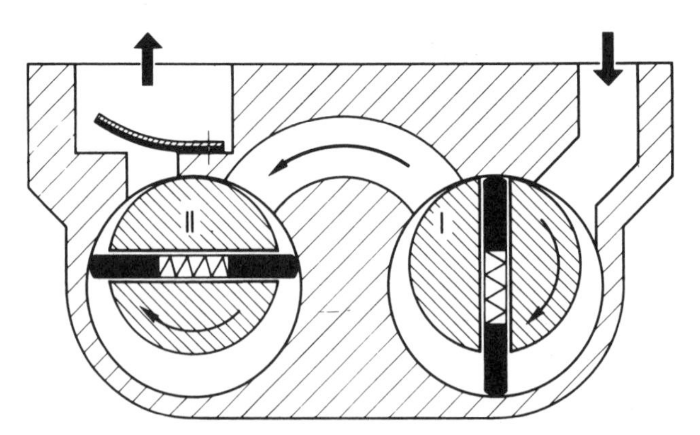

Figure 5.2. Cross section of a two-stage rotary vane pump. (Copyright © 1985 by Leybold–Hereaus Vacuum Products Inc. Used with permission.)

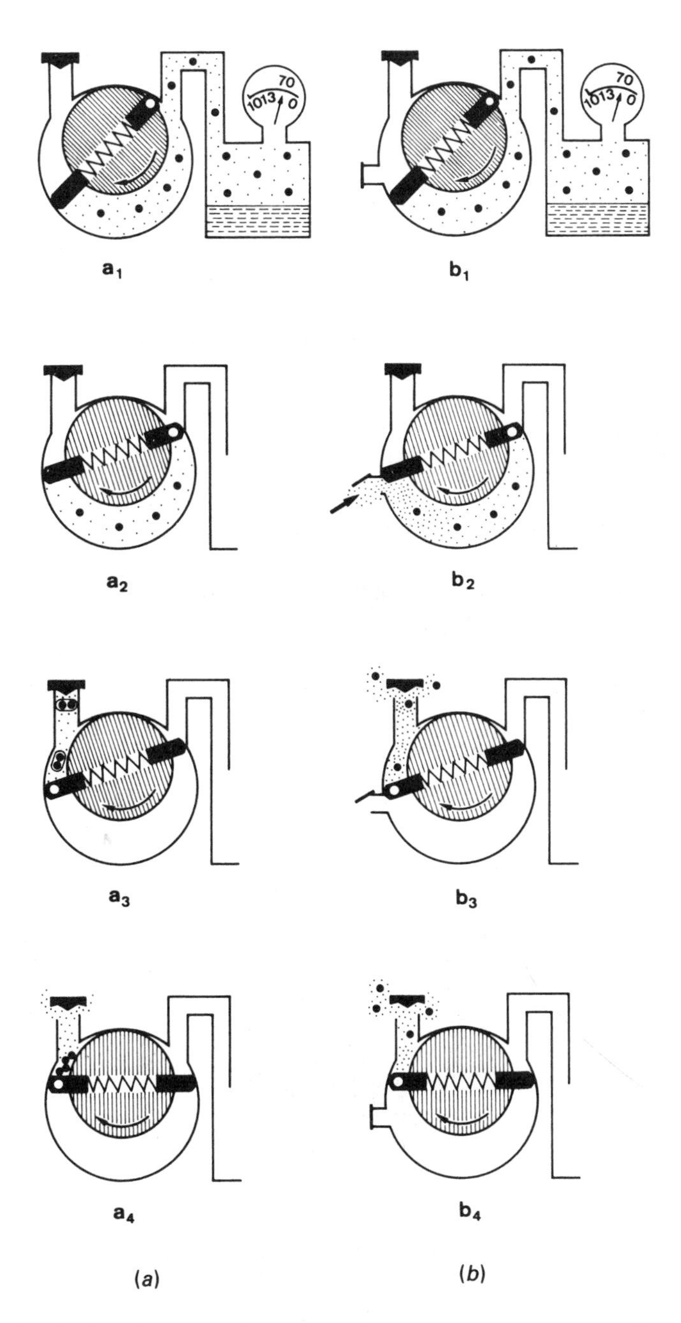

Figure 5.3. Illustration of the pumping process in a rotary vane pump (*a*) without the gas ballast valve in use and (*b*) with the gas ballast valve in use. (Copyright © by Leybold–Hereaus Vacuum Products Inc. Used with permission.)

Rotary plunger pumps. Rotary plunger pumps are similar to rotary vane pumps except that the rotor rotates on a shaft that is concentric with the inside of the chamber but not with the rotor. The valve is attached to the chamber and is forced against the outside of the rotor by a spring mechanism. Again seals are accomplished by means of oil. Figure 5.4 shows the general design. Seals are made between the end of the plunger and the rotor, and between the edge of the rotor and the inside of the chamber. Valving is performed by a hole on the side of the plunger mechanism.

A similar type of pump is shown in Figure 5.5. This is the type manufactured by Cenco (a two-stage pump is shown). The operation is similar to that of the pump shown in Figure 5.4 except that the valving is performed by a check valve (indicated D in the figure).

Trochoid pump. The trochoid pump has an elliptically shaped rotor that turns by means of the gear arrangement shown in Figure 5.6. The chamber is roughly heart shaped and the rotor remains in contact with it at either end. This may seem like a rather elaborate arrangement but it is better balanced and vibrates less than the two previous kinds of pumps.

All three of these pumps are similar in their uses. They will all produce vacuums in the 1–100 μg Hg range and will all exhaust to the atmosphere. Figure 5.7 shows the operating ranges of these and other pumps. The first two types discussed are the most commonly encountered and they are all

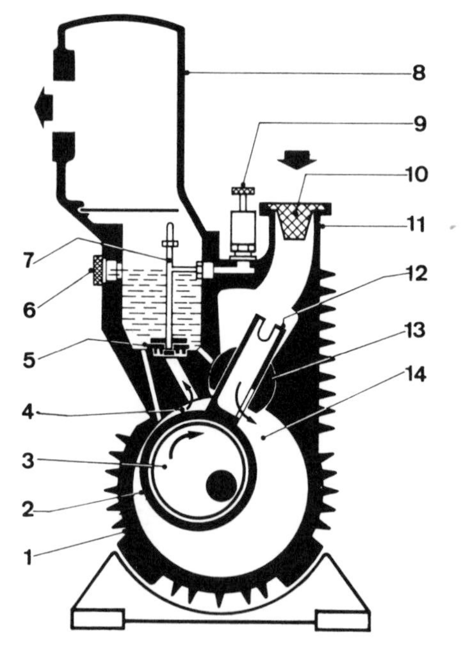

Figure 5.4. Cross-sectional view of a rotary plunger pump. (Copyright © 1985 by Leybold–Hereaus Vacuum Products Inc. Used with permission.)

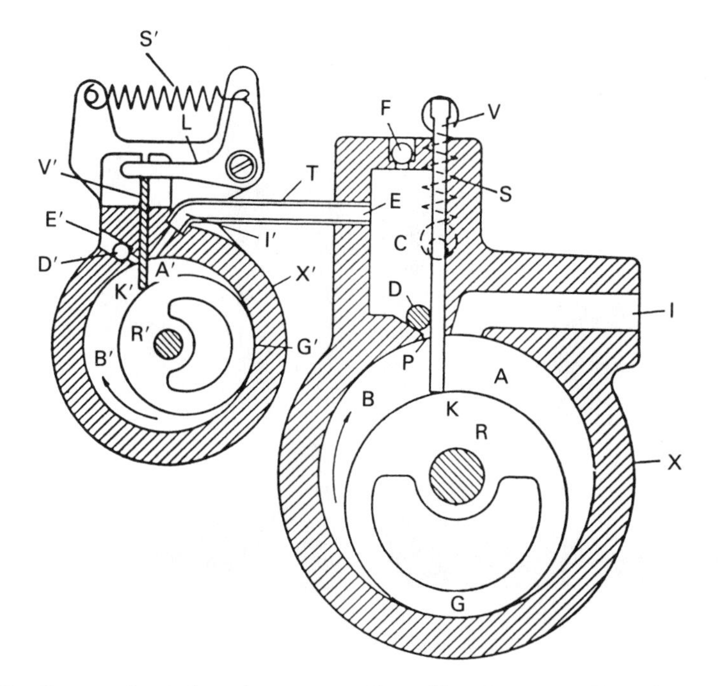

Figure 5.5. Cross-sectional view of a two-stage Cenco Hypervac pump. The valves are shown by D and D′ in the figure. (From S. Dushman, *Vacuum Technique*. Copyright © 1949 by John Wiley & Sons. Used with permission.)

frequently referred to as "roughing pumps", "backing pumps" or "fore-pumps" since they are often used in conjunction with another kind of pump to produce a better vacuum.

Roots blower. The roots blower consists of an oval chamber with two "figure-of-eight" shaped rotors as shown in Figure 5.8. The two rotors rotate in opposite directions and are synchronized by a gear arrangement. The rotors do not touch each other or the walls of the chamber; the clearance is a fraction of a millimeter. Hence the pump can be run at high speed without mechanical wear.

Although the roots blower can displace large volumes of gas per unit time, the fact that there are no real "seals" between the parts means that it cannot operate with a large pressure difference between the intake and the exhaust. It must therefore be used in conjunction with a roughing pump.

d. Kinetic pumps

Oil vapor diffusion pumps. Diffusion pumps are not mechanical devices in the same sense as the previously described pumps, since the only moving part is the oil.

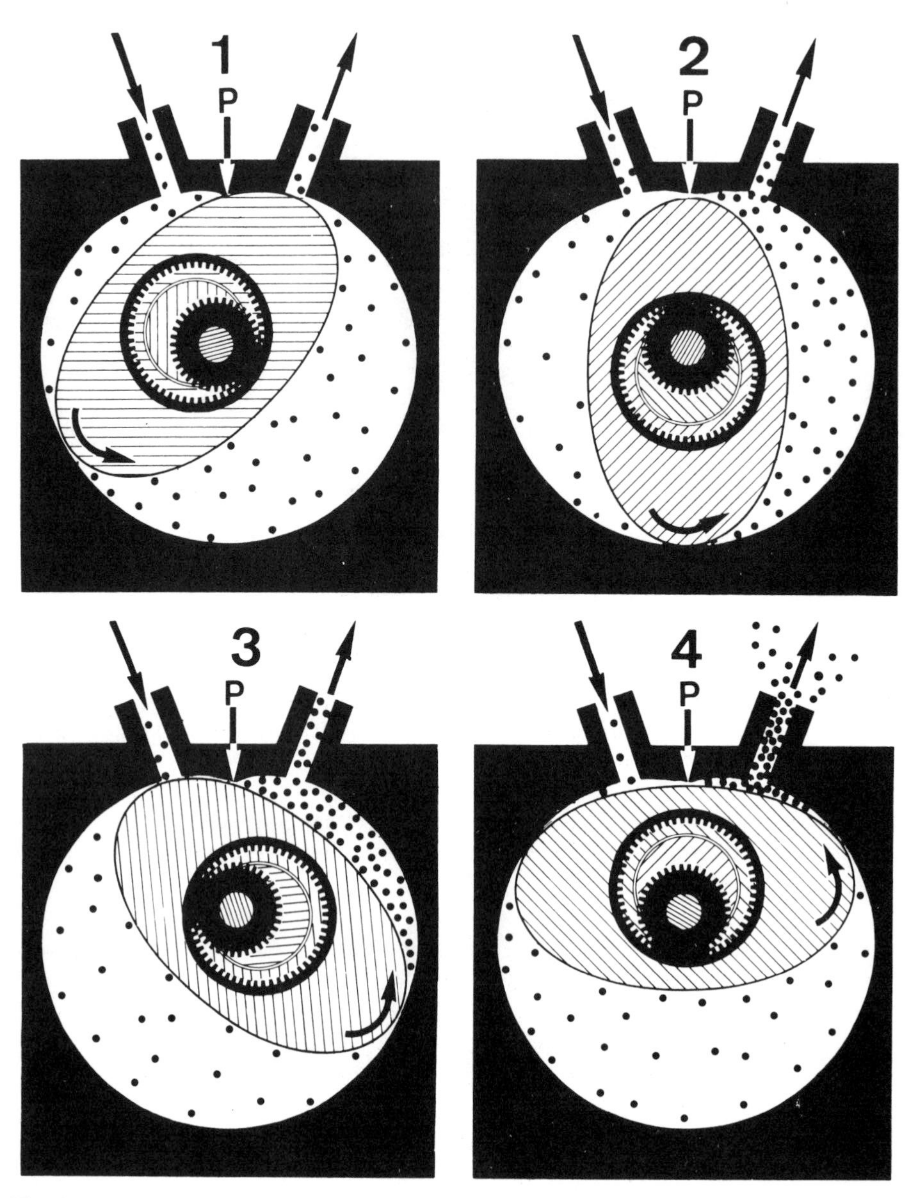

Figure 5.6. The operation of the trochoid pump. (Copyright © by Leybold–Hereaus Vacuum Products Inc. Used with permission.)

The general design is shown in Figure 5.9. It consists of an outer cylinder that is usually metal but can be glass. There is a heater at the bottom of the cylinder and a layer of oil (usually a couple of centimeters or so in a small pump). Inside is a metal construction that we will discuss in detail later. Outside there are usually water pipes to keep the walls of the cylinder cool;

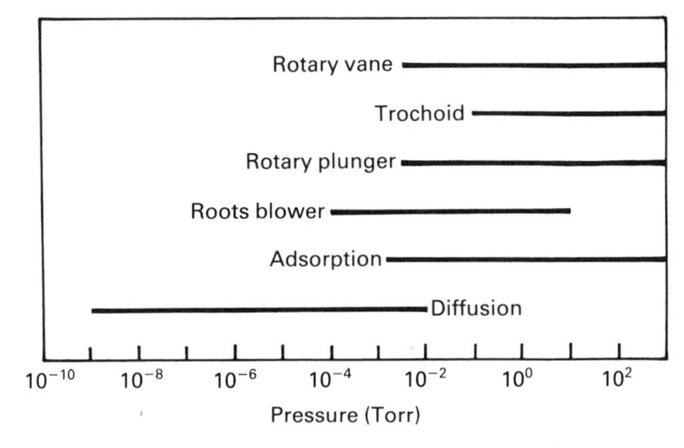

Figure 5.7. Operating range for various types of pump.

in smaller pumps there may simply be metal fins and a fan to keep it cool. The intake is at the top of the cylinder and the exhaust is through a pipe on the side, not too far from the bottom but high enough to be above the oil level.

The heater heats the oil. This vaporizes and travels up the inside of the metal construction in the center of the pump. Figure 5.10 shows in more detail what this metal insert, called the *nozzle system,* looks like. The movement of the oil vapor up through the nozzle system is very much like the way a coffee percolator works. Flanges direct the flow of vapor outwards towards the pump wall. At the same time the flow is directed somewhat downwards. A pump would usually have 2–4 nozzles or flanges. These are referred to as stages, so Figure 5.9 represents a four-stage pump. Gas molecules are carried downward along with the flow of oil vapor. The oil

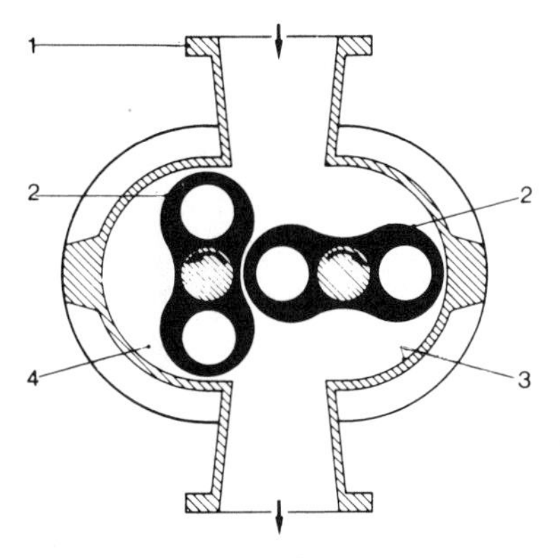

Figure 5.8. The roots blower. (Copyright © by Leybold–Hereaus Vacuum Products Inc. Used with permission.)

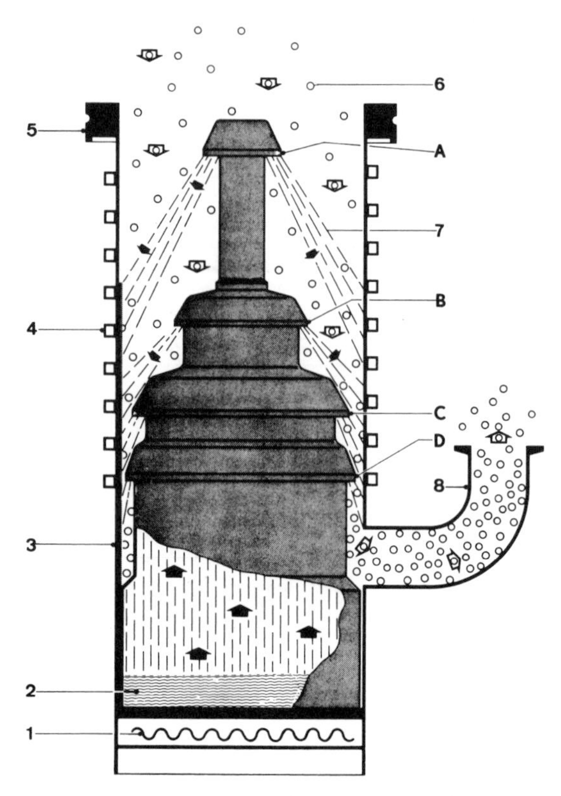

Figure 5.9. The operation of the oil vapor diffusion pump. (Copyright © by Leybold–Hereaus Vacuum Products Inc. Used with permission.)

condenses on the cooled walls and runs back into the reservoir at the bottom of the pump.

The gas molecules are pushed downward and out through the exhaust port. In a fractionating type of pump, divisions between the nozzle elements extend down into the oil as shown in Figure 5.10*a*. Thus the quantity of oil vapor traveling to each nozzle is controlled. This yields a somewhat lower ultimate pressure than the nonfractionating type, shown in Figure 5.10*b*, in which the inside of the nozzle system is not divided.

As you can see in Figure 5.7, oil vapor diffusion pumps can pump to exceedingly low pressures. They can also pump very large volumes. However, they cannot operate with their exhaust at above a few hundred mTorr. They must therefore be used with a backing pump. An oil vapor diffusion pump along with a rotary vane or rotary plunger backing pump forms the most common high-vacuum system. The construction and operation of such a system will be discussed later.

Mercury vapor diffusion pumps. The construction and operation of a mercury diffusion pump is essentially the same as for the oil vapor diffusion

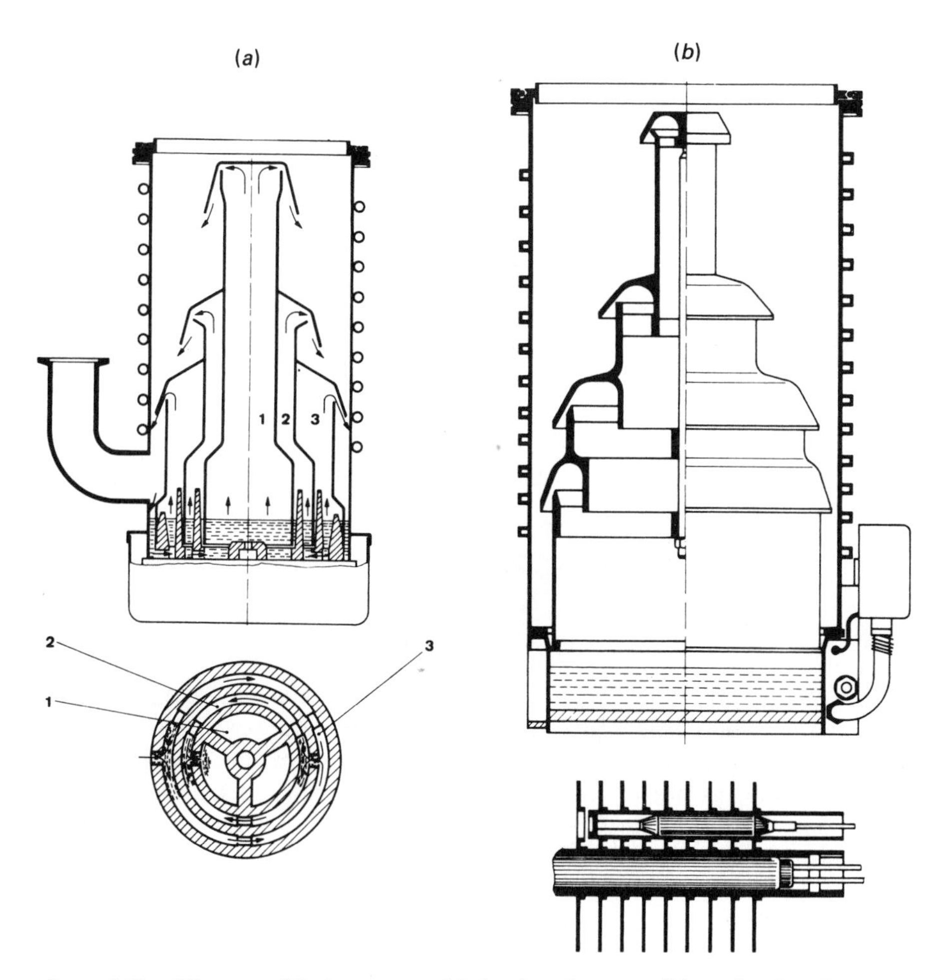

Figure 5.10. Oil vapor diffusion pumps: (*a*) fractionating type; (*b*) nonfractionating type. (Copyright © by Leybold–Hereaus Vacuum Products Inc. Used with permission.)

pumps with one important difference. The vapor pressure of mercury is several orders of magnitude higher than that of most oils. As a result, pumps of this sort are most commonly cooled with liquid nitrogen to facilitate the condensation of the vapor on the pump walls. The use of these pumps is becoming increasingly less common because of the toxicity of mercury vapor.

Adsorption pumps. Adsorption pumps remove gas by adsorbing the gas molecules on to the surface of an adsorption material, commonly a zeolite. Zeolites are alkali aluminosilicates that have a surface area of about 10^3 m^2 g^{-1}. All of the gas contained in 1 litre at atmospheric pressure can be adsorbed on about 10 g of zeolite. The adsorption process is only effective

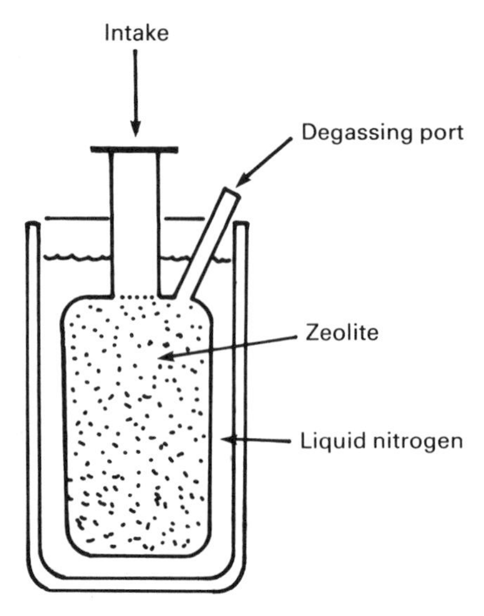

Figure 5.11. The adsorption pump.

at low temperature, so these kinds of pumps (sometimes called sieve pumps) are immersed in liquid nitrogen. The general design is illustrated in Figure 5.11. These pumps contain a few kilograms of zeolite and can adsorb a considerable quantity of gas. However, the zeolite must eventually be replaced or degassed. The lower pressure limit on these pumps is somewhat better than for roughing pumps but not as good as for oil vapor diffusion pumps. These are of particular use in cases where small quantities of oil impurities (from oil-type pumps) could cause problems in an experiment: surface physics is a field in which this could be the case.

5.2 Vacuum and pressure gauges

We consider several of the more commonly used types of pressure and vacuum gauges.

a. Manometers

Manometers are devices that measure pressure by determining the height of a column of liquid. In the interest of conservation of space, mercury is commonly used because of its high density. These types of gauges have essentially been replaced by more "modern" types and are now rarely used as they are less convenient and are vulnerable to breakage. However, the principle of their operation is of interest. A simple type is pictured in Figure 5.12. The "U" shaped tube is partially filled with mercury. On one end a

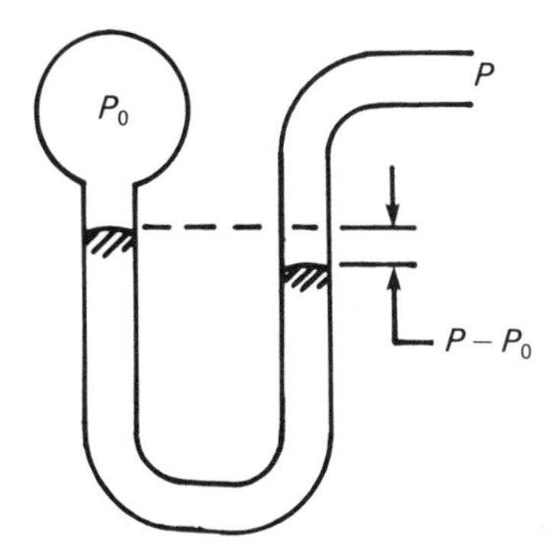

Figure 5.12. A simple mercury manometer.

bulb is evacuated to a known pressure (generally the room-temperature vapor pressure of mercury, $\sim 10^{-3}$ Torr. The difference, in millimeters, in the levels of the mercury on the two sides is equal to the difference in the pressures P and P_0 in millimeters of mercury (mm Hg). In this way the pressure can be measured from ~0.1 mm Hg to atmospheric pressure with an accuracy of a fraction of a millimeter of mercury.

b. McLeod gauge

A device related to the mercury manometer but that is more useful at low pressures is the McLeod gauge. The principal of measurement is similar to that of the manometer except that a low pressure is measured by compressing a large volume of the low-pressure gas into a smaller volume of high-pressure gas by the use of the column of mercury. The device is illustrated in Figure 5.13 and it works as follows. The reservoir is lowered so that the level of mercury is below point 1. In this case the pressure P is connected to the bulb of volume V and this bulb is then full of gas at this pressure. The volume of gas trapped is then $\sim V$. The reservoir is then raised above point 2. The gas trapped in V is compressed in the capillary (a). This capillary has a diameter d. The reservoir is raised so that the level of the mercury in capillary (b) is level with the top of capillary (a). At this point the difference in the levels is given by h. Therefore the pressure of gas in capillary (a) is given by

$$P_a V_a = nRT \tag{5.6}$$

or

$$\tfrac{1}{4} P_a \pi d^2 h = nRT \tag{5.7}$$

The number of moles of gas is given by

$$n = \frac{PV}{RT} \tag{5.8}$$

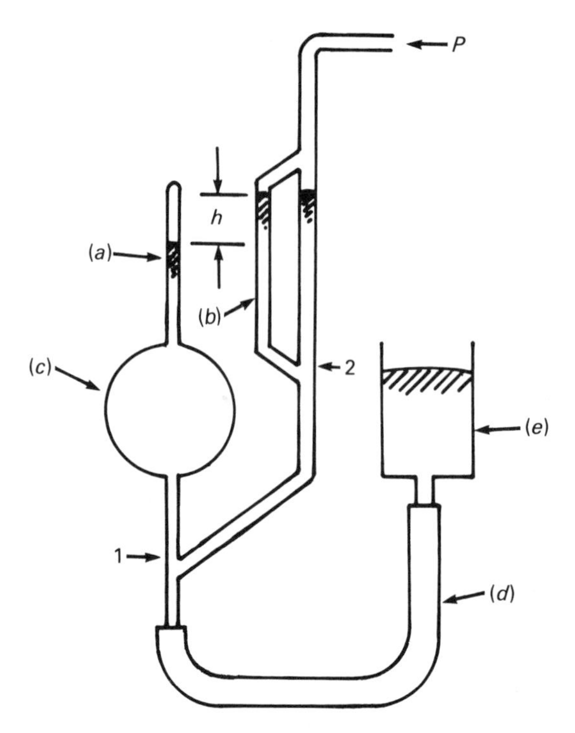

Figure 5.13. The McLeod gauge: (*a*) measurement capillary; (*b*) comparison capillary; (*c*) bulb of volume *V*; (*d*) flexible tube; (*e*) mercury reservoir; (*P*) pressure to be measured.

and equation (5.8) becomes

$$\tfrac{1}{4}P_a \pi d^2 h = \frac{PV}{RT}RT \tag{5.9}$$

We know that the difference in the levels of the mercury on the two sides is related to the difference in the pressures,

$$(P_a - P) = h \tag{5.10}$$

when the pressures are measured in millimeters of mercury and h is in millimetres. Now we write

$$P_a = h + P \tag{5.11}$$

and equation (5.9) becomes

$$\tfrac{1}{4}(h + P)\pi d^2 h = PV \tag{5.12}$$

Solving for P, we find

$$P = h\left(\frac{4V}{\pi d^2 h} - 1\right)^{-1} \tag{5.13}$$

It is often the case that $4V/(\pi d^2 h) \gg 1$ and (5.13) is then written as

$$P = \frac{\pi h^2 d^2}{4V} \tag{5.14}$$

There are variations on this method: often an arrangement in which the whole device rotates is used in place of the raising and lowering of the reservoir.

c. *Bourdon gauge*

The Bourdon gauge consists of a spiral shaped tube attached to an indicator needle. As the pressure in the tube changes relative to the pressure outside of the tube, the spiral winds or unwinds and the needle indicates the pressure on a circular scale. Figure 5.14 shows a fairly elaborate Bourdon gauge that uses two spiral tubes, and a gear arrangement for driving the needle. These devices can range from very inexpensive ones that are accurate to no better than 50 mm Hg or so, to fairly expensive ones that are accurate to better than 1 mm Hg. There are devices of this kind that are useful for measuring pressures above atmospheric pressure. One fault is that, unless the volume around the spiral tube is sealed at a known and fixed pressure, or unless two tubes are used as in Figure 5.14, then the reading obtained with such a gauge depends upon the atmospheric pressure.

All of the above gauges use direct methods to measure the pressure; that is, the force caused by the pressure is measured directly. They also have

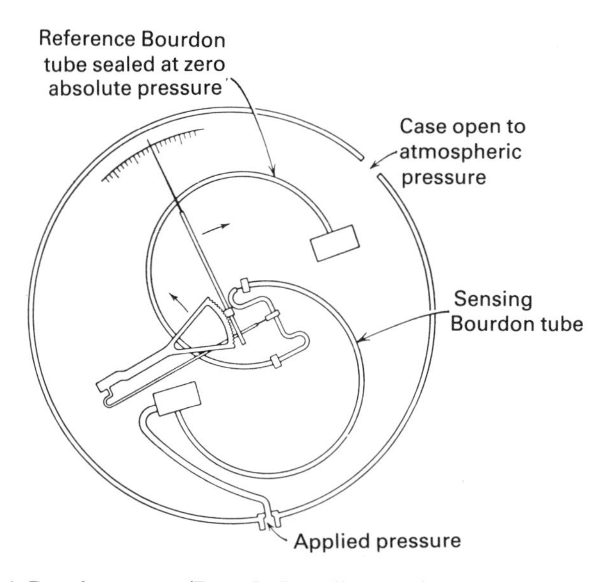

Figure 5.14. A Bourdon gauge. (From P. Benedict, *Fundamentals of Temperature, Pressure and Flow Measurements*. Copyright © 1969 by John Wiley & Sons. Used with permission.)

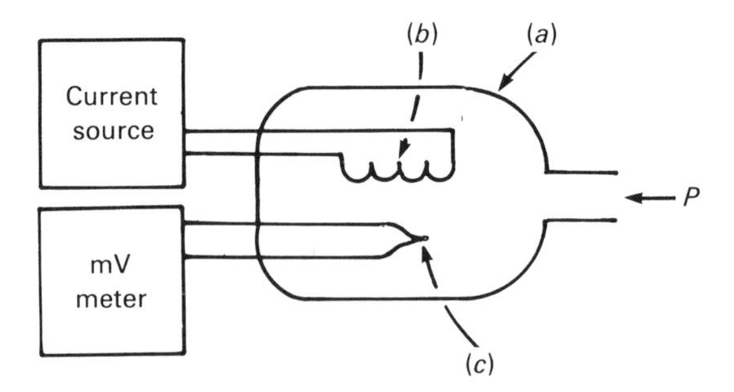

Figure 5.15. The thermocouple gauge: (*a*) enclosure; (*b*) filament (heater); (*c*) thermocouple; (*P*) pressure to be measured.

something else in common—the pressure they measure is independent (as it should be) of the kind of gas present. The remaining types of gauges are all of the "electrical" type and do not measure the pressure itself, but rather something related to the pressure. They are, therefore, indirect gauges. They give a result that must be calibrated for the type of gas present. However, they are the types of gauges in most common use in current vacuum technology.

d. *Thermocouple gauges*

Thermocouple gauges (also called Pirani gauges) measure the pressure of a gas by measuring its thermal conductivity. The thermal conductivity is not only a function of the pressure but also depends on the nature of the gas molecules. A typical arrangement is illustrated in Figure 5.15. A constant current is supplied to the filament. The temperature the thermocouple measures is related to the conductivity of the medium between the filament and the thermocouple. Figure 5.16 shows a plot of the heat transferred

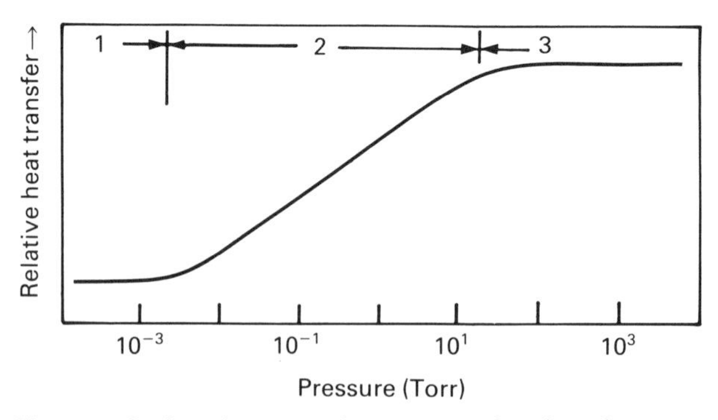

Figure 5.16. Heat transfer in a thermocouple gauge as a function of pressure. The dominant mechanism for heat transfer in the three regions is (1) conductivity of the wires, (2) conductivity of the gas and (3) convection in the gas.

between the filament and the thermocouple as a function of pressure. In region 1 the heat transfer is dominated by conduction through the electrical wires; in region 2 it is dominated by the thermal conductivity of the gas; and in region 3 it is dominated by convection. It is only in region 2 that there is a direct relationship between heat transfer and pressure and the thermocouple gauge is useful only in this region. This region is between about 1 mTorr and 50 Torr. This is probably the most commonly used gauge for this pressure range. Gauges are usually calibrated for nitrogen (essentially same as for air). Manufacturer's specifications will tell you how to convert to the correct pressure from the measured pressure if the system contains some other gas.

e. *Ionization gauges*

Ionization gauges measure a current of ions formed by the ionization of gas molecules. This ionization is caused by either large electric fields, or electrons emitted by a hot cathode. These are referred to as cold-cathode and hot-cathode gauges, respectively.

Cold-cathode gauges. Cold-cathode gauges (also called Penning gauges) contain a cathode made of an active material (such as thorium or zirconium) that emits electrons as a result of a large potential (~2 kV) between it and the anode. A magnetic field is supplied by a permanent magnet in such a way that the electrons spiral their way between the cathode and anode; their total path is several hundred times the actual distance between the two electrodes. In this process they cause considerable ionization. This ion current is proportional to the pressure. A simple arrangement is shown in Figure 5.17.

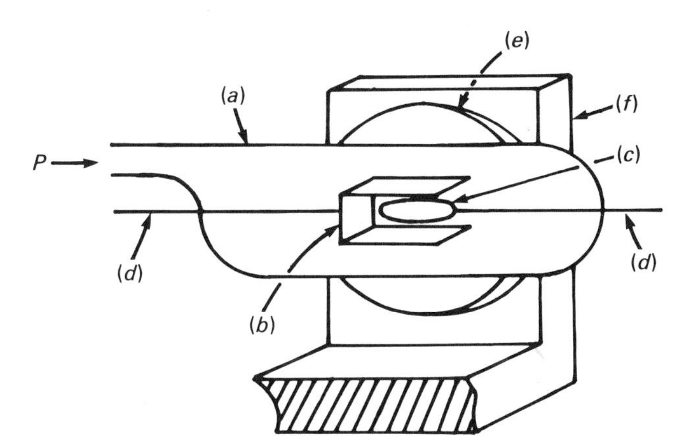

Figure 5.17. The cold-cathode ionization gauge (Penning gauge): (*a*) enclosure; (*b*) anode plates; (*c*) ring-shaped cathode; (*d*) electrical connections; (*e*) magnet (1 of 2 shown); (*f*) magnetic yoke (half shown).

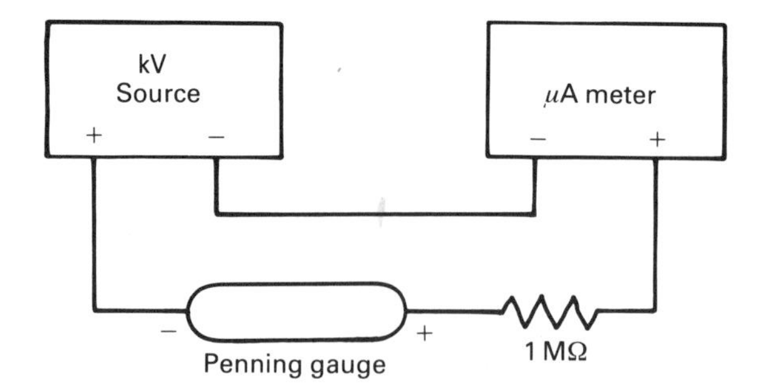

Figure 5.18. Wiring diagram for the cold-cathode ionization gauge.

The wiring diagram is shown in Figure 5.18. The gauge has good sensitivity in the range of 10^{-5} to 10^{-3} Torr, this is illustrated in Figure 5.19. The advantages of this kind of gauge are that (1) it is inexpensive; (2) it is fairly insensitive to mechanical shock; and (3) there is nothing to burn out if it is left on at too high a pressure. Unfortunately, it has the disadvantage that it will not measure pressures as low as we might want to if we are using a diffusion pump.

Hot-cathode gauge. In the hot-cathode gauge a filament (the cathode) provides large quantities of electrons, and the need for high voltages and electric fields is eliminated (Figure 5.20). The electrons cause ionization as they are accelerated from the cathode to the anode, not so much because of the large accelerating voltages (here we have only a hundred volts as so) but because of the large quantity of them. Ions that are formed are further accelerated towards the ion collector, which is at ground potential, and the current is measured by an appropriate meter. These gauges will read from about 10^{-10} Torr (10^{-13} Torr in some cases) up to about 10^{-3} Torr. There is

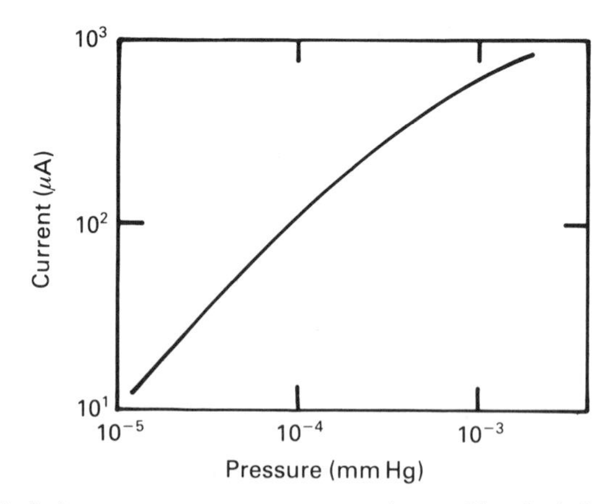

Figure 5.19. Typical current versus pressure curve for a cold-cathode ionization gauge.

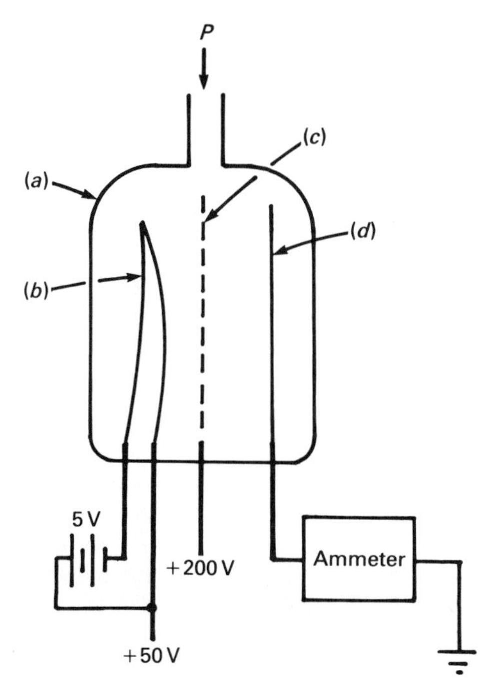

Figure 5.20. The hot-cathode ionization gauge: (*a*) enclosure; (*b*) cathode (filament); (*c*) anode; (*d*) ion collector; (*P*) pressure to be measured.

one danger: the filament must never be left on when the pressure exceeds 10^{-3} Torr, as there is a possibility of burning it out.

The current measured in both these types of ionization gauges is related not only to pressure of the gas but also to its ionization potential. The conversion of the measured pressure to the actual pressure if the gauge is

Table 5.2. Ionization gauge calibration factors for different gases

Gas present	Multiply reading on air calibrated gauge by this factor
He	6.0
Ne	3.7
Ar	0.71
Kr	0.50
Xe	0.33
Hg	0.27
H_2	1.8
CO	0.85
CO_2	0.59
CH_4	0.7

not calibrated for the kind of gas present is not straightforward. Besides, it is to some extent geometry dependent and varies from gauge to gauge. However, a reasonable estimate of the pressure can be obtained using the conversion factors given in Table 5.2.

5.3 Design and operation of vacuum systems

For applications in which a rough vacuum is required, a roughing pump may be used by itself and the gas may be exhausted to the atmosphere. In many cases, however, a better vacuum is needed; it is then necessary (usually) to resort to using two pumps in conjunction. The most common arrangement is an oil diffusion pump backed by a rotary vane or rotary plunger pump. A typical arrangement is illustrated in Figure 5.21.

The roughing pump will pump to about 10^{-2} Torr or so and the diffusion pump with its exhaust at this pressure will pump the chamber down to 10^{-6} Torr or so.

Turning the system on and off requires a particular procedure. The following is the method of turning on a system that is cold and at atmospheric pressure.

I. Turning on roughing pump.
   1. Make sure all valves are closed.
   2. Turn on roughing pump switch.

II. Diffusion pump heat up.
   3. Turn on TC #1 and make sure pressure goes down to $\sim 10^{-2}$ Torr.
   4. Open valve #1 make sure pressure returns to $\sim 10^{-2}$ Torr.

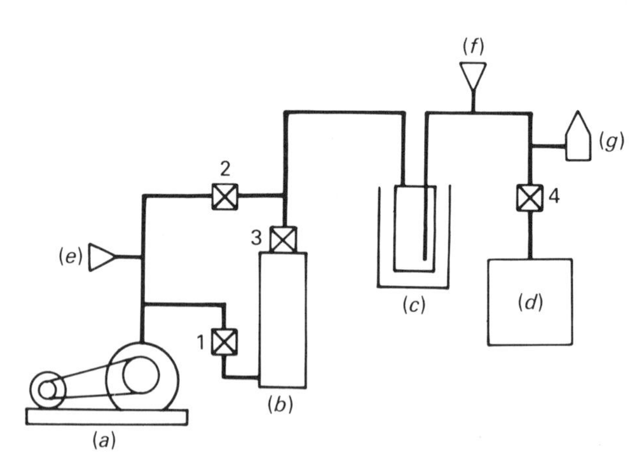

Figure 5.21. A conventional high-vacuum system: (*a*) roughing pump, (*b*) diffusion pump; (*c*) cold trap; (*d*) chamber to be evacuated; (*e*) thermocouple (TC) gauge #1; (*f*) thermocouple gauge (TC) #2; (*g*) ionization gauge. ⊠ = valve.

5. Turn on diffusion pump. This will take 10–30 min to warm up. If it is a water-cooled pump, make sure the water is ON. If it is an air-cooled pump, make sure the fan is on.

III. Rough out chamber.

6. Close valve #1.
7. Open valves #2 and #4.
8. Turn on TC #2 and make sure pressure goes down to $\sim 10^{-1}$ Torr at TC #2.

IV. Evacuate chamber.

9. Close valve #2.
10. Open valve #1.
11. Make sure pressure at TC #1 $\sim 10^{-2}$ Torr.
12. Make sure pressure at TC #2 is not rising too fast.
13. Open valve #3.
14. Ensure that pressure at TC #2 ~ 0 Torr.
15. Turn on ionization gauge.

V. Cold trap.

16. Cold trap may now be filled with liquid nitrogen. This helps to condense out heavier molecules that are in the chamber. This has two effects:
    (1) it lowers pressure in chamber by about an order of magnitude; and
    (2) it preserves the pump oil and mechanicals by eliminating water vapor that can damage both.

The procedure to shut down a system is more or less the opposite of the above. It is as follows.

1. Turn off ionization gauge and TC #2.
2. Close all valves except #1.
3. Switch off diffusion pump and allow this to cool for an half hour or so.
4. Close valve #1.
5. Switch off roughing pump and TC #1.

If the system is hot and evacuated and valve #4 is closed and you do something to the chamber D so that gas is admitted, you can then re-evacuate it according to the turn-on procedure, beginning at III.

SOME IMPORTANT CAUTIONS

1. If the pressure at TC #1 exceeds $\sim 10^{-1}$ Torr at any time when the diffusion pump is on, then something is wrong. Begin by closing valve #4. If this improves matters, there is a leak in the chamber D and you will have to fix it. If this does not change the pressure, there is a problem elsewhere and you will have to go through the shut-down procedure immediately.
2. If the pressure at TC #2 exceeds $\sim 10^{-3}$ Torr, this is an indication of a

similar problem to the above and the same procedure should be followed.

3. It is of utmost importance that neither the intake nor the exhaust pressure of the diffusion pump should exceed $\sim 10^{-1}$ Torr when the oil is hot. This can damage the oil or even the diffusion pump. Since diffusion pump oil costs from a few hundred to over a thousand dollars a litre, more than \$100 worth of oil can be contained in even a small pump. While synthetic oils tend to be less expensive and are not so susceptible to damage in this way, it is still not advisable to allow hot oil to come into contact with gas at higher pressures.

6 Thermometry

In this chapter we discuss several methods of measuring temperature. We concentrate on methods in which the temperature is measured electrically (we also discuss optical pyrometry, which is a nonelectrical method). There are also more "conventional" methods—mercury thermometers, vapor-pressure thermometers, gas thermometers, etc. These fall into two categories—those, like the mercury thermometer, that are straightforward to use but are not especially accurate; or those, like vapor-pressure thermometers, that can be accurate but are sufficiently complicated to use that their applications are limited. A good discussion of the details of these methods is given by G. K. White in *Experimental Techniques in Low Temperature Physics* (Oxford, 1959; or later editions) but we will not deal further with them here.

6.1 Thermocouples

Consider a bar of metal that has a thermal gradient between the two ends. This thermal gradient causes a flow of heat in the bar from the hot end to the cold end. Part of this heat is carried by a flow of conduction electrons. So there is a flux of electrons, call it Q_1, from the hot end to the cold end, and these electrons carry both charge and excess heat.

When the bar is not electrically connected to anything, there can be no net flow of current out of the material. This means there must be a flux of electrons Q_2, where $Q_2 = Q_1$, from the cold end to the hot end. These electrons carry charge but no excess heat. The driving force for Q_1 is the thermal gradient. The driving force for Q_2 is an electric field caused by a charge distribution inside the material that results from Q_1. This field is just sufficient to produce a flux Q_2 that cancels Q_1. The net result of this is that when a thermal gradient exists in a metal, then a potential difference is produced between the ends. Figure 6.1*a* shows this.

The potential difference is a function of ΔT and is given by

$$\Delta V = S(T)\,\Delta T \tag{6.1}$$

The quantity $S(T)$ is called the *absolute thermoelectric power*. As you can see, it is a function of T. It is also a function of the intrinsic properties of the

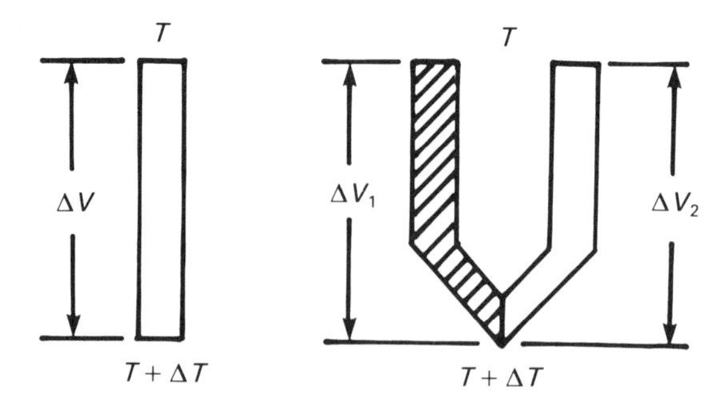

Figure 6.1. (*a*) Potential difference between the two ends of a metal bar with a thermal gradient between the ends. (*b*) Two bars of different metals with a thermal gradient between the ends and with one end in electrical contact.

metal, such as the conduction-electron density and the conduction-electron mobility. Figure 6.1*b* shows two different metals with a temperature difference ΔT between the ends. The hot ends are in electrical contact with each other. We can write

$$\Delta V_1 = S_1(T)\,\Delta T \qquad \text{and} \qquad \Delta V_2 = S_2(T)\,\Delta T \tag{6.2}$$

ΔT and T are the same for the two metals, but the $S_i(T)$ are not. Now we have a potential between the two cold ends that is given by

$$\Delta V_T = \Delta V_2 - \Delta V_1 = [S_2(T) - S_1(T)]\,\Delta T \tag{6.3}$$

That is, it is related to ΔT but also to the difference in the thermopowers of the two materials. (If the two materials are the same, then $S_1 = S_2$ and $\Delta V_T = 0$). This device is a thermocouple and we could in principle connect it directly to a voltmeter to measure ΔV_T. If we knew S_1 and S_2 as well as the temperature of the meter, we could determine the temperature of the thermocouple junction $T + \Delta T$. $S_i(T)$ is known for a number of metals and is tabulated in a number of references. The problem here is that in order to measure $T + \Delta T$ we have to know T. The junction may be attached to an experiment in which we want to measure the temperature while the meter is at the temperature of the laboratory. So in order to measure the temperature of the experiment, we have to monitor the temperature of the laboratory as well. This can be inconvenient. There is, however, a simple way of getting around this problem.

Let us consider the experiment in Figure 6.2. The thin wires are made of one material and the heavy wires are made of another material with a different value of $S_i(T)$. The total potential drop we measure at the meter is the sum of three terms:

$$\begin{aligned}\Delta V_T = \Delta V_1 + \delta V + \Delta V_2 &= S_1(T)[\Delta T + \delta T] - S_2(T)[\delta T] - S_1(T)\,\Delta T \\ &= [S_1(T) - S_2(T)]\,\delta T \end{aligned} \tag{6.4}$$

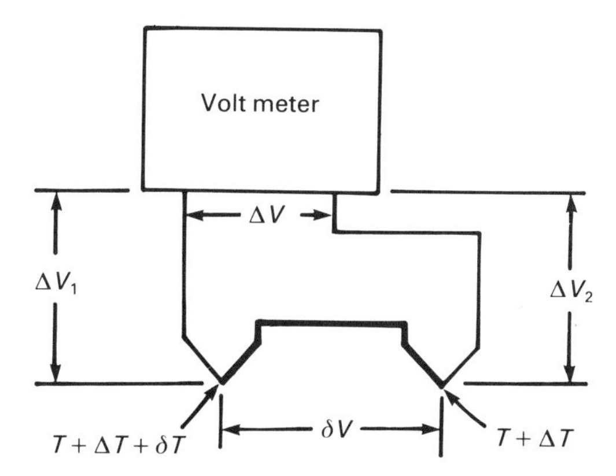

Figure 6.2. The measurement of temperature using a thermocouple with a reference junction.

So we measure a potential that is a function of the difference of the $S_i(T)$ and the temperature difference between the two thermocouples but that is independent of the temperature of the meter. Now we can place one junction (the right-hand one) some place where we know the temperature is fixed and we can measure the relative temperature of the other junction. Commonly the reference junction is placed in a beaker of ice–water so the temperature is measured relative to 0°C.

It is possible to buy digital temperature indicators that consist of a meter with the reference built-in. In this case the reference is not a thermocouple held at a fixed temperature (like 0°C) but is a constant-voltage source that produces the same voltage that this reference thermocouple would.

While we could, in principle, make a thermocouple out of any two dissimilar metals, there are several points to consider.

1. The materials must not melt at the temperatures of interest.
2. Some metals oxidize or react with certain environments at higher temperature, so we must take into account this possibility.
3. The difference $[S_1(T) - S_2(T)]$ must be large enough to give the thermocouple the desired sensitivity.

This last point requires careful consideration. The value of $S_1 - S_2$ tells us the change in voltage per unit change in temperature. So from (6.4)

$$\frac{\partial V}{\partial T} = [S_1(T) - S_2(T)] \tag{6.5}$$

This is not normally a problem at or above room temperature but $S_i(T)$ for all materials goes to zero at $T = 0$ K. Thermocouples are therefore useless in the limit of zero temperature. Even in the range of $T < 100$ K we have to choose carefully the two metals to keep $S_1 - S_2$ as large as possible, and there are only a few combinations of materials that can be used at liquid-helium temperatures.

Table 6.1. Some fixed temperature points

Fixed point	Value
Ice point	0°C
Steam point	100.000°C
Boiling point of sulfur	444.600°C
Freezing point of silver	960.8°C
Freezing point of gold	1063.0°C
Boiling point of oxygen	90.18 K
Triple point of oxygen	54.36 K
Boiling point of hydrogen	20.39 K
Triple point of hydrogen	13.95 K
Boiling point of helium	4.214 K
λ Point of helium	2.173 K

One problem with thermocouples, as well all temperature measuring devices, is that of calibration. A number of fixed temperature references have been established to aid in the calibration of temperature-measuring devices (see Table 6.1). We could do this for the thermocouple, for example, by fixing our reference temperature to (say) 0°C and measuring ΔV for several fixed temperatures. Intermediate values can be obtained by fitting the ΔV to a power series of the form,

$$\Delta V = aT + bT^2 + cT^3 + \cdots \tag{6.6}$$

It would be preferable if we used two materials for which $\Delta V(T)$ or at least $S_1(T)$ and $S_2(T)$ were known for some convenient reference temperature. To this end, several "standard" thermocouple pairs have been established. Information on standard thermocouple materials is given in Tables 6.2 and 6.3.

Table 6.2. Thermocouple properties

Thermocouple type designation	Connector color code	Material[a]	Useful temperature range
T	Blue	$\underline{\text{Copper}}$/constantan	−200 to 350°C
E	Violet	$\underline{\text{Chromel}}$/constantan	−200 to 900°C
J	Black	$\underline{\text{Iron}}$/contantan	0 to 750°C
K	Yellow	$\underline{\text{Chromel}}$/alumel	−200 to 1250°C
G	Red/green	$\underline{\text{W}}$/W + 26%Re	0 to 2300°C
C	Red	$\underline{\text{W + 5\% Re}}$/W + 26%Re	0 to 2300°C
D	Red/white	$\underline{\text{W + 3\% Re}}$/W + 25%Re	0 to 2300°C
R	Green	Pt/$\underline{\text{Pt + 13\%Rh}}$	0 to 1450°C
S	Green	Pt/$\underline{\text{Pt + 10\%Rh}}$	0 to 1450°C
B	White[b]	Pt + 6%Rh/$\underline{\text{Pt + 30\%Rh}}$	0 to 1700°C

[a] The (+) wire is indicated by the underline.

[b] White is uncompensated (type U) and is a copper–copper connector. This is used for type B thermocouples.

Table 6.3. Properties of thermocouple materials

Material	Composition	Designation[a]	Other names	T_{melt}(°C)
Iron	Fe	JP	—	1536
Copper	Cu	TP	—	1083
Gold	Au	—	—	1063
Gold–iron	Au + 0.03%Fe	—	—	1063
	Au + 0.07%Fe	—	—	1063
Constantan	55%Cu–45%Ni	JN, EN or TN	Cupron, Advance	1270
Chromel	90%Ni–10%Cr	KP or EP	Tophel, T_1, Thermokanthal KP	1430
Alumel	95%Ni–2%Mn–1%Si–2%Al	KN	Nial, T_2, Thermokanthal KN	1400
Platinum	Pt	RN or SN	—	1769
Tungsten	W	—	—	3410
Tungten–rhenium	97%W–3%Re	—	—	3410
	95%W–5%Re	—	—	3350
	75%W–25%Re	—	—	3150
	74%W–26%Re	—	—	3120
Platinum–rhodium	94%Pt–6%Rh	BN	—	1810
	90%Pt–10%Rh	SP	—	1830
	87%Pt–13%Rh	RP	—	1840
	70%Pt–30%Rh	BP	—	1910

[a] The first letter refers to the thermocouple type, the second letter to the polarity.

Table 6.4. Thermocouple lead compensating alloys

Thermocouple	(+) lead	(+) compensating[a]	(−) lead	(−) compensating[a]	Type codes
Pt/Pt–Rh	Pt + 10–13%Rh	Cu[1083]	Pt	#11[1090]	R & S
W/W–Re	W	#200[1430]	W + 26%Re	#226[1450]	G
	W + 5%Re	#405[1410]	W + 26%Re	#426[1520]	C
	W + 3%Re	#203[1400]	W + 26%Re	#225[1370]	D

[a] The melting temperature is given in degrees Celsius in square brackets.

While materials such as copper or iron, or even Chromel or Constantan, are reasonably inexpensive, you can probably guess that Pt–Rh or W–Re alloys are costly. This is an important factor if the meter is on one side of the laboratory and the experiment whose temperature you want to measure is on the other side. As a result, *compensating alloys* have been developed. These have the same, or nearly the same, room-temperature thermopower as the expensive metals. These are listed in Table 6.4. Figure 6.3 shows a typical set up for a Pt *vs* Pt–10%Rh thermocouple with compensating leads. All junctions labeled "J" are held at room temperature (more or less). You may ask: Why not just use the compensating alloys instead of the expensive metals? There are two considerations.

1. The melting temperature of the compensating alloys is much lower than those of the thermocouple materials (see Tables 6.3 and 6.4). One reason for using such exotic materials is because they can be used to very high temperatures.
2. Although $S(T)$ is the same for the thermocouple material and the compensating alloy at room temperature, this is not the case at all other temperatures. In fact $S(T)$ for the compensating alloys is probably such that they would not make very good thermocouples over a very large temperature range.

A few comments might be helpful on the use of thermocouples below ~100 K. Most pairs of materials do not have $S_1 - S_2$ values at low temperatures that are large enough to make useful thermocouples. Iron-

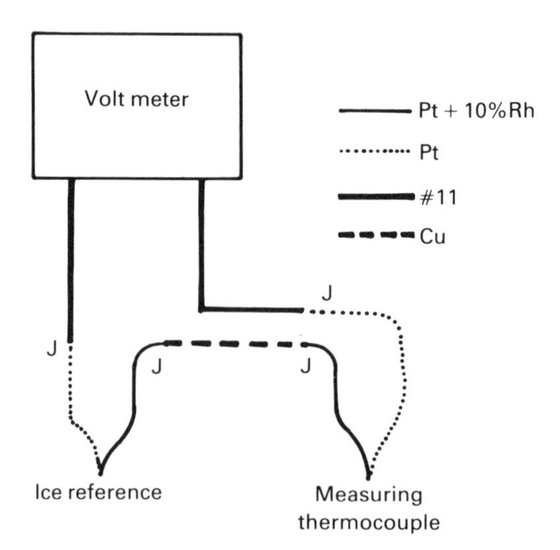

Figure 6.3. Compensated Pt *vs* Pt + 10%Rh thermocouple.

doped gold alloys have been most commonly used. Some examples are

Ag *vs* Au + 0.07 at.% Fe
Cu *vs* Au + 0.07 at.% Fe
Chromel *vs* Au + 0.07 at.% Fe
Chromel *vs* Au + 0.03 at.% Fe

Figure 6.4 shows why Chromel *vs* Au/Fe is a better thermocouple at low temperatures than the standard types. In all cases with these thermocouples, the doped-gold side is negative.

In practice, after measuring ΔV we would normally obtain the temperature by looking it up in a thermocouple table. Each table is for a particular

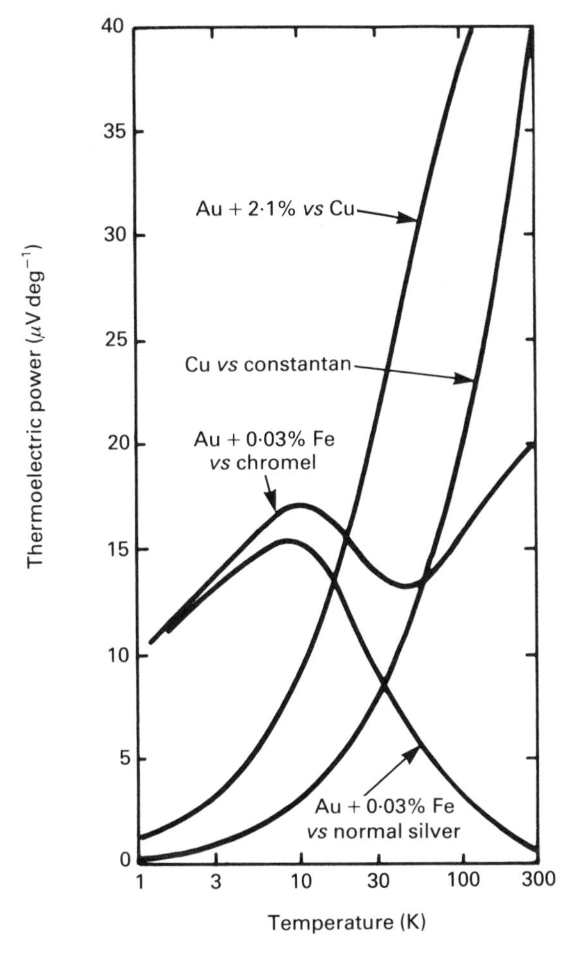

Figure 6.4. Thermoelectric powers for some thermocouple pairs at low temperatures. (From G. K. White, *Experimental Techniques in Low Temperature Physics,* 2nd ed. Copyright © 1968 by Oxford University Press.)

pair of thermocouple materials and the voltages are measured relative to the voltage at some reference temperature (usually 0°C except for low-temperature types, which are often referred to 4.2 K). Numerous books of thermocouple tables are available and some are given in the list of references.

6.2 Metal resistance thermometers

Since, as we have seen previously, the electrical resistivity of a metal is a function of temperature, it provides a suitable parameter for determining the temperature. In an actual metal there are two contributions to the resistivity: first, as we have discussed before, a contribution ρ_t due to thermal vibrations; additionally, there is a contribution ρ_i that results from the scattering of electrons from impurity atoms. Thus we write

$$\rho = \rho_t + \rho_i \tag{6.7}$$

As we know, ρ_t is roughly linear in temperature while ρ_i is independent of temperature. Since all actual materials contain impurities, ρ_i is always present. Hence, the ρ_t term will dominate at higher temperatures, but at lower temperatures ρ will not approach zero but will be limited by ρ_i.

In choosing an appropriate material for a resistance thermometer there are several criteria to consider.

1. The ρ_t should be reasonably linear in the high temperature region—this improves the accuracy of interpolating between fixed points.
2. The $\partial\rho/\partial T$ should be fairly large—this determines the sensitivity.
3. The linear region should extend to low temperatures—this is important for low-temperature measurements, since, when ρ_i dominates, then $\partial\rho/\partial T \rightarrow 0$. This criterion requires the preparation of samples of high purity in order to make ρ_i small.
4. The material must be able to be formed into an appropriate shape—usually wire.
5. If high temperatures or reactive atmospheres are to be used the material must be resistant to degradation.

Platinum has become the most commonly used metallic material for resistance thermometers, although copper has been used as well. The resistance of copper is shown in Figure 6.5 as a function of temperature. The sensitivity and linearity of the thermometer are easily seen by looking at the derivative of the resistivity. The resistance thermometer made of platinum is probably the most accurate and simplest method of measuring temperatures over the largest range. These can measure to better than 50 mK from 4.2 K to above room temperature. They are highly insensitive to temperature cycling. This means that we can warm them and cool them many times without changing the calibration.

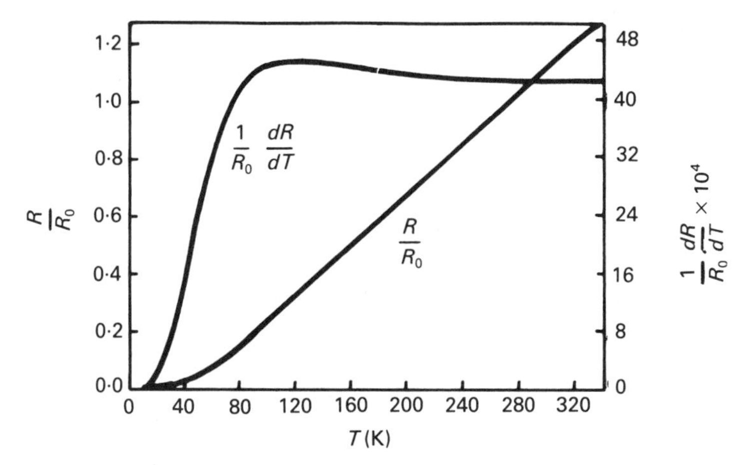

Figure 6.5. Temperature dependence of the resistivity and its derivative for copper. (From G. K. White, *Experimental Techniques in Low Temperature Physics,* 2nd ed. Copyright © 1968 by Oxford University Press.)

When we make a measurement, we actually measure the resistance R of the thermometer, not the resistivity ρ. The common method of calibration is as follows. Define a function $Z(T)$ as

$$Z(T) = \frac{R(T) - R(4.2\,\mathrm{K})}{R(273\,\mathrm{K}) - R(4.2\,\mathrm{K})} \tag{6.8}$$

Since we have taken ratios of resistances, these should be the same as ratios of resistivities and the geometrical factor should be eliminated. Hence, $Z(T)$ is a function only of the temperature and the material used and not of its size and shape. However, the purity of the material is important. Fortunately, platinum can be made very pure. A measure of its purity can be obtained by measuring $R(273\text{ K})/R(4.2\text{ K})$, as this is a measure of the ratio of ρ_t to ρ_i. For samples of platinum (and copper) with $R(273\text{ K})/R(4.2\text{ K}) > 300$, values of $Z(T)$ have been tabulated in Table 6.5. For purposes of interpolation, values of Z can be fitted to a polynomial of the form

$$Z(T) = A_1 + A_2T + A_3T^2 + A_4T^3 + \cdots \tag{6.9}$$

Since $R(T)$ is actually nearly linear, terms in (6.9) of higher order than A_3T^2 are very small. Even in cases where accuracy in the order of a few millikelvin is required, it is generally not necessary to take terms of higher order than T^4.

In a given thermometer we measure $R(T)$ and obtain T from the tabulated (and interpolated) values of $Z(T)$. It is necessary to know $R(4.2\text{ K})$ and $R(273\text{ K})$ as well. For highest accuracy, it is customary to buy a platinum thermometer with values of these calibration constants provided by the manufacturer. If accuracy is not so crucial, we can easily measure

Table 6.5. Measured $Z(T)$ function for high-purity platinum and copper, where $Z = [R(T) - R(4.2\ \mathrm{K})]/[(R(273\ \mathrm{K}) - R_1(4.2\ \mathrm{K})]$

T (K)	Copper		Platinum	
	Z	$\Delta T/\Delta Z$	Z	$\Delta T/\Delta Z$
295	1.09467	230.8	—	—
273.15	1.00000	236.6	1.00000	—
260	0.94295	229.7	—	—
240	0.85589	228.5	—	—
220	0.76839	227.0	—	—
200	0.68030	225.1	—	—
180	0.59146	223.0	—	—
170	0.54662	221.5	0.581609	241.1
160	0.50148	221.1	0.540114	239.8
150	0.45626	219.2	0.498400	238.5
140	0.41064	218.2	0.456456	237.1
130	0.36482	217.7	0.414269	235.7
120	0.31888	217.6	0.371819	234.1
110	0.27292	218.6	0.329090	232.6
100	0.22718	222.9	0.286063	231.2
90	0.18231	227.8	0.242719	229.8
85	0.16036	233.1	0.220947	230.2
80	0.13891	245.0	0.199215	231.4
75	0.11850	253.0	0.177599	233.2
70	0.09874	269.8	0.156140	235.8
65	0.08021	292.4	0.134928	240.7
60	0.06311	328.1	0.114147	249.2
55	0.04787	379.1	0.094083	261.9
50	0.03468	432.0	0.074982	274.6
48	0.03005	472.8	0.067694	284.1
46	0.02582	497.5	0.060651	295.6
44	0.02180	570.0	0.053881	309.4
42	0.01829	647.2	0.047414	326.2
40	0.01520	743	0.041280	346.6
38	0.01251	816	0.035508	372.3
36	0.01006	943	0.030132	404.4
34	0.00793	1123	0.025183	445.1
32	0.00615	1371	0.020688	497.5
30	0.004700	1668	0.016666	565.6
28	0.003501	2045	0.013129	656.2
26	0.002523	2618	0.010079	779.4
24	0.001759	3484	0.007512	951.5
22	0.001185	4556	0.005409	1203
20	0.000746	—	0.003745	1472
19	—	—	0.003066	1705
18	—	—	0.002479	2000
17	—	—	0.001979	2381
16	—	—	0.001558	2862
15	—	—	0.001208	3495
14	—	—	0.000922	4357
13	—	—	0.000692	5447
12	—	—	0.000509	—

Source: From G. K. White, *Experimental Techniques in Low Temperature Physics* (New York: Oxford University Press, 1959).

these constants for ourselves. The accuracy required for the resistance measurement is determined by how accurately we want to know the temperature. Consider an example: for a thermometer (platinum) with $R(273\text{ K}) = 100\ \Omega$, we might have $R(4.2\text{ K}) = 0.1\ \Omega$. So

$$Z(T) = [R(T) - 0.01]/99.9 \tag{6.10}$$

At 100 K, for example, we have from Table 6.5

$$\frac{\partial Z(T)}{\partial T} = 4.3 \times 10^{-3}\ \text{K}^{-1} \tag{6.11}$$

(this is the inverse of $\Delta T/\Delta Z$ given in the table). We see from (6.10) that

$$\frac{\partial R(T)}{\partial T} = 99.9 \frac{\partial Z(T)}{\partial T} = 0.43\ \Omega\ \text{K}^{-1} \tag{6.12}$$

So a typical thermometer might have a resistance that changes by about half an ohm per degree of temperature. Results are given in Table 6.6 for a 100 Ω thermometer.

The resistance of a couple of meters of copper hook-up wire might be in the order of 0.1 Ω, and an inexpensive digital multimeter might read to $3\frac{1}{2}$ digits. So you see that a very simple measurement might give an accuracy of ±1 K. If we want to do better than that, we need to improve our measurement of *R*. The first thing we must do is to make a four-terminal resistance measurement.

Recently, digital multimeters have become available at reasonable prices ($1000–2000) that can measure to $5\frac{1}{2}$ or even $6\frac{1}{2}$ digits and have built into them the capability of performing four-terminal resistance measurements. With a well-calibrated platinum resistance thermometer this will allow for temperature measurements accurate to a few millikelvin. It is important to realize, however, that great care must be taken to ensure that temperature gradients much larger than this do not occur over the dimensions of the sensor.

Table 6.6. Required accuracy of measurement

Desired *T* accuracy (K)	Required *R* accuracy (Ω)	No. of digits to be measured
10 K	5 Ω	$2\frac{1}{2}$
1 K	0.5 Ω	$3\frac{1}{2}$
100 mK	50 mΩ	$4\frac{1}{2}$
10 mK	5 mΩ	$5\frac{1}{2}$
1 mK	0.5 mΩ	$6\frac{1}{2}$

6.3 Semiconductor temperature sensors

In this section we consider the use of conventional semiconducting materials (germanium, silicon, gallium arsenide, etc.) as well as carbon and carbon–glass as temperature sensors. These latter are not strictly semiconducting materials but are similar in several respects.

a. Intrinsic semiconductors

In Chapter 1 we developed a model for the temperature dependence of the resistivity $\rho(t)$ of an intrinsic semiconductor. Recall that

$$\rho(T) = AT^{-3/2} \exp\left(\frac{\mathscr{E}_g}{2k_B T}\right) \tag{6.13}$$

The resistivity will be a relatively slowly varying function of T at higher temperatures but will be a very steep function of T at lower temperatures. Let us consider the case of germanium. We recall that $\mathscr{E}_g = 0.7$ eV, and find in the literature the room temperature resistivity of germanium; this is $\rho(295\text{ K}) = 46\ \Omega$ cm. Consider a piece of germanium 1 mm × 1 mm × 2 mm; its resistance (along the long axis) will be 920 Ω. We find the value of A and make a table of $R(T)$ (See Table 6.7). You can see that the resistance becomes too large to measure for temperatures below about 200 K. Even if we changed the dimensions of the sample so that the room temperature resistance as two or three orders of magnitude lower, it would not help much, because $\Delta R/\Delta T$ becomes so large at low temperatures. The fractional change in the resistivity at room temperature is given by

$$\left.\frac{1}{R}\frac{\partial R}{\partial T}\right|_{295\text{ K}} = 0.26\text{ K}^{-1} \tag{6.14}$$

Compare this result with the value for the 100 Ω platinum thermometer, 0.005 K^{-1}.

A difficulty with this type of sensor is that ρ is very sensitive to the

Table 6.7. Resistance of a piece of intrinsic germanium.

$T(K)$	$R(\Omega)$
295	920
250	1.40×10^4
200	1.13×10^6
150	1.5×10^9
100	2.1×10^{15}
80	7.4×10^{19}

purity—much more so than in metals like platinum. One thing we can do is intentionally add impurities. This is discussed below.

b. Doped semiconductors

The advantage of doped semiconductors is that (1) we know what the impurity level is, and (2) we can decrease the low-temperature resistance to make the device more usable. We ask: What will the temperature dependence of the doped semiconductor look like? Consider the n-type material shown in Figure 6.6. The Fermi energy has shifted by $\mathscr{E}_d$ from the center of the gap at low temperature. So the carrier concentration is dominated by the donor electrons (majority carriers) and is

$$n = n_e = 2UT^{3/2} \exp\left(-\frac{\mathscr{E}_g - 2\mathscr{E}_d}{2k_B T} \right) \tag{6.15}$$

As the temperature increases, the Fermi level shifts towards the center of the energy gap so that at high temperature the number of carriers per unit volume becomes

$$n = n_e + n_h = 4UT^{3/2} \exp\left(-\frac{\mathscr{E}_g}{2k_B T} \right) \tag{6.16}$$

So the material behaves as though it is an intrinsic semiconductor. Figure 6.7 shows the carrier concentration as a function of temperature. The linear regions for $\log n$ versus $1/T$ show the regions dominated by impurity effects and by intrinsic effects. The temperature at which the transition occurs is a function of the impurity concentration: the more impurities, the higher the temperature; that is, the larger $\mathscr{E}_d$ in Figure 6.6 and the more T has to be

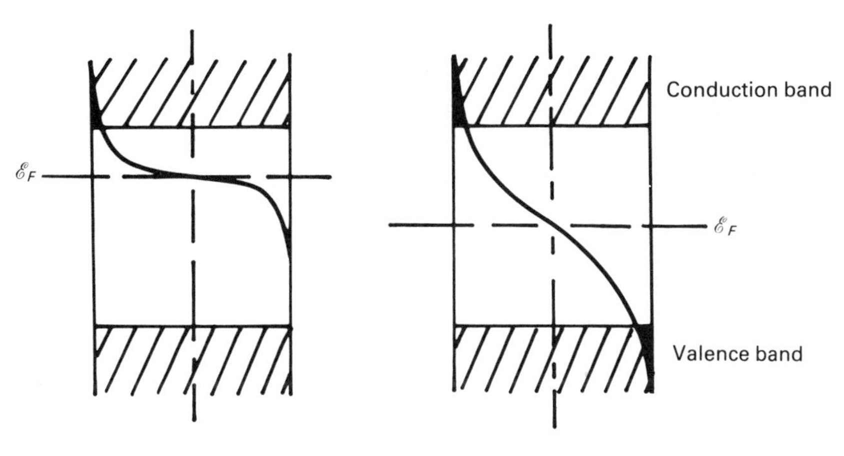

Figure 6.6. The shift of the Fermi energy towards the center of the energy gap with increasing temperature for an n-type semiconducting material.

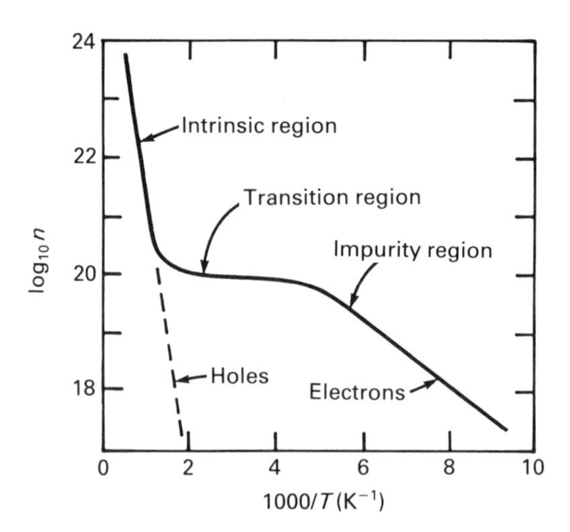

Figure 6.7. Carrier concentrations in a typical n-type semiconductor as a function of temperature.

raised for $\mathscr{E}_F$ to return to near the center of the gap. The resistivity of the doped semiconductor follows from equation (1.32) and Figure 6.7 as

$$\rho(T) = \frac{1}{\sigma(T)} = [e(n_e(T)\mu_e + n_h(T)\mu_h)]^{-1} \tag{6.17}$$

Figure 6.8 shows typical results for $\rho(T)$ for some doped semiconductors. This type of thermometer has several advantages as well as some disadvantages. The advantages are that

1. The doping allows use at low temperatures.
2. The large $\Delta\rho/\rho T$ provides very good sensitivity.

And the disadvantages are that

1. The function $\rho(T)$ is not very manageable and it is not a simple matter to fit it to the calibration values.
2. The large change in ρ as a function of temperature means that a particular thermometer may not cover a very large temperature range and we may need several thermometers with different doping levels to cover a range from (say) 4.2 K to room temperature. Each of these would have to be calibrated independently over the temperature range for which it was to be used.
3. The large change in ρ with doping level makes it necessary to calibrate each thermometer separately. Two thermometers of the same nominal doping level could be vastly different.

We should point out that at any fixed temperature the electrical

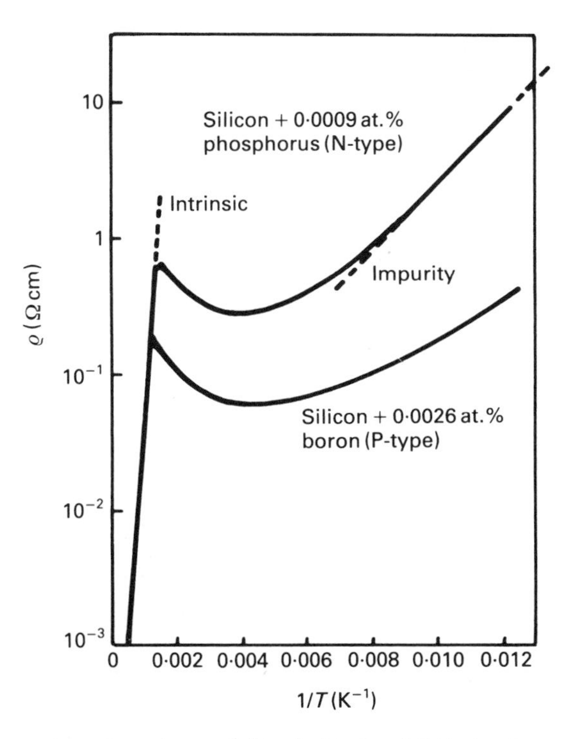

Figure 6.8. Temperature dependence of the electrical resistivity for two samples of doped silicon. (From G. K. White, *Experimental Techniques in Low Temperature Physics,* 1st ed. Copyright © 1959 by Oxford University Press.)

properties of the pure and doped semiconductors, as well as of the metal oxide, carbon, and carbon–glass thermometers to be discussed, are ohmic. This is not true for the diode thermometer that is discussed next.

c. *Diode thermometers*

The use of a forward-biased diode eliminates most of the problems encountered with the doped-semiconductor thermometer. The physics of this device is very complicated and we will merely describe the behavior phenomenologically. We measure the forward bias voltage necessary for a certain forward current to flow as a function of temperature. We can measure this by forward biasing the diode using a constant-current source and measuring the voltage drop across it. For gallium arsenide we find, for T greater than a few kelvin, that $V(T)$ decreases nearly linearly: we see this in Figure 6.9. The rate of decrease is small enough to make the thermometer suitable for use over a large temperature range. A fairly simple empirical relation can be used to fit the temperature dependence of the resistivity to values at a few fixed points. Silicon, as shown by the figure,

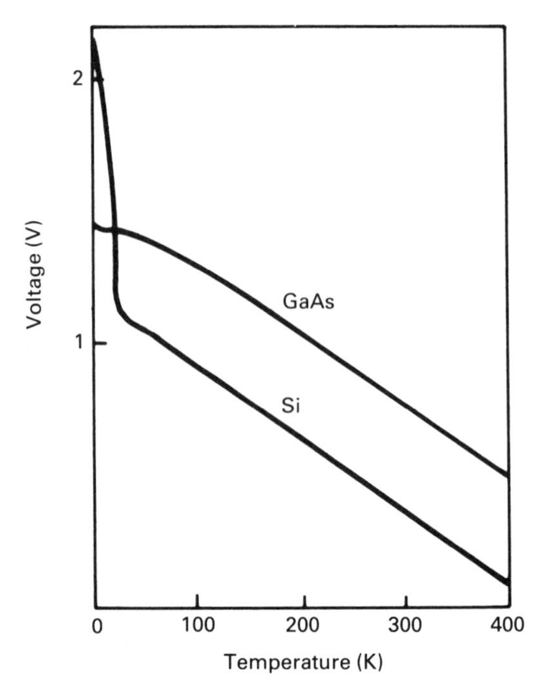

Figure 6.9. Temperature dependence of the forward bias voltage at constant current for silicon and gallium arsenide diode thermometers. (From T. J. Quinn, *Temperature*. Copyright © 1983, Academic Press. Used with permission.)

has a $V(T)$ that is more nearly linear above ~30 K but departs from linearity below this temperature.

d. Metal oxide thermometers

These have been used for a number of years and are commonly called thermistors. These are made of various metal oxides, usually of a nonstoichiometric composition, that have electrical properties similar to those of a semiconductors. The actual sensor is generally made by the sintering of micrometer-sized particles of the material. An example of a thermistor material is $MgTiO_3$. The resistivity of this material is easily adjusted by adjusting its stoichiometry, that is, primarily by changing the proportion of oxygen. This can be done in one of two ways:

1. Mix a little MgO into the material.
2. Sinter in either an oxidizing atmosphere (oxygen) or a reducing atmosphere (hydrogen).

In the sense of semiconductors, the ratio of electron carriers to hole carriers, and the resistivity of the material, can be adjusted by controlling the quantity of oxygen. That is, it is like doping. Sintering in a reducing or oxidizing atmosphere can result in particles that are, say, n-type on the surface and p-type inside—or the other way around. The electrical properties of the sensor depend not only on the carrier concentrations in the particles but on the quality of the electrical contact between the sintered particles. As a result, the electrical properties of these materials can be varied over a large range by the manufacturing process. One advantage of this type of sensor is that it can be made very small. Spherical sensors less than 0.1 mm diameter with leads of 0.01 mm diameter have been produced. It is customary to coat the sensors with glass in order to eliminate changes in the electrical properties that could result from reactions with atmospheric oxygen. The most commonly used materials for the manufacture of these devices are mixtures of nickel and manganese oxides or nickel, manganese, and cobalt oxides.

In view of their semiconductor-like behavior we expect a temperature dependence that is somehow exponential. A semiempirical relationship that is often used is

$$R(T) = R_0 \exp\left[-B\left(\frac{1}{T} - \frac{1}{T_0}\right)\right] \tag{6.18}$$

where R_0 is the resistance at T_0.

Most commonly produced thermistors have room-temperature resistances of 5–30 kΩ. These are useful from around zero to a few hundred degrees Celsius, their resistances becoming too high at lower temperatures to make them usable. Low-resistance sensors made of nonstoichiometric iron oxides have been used successfully down to 4.2 K, although other sensors are generally more suitable for work in this temperature range.

e. *Carbon thermometers*

Although, like a semiconductor, carbon shows a decreasing resistivity as a function of increasing temperature, it is not, in a strict sense, a semiconductor as are germanium or silicon. Carbon is highly anisotropic and a single crystal of carbon would show a room-temperature resistivity of about 1 Ω cm parallel to the hexagonal axis but a resistivity of about 10^{-4} Ω cm perpendicular to this axis. Understandably, the resistivity of polycrystalline carbon is rather complex. While we cannot undertake a rigorous derivation of $\rho(T)$ we can, as in the case of the metal oxide thermometer, state some semiempirical expressions that seem to work.

The most common carbon thermometer is the standard, commercially available carbon composition resistor. Those manufactured by Allen–Bradley of the $\frac{1}{8}$ W, $\frac{1}{4}$ W, $\frac{1}{2}$ W or 1 W size are usually preferred, but those made by Speer (grade 1002) and I.R.C. (Ohmite) have been used as well. Typical resistance curves for some carbon resistors are shown in Figure 6.10. The resistors are referred to by their nominal room temperature

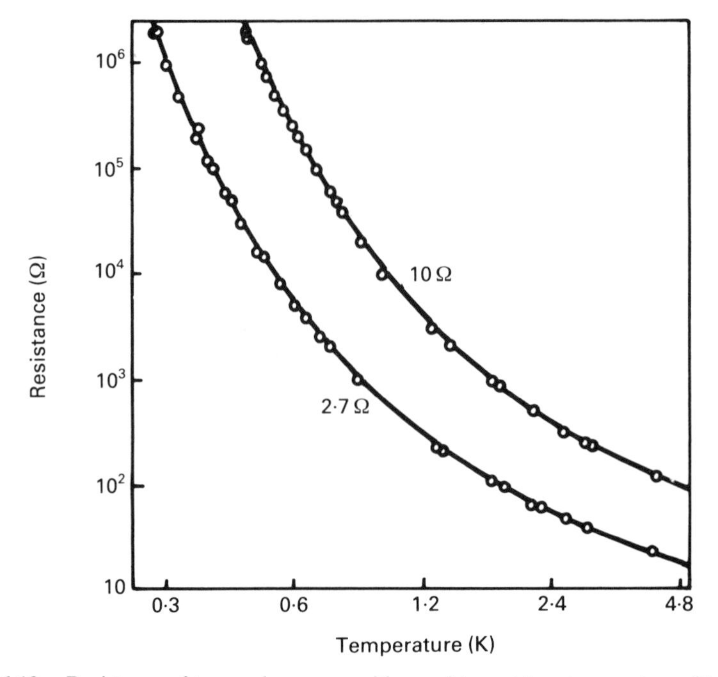

Figure 6.10. Resistance of two carbon composition resistors at low temperature. (From G. K. White, *Experimental Techniques in Low Temperature Physics,* 2nd ed. Copyright © 1968, Oxford University Press.)

resistance. You can see that the resistance remains in the measurable range over a very large temperature range.

An expression that is commonly used to fit these curves is

$$\log_{10} R + \frac{K}{\log_{10} R} = A + \frac{B}{T} \tag{6.19}$$

where $K, A,$ and B are determined for each thermometer from resistance measurements at at least three fixed points. Figure 6.11 shows that the data fit fairly well to this expression over a fairly large temperature range. Another expression that is sometimes used is:

$$\ln R = AT^{-m} + B \tag{6.20}$$

where again three or more fixed points are needed to fit the parameters.

Advantage of the carbon thermometer include the following.

1. It is very inexpensive.
2. It operates over a large temperature range.
3. It has good sensitivity at low temperatures.

However, there are, of course disadvantages too.

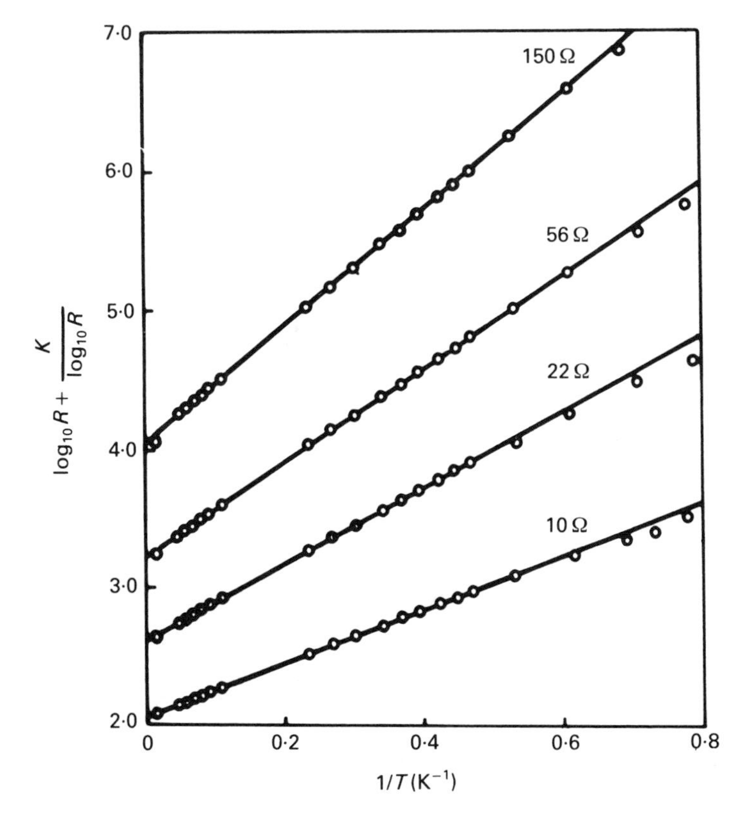

Figure 6.11. Graphical test of equation (6.18) for calibrating four carbon thermometers. (From G. K. White, *Experimental Techniques in Low Temperature Physics,* 2nd ed. Copyright © 1968, Oxford University Press.)

1. Each has to be calibrated individually.
2. Calibration from equation (6.19) or (6.20) may not give sufficient accuracy over large temperature ranges.
3. Cyclability is not as good as for platinum or diode thermometers.
4. Resistance tends to drift even if temperature is constant.

f. Carbon–glass

Carbon glass thermometers are made by first preparing a porous glass (in this case an alkali borosilicate glass); the pores are in the order of a few nanometers in size. The glass is then impregnated with carbon. The physics of these devices is the same (more or less) as it is for the carbon thermometer. Hence, calibration can be done using equation (6.19) or (6.20). They are considerably more expensive than carbon thermometers but they are better in several respects in that they have

1. Less drift
2. Better cyclability
3. Higher sensitivity
4. Better uniformity between devices

6.4 Optical pyrometry

Optical pyrometry is used only for measuring very high temperatures. A black body at temperature T emits radiation with an energy spectrum defined as

$$\mathscr{E}(\omega)\,d\omega = \frac{\hbar\omega^3\,d\omega}{4\pi^2c^2(e^{\hbar\omega/k_BT}-1)} \tag{6.21}$$

This gives the energy per unit frequency. The total energy radiated is given by

$$\int \mathscr{E}(\omega)\,d\omega = \sigma T^4 \tag{6.22}$$

Equation (6.21) is the Planck radiation law and equation (6.22) is the Stefan–Boltzmann law. The higher the temperature the more energy radiated; but something else happens as the temperaure increases. Figure 6.12 is a plot of equation (6.21). As the temperature of the black body increases, the peak in $\mathscr{E}(\omega)$ shifts to higher frequencies. At 1200 K only the tail of $\mathscr{E}(\omega)$ overlaps the red end of the visible spectrum. Hence an object at

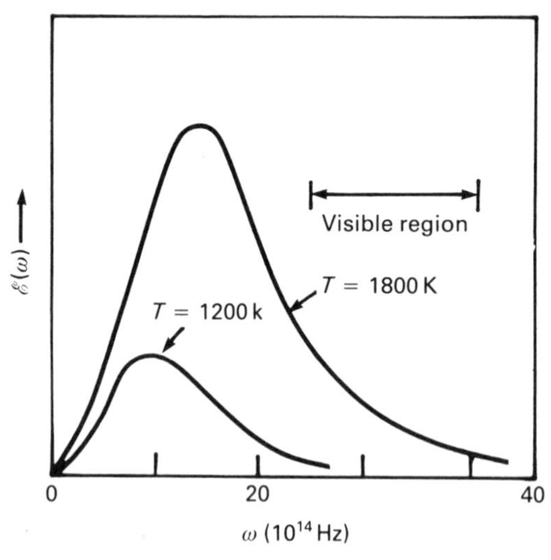

Figure 6.12. The Planck radiation law for objects at two different temperatures.

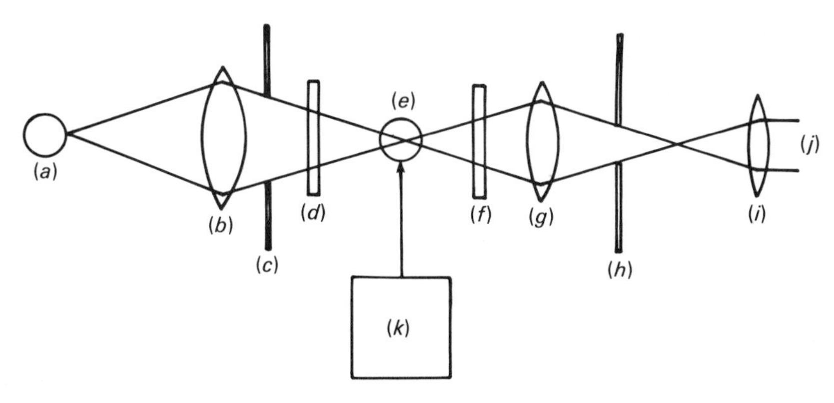

Figure 6.13. Schematic diagram of the optical pyrometer: (*a*) object; (*b*) lens; (*c*) aperture; (*d*) filter for high-temperature operation; (*e*) tungsten lamp; (*f*) red filter; (*g*) lens; (*h*) aperture; (*i*) eyepiece; (*j*) observer; (*k*) current source.

this temperature appears to glow red. As T increases, the average colour shifts towards yellow. If T is high enough, we see a distribution of frequencies over the entire visible region and the object appears white. So the apparent color of a hot object that is giving off visible radiation is related to its temperature. This is the principle on which the optical pyrometer works.

The optical pyrometer consists of an optical system for viewing the test material and also for simultaneously viewing a standard light source. A diagram of a typical system is shown in Figure 6.13. The temperature of the test material is determined by adjusting the current to the lamp until the brightness of the filament just matches the brightness of the test material. In order to simplify matters somewhat, the brightness of the object is compared only at a single frequency.

Thus we want to look at the image of the filament and the test body through a optical band pass filter. This filter comprises two elememts: (1) the *red filter* (*f* in Figure 6.13) and (2) the *human eye.* The red filter is a low-pass filter with a cutoff at about 630 nm (see Figure 6.14). The high-frequency cutoff is proved by the diminishing sensitivity of the human eye for red light. The mean wavelength used for most pyrometers is 655 nm. The preferred filament for pyrometer lamps is tungsten enclosed in an evacuated glass tube. The filament temperature is then determined as a function of lamp current. Although the melting temperature of tungsten is quite high (~3400°C), pyrometer lamps filaments are not usually used above 1350°C. Operation at higher temperatures can cause the calibration to change. For the measurement of test materials at higher temperatures, a filter (*d* in Figure 6.13) is inserted between the object and the filament to reduce the apparent brightness of the source. Thus the test material over 1350°C appears in the pyrometer as a material under 1350°C and a different calibration scale for temperature as a function of filament current is used.

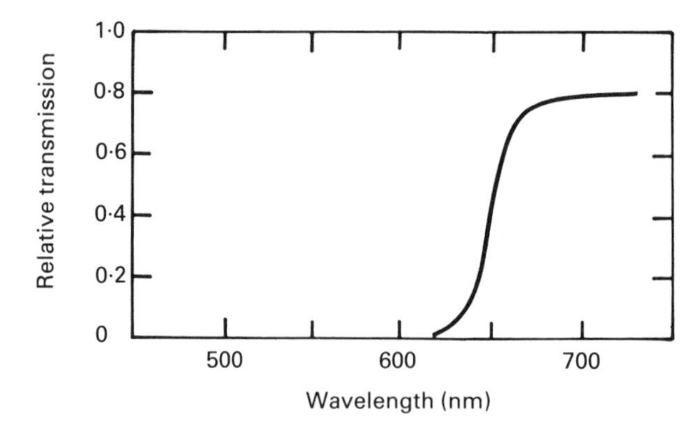

Figure 6.14. Relative transmission of a red filter as a function of wavelength.

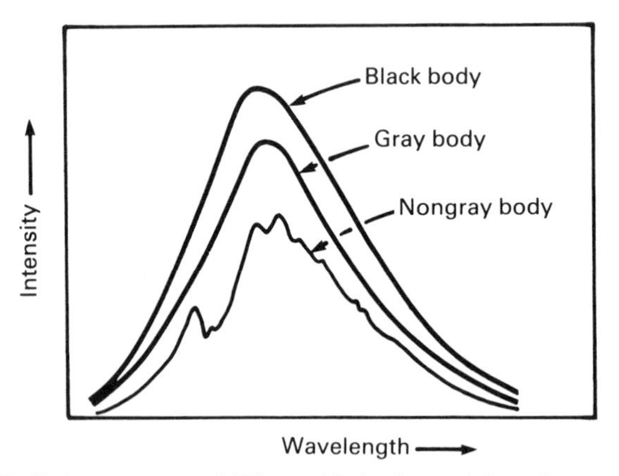

Figure 6.15. Radiation spectrum of different kinds of materials at the same temperature.

Table 6.8. Emissivities of some materials at 6550 nm in the solid and liquid states

Material	ϵ(solid)	ϵ(liquid)
Cr	0.34	0.39
Co	0.36	0.37
Cu	0.10	0.15
Au	0.14	0.22
Fe	0.35	0.37
Mn	0.59	0.59
Mo	0.37	0.40
Ni	0.36	0.37
Pd	0.33	0.37
Pt	0.30	0.38
Rh	0.24	0.30
Ag	0.07	0.07
Ti	0.63	0.65
W	0.43	—

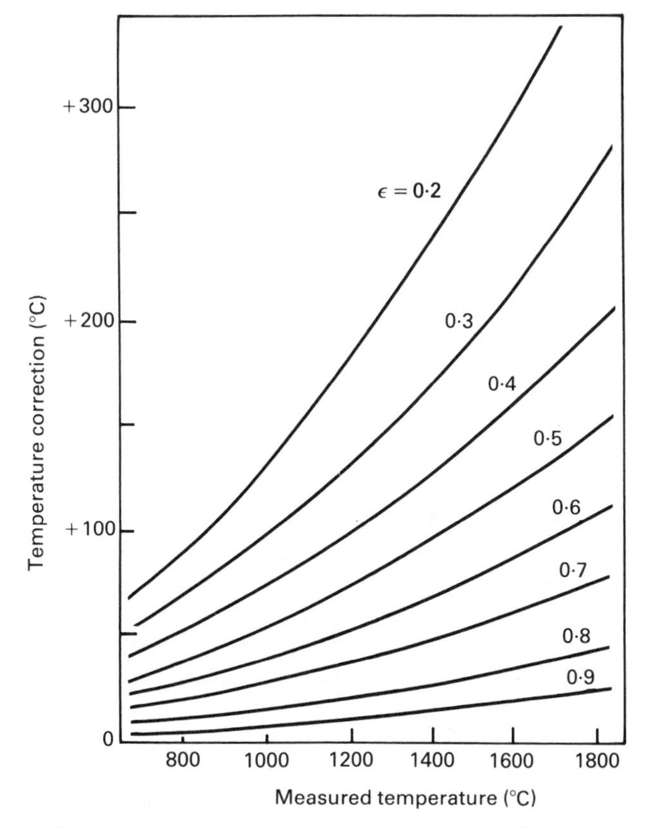

Figure 6.16. Temperature corrections due to emissivity differences at 655 nm.

One problem is that not many objects are actually black bodies. At a given temperature, no object produces as much radiation at particular frequency as does a black body (see Figure 6.15.) The ratio of the brightness of one object to the brightness of a black body as the same temperature is called the emissivity ϵ. If ϵ is independent of wavelength, then the material is referred to as a gray-body if ϵ is a function of wavelength, then this is a nongray body (see Figure 6.15). Since for the optical pyrometer we are concerned only with what is happening at one particular wavelength (655 nm), we are really only concerned with the emissivity at 655 nm and not whether the material is a gray body or a nongray body. Table 6.8 gives the emissivities of some common materials. Since ϵ is always less than 1, the brightness is always less than that of a black body at the same temperature, and so the temperature we measure is always less than the actual temperature. Figure 6.16 shows the amount that has to be added to the measured temperature to obtain the actual temperature for different emissivities.

7
Cryogenic technology

7.1 Cryogenic fluids

The cooling of experimental apparatus to low temperatures is generally done by the use of cryogenic fluids. In this section we describe the properties of the fluids in common use. Continuous-flow helium gas refrigerators can be used for similar purposes. Although their use in research laboratories is becoming more common, they are not generally used in undergraduate teaching laboratories and their use will not be described here. We will also not delve into any great detail concerning the processes by which cryogenic fluids are produced. For interested readers, detailed descriptions of these processes are given in *Advanced Cryogenics* by L. A. Bailey.

We begin by looking at the phase diagram of a typical material. We see this in Figure 7.1. Some interesting features of this diagram are the critical point (labeled CP) and the triple point (labeled TP). For temperatures and/or pressures above those of the CP there is no distinction between the liquid and gas phases, and at the TP all three phases coexist.

A liquid at a particular pressure that is an equilibrium with the gas phase will be at some temperature along the line TP–CP. A substance at an appropriate pressure (as shown in line A) will undergo transitions from gas to liquid and from liquid to solid as the temperature is lowered. Water at 1 atmosphere pressure shows this behaviour. The TP and CP values for various substances that are of cryogenic interest are given in Table 7.1.

The experiment that we desire to cool to a particular temperature is thermally coupled to a bath of some appropriate cryogenic fluid. Normally the pressure would be 1 atmosphere and the temperature the boiling point (BP) of the fluid at this pressure as given in Table 7.1. The temperature of the bath may be adjusted by changing the vapor pressure over the liquid. This generally consists of lowering the vapor pressure by pumping on the liquid with a vacuum (roughing) pump, but it is possible to pressurize the vapor over the liquid. As the pressure is raised or lowered the temperature of the liquid follows the CP–TP line in Figure 7.1. The usual range of accessible temperatures for a particular fluid is between the CP and TP. Although in practice it is possible in some cases to obtain somewhat lower temperatures than that corresponding to TP. An investigation of Table 7.1

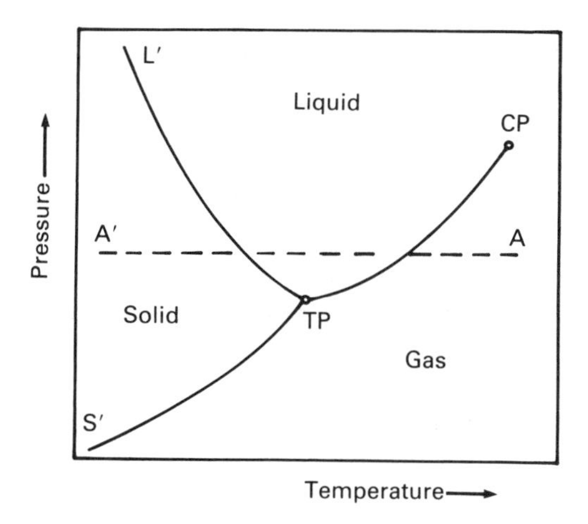

Figure 7.1. A typical phase diagram.

shows that temperatures in the range of 5–15 K and 30–60 K are difficult to access directly. We shall see in the following sections on cryostat design and temperature control how this problem can be overcome.

The various cryogenic fluids are discussed in turn below.

a. Nitrogen

Nitrogen is a readily accessible cryogenic liquid. It is inexpensive, safe, and easy to use for experiments that can be performed at ~77 K. Temperatures down to about 54 K may be obtained by pumping on nitrogen. Down to the TP, the TP–CP line in Figure 7.1 is followed. The cooling can be thought of as resulting from the vaporization of some of the liquid, the vapor carrying with it the latent heat of vaporization (see Table 7.1). Below the TP the S′–TP line is followed; cooling results from sublimation. Below the TP, however, the nitrogen bath is a solid and because of thermal expansion it is

Table 7.1. Properties of some materials of cryogenic interest

Substance	BP (K) at 1 atm	CP T(K)	CP P(atm)	TP T(K)	TP P(mm Hg)	Latent heat (kJ kg^{-1})
Oxygen	90.2	154.8	50.1	54.4	1.14	213
Hydrogen	20.4	33.25	12.8	14.0	54.0	451
^{3}He	3.20	3.35	1.2	—	—	8.2
^{4}He	4.18	5.25	2.26	—	—	20.5
Neon	27.2	44.5	25.9	24.9	324	89.8
Nitrogen	77.3	126.2	33.5	63.1	96.6	199

not easy to retain good thermal coupling between this bath and the experiment. Liquid nitrogen is prepared by first liquefying air (79 percent nitrogen) and distilling this to separate the nitrogen and oxygen.

In addition to providing a common method for cooling experiments to moderate temperatures, liquid nitrogen has a number of other common uses in the laboratory:

1. maintaining an intermediate temperature bath in helium cryostats (better-insulated cryostats are making this use out-dated)
2. precooling cryostat parts that are to be subsequently cooled to liquid helium temperature
3. for use in cold traps and adsorption pumps

b. ***Helium*-4**

^{4}He is the "common," naturally abundant isotope of helium, and the term helium or liquid helium without further specification refers to this isotope. This is the most common cryogenic fluid for use at temperatures below that of liquid nitrogen.

The temperature can be lowered to ~1 K or so by pumping. In principle, lower temperatures can be attained, but the pressure necessary is impractical to obtain (see Table 7.2).

The use of helium is the easiest and best method of cooling an experiment down to lower than 5 K. It is obvious, however, from Table 7.1 that ^{4}He (and ^{3}He as well) has an unusually low latent heat of vaporization. This means that a given quantity of liquid helium will not cool very much material to 4.2 K. However, the gas from evaporated liquid helium is very cold and is capable of absorbing a good deal of heat. Table 7.3 shows the savings that can be obtained if the heat capacity of the cold helium gas as well as the latent heat of vaporization is used to cool an experiment. This stresses the necessity of not cooling the experiment too quickly.

c. ***Helium*-3**

^{3}He has the lowest known BP at any given pressure; we see this from Table 7.1. ^{3}He occurs naturally in only very small quantities—10^{-4} percent of all

Table 7.2. Vapor pressure of helium (mm Hg)

	T(K)		
Isotope	1.0	0.6	0.3
^{3}He	8.56	0.5	1.5×10^{-3}
^{4}He	0.12	2.8×10^{-4}	3×10^{-10}

Table 7.3. Quantity of the liquid necessary to cool 1 kg of copper from 77 K to 4.2 K

Method	Amount of liquid He needed (litres)
Cooling using latent heat of vaporization only	2.3
Cooling using heat capacity of gas as well as latent heat	0.14

helium. It is produced artifically as a byproduct of the radioactive decay of tritium [${}^3_1H \rightarrow {}^3_2He + e^-$], again only in very small quantities. As a result, this isotope is particularly expensive. It is therefore used only when absolutely necessary and then only in a closed experiment so that it is not lost. As you can see from Table 7.2, lower temperatures can be attained with the use of ^3He than with ^4He. Its principal use at present is in dilution refrigerators, which can produce temperatures of about 10 mK.

d. Hydrogen

Liquid hydrogen is inexpensive, since it is produced from a common gas. Furthermore, it is readily available commercially because it is in common use as a rocket fuel. It is of cryogenic interest because it fills in the temperature gap between liquid helium and liquid nitrogen. It has a large latent heat of vaporization, as shown in Table 7.1, despite its low density. Its main drawback is the danger involved in its use. It is very flammable and one must take extreme care, particularly in cases where electrical connections are made in the vicinity of the hydrogen, to avoid explosion. For this reason alone it has largely been replaced by the use of ^4He in cryostats that allow operation at higher temperatures (see section on cryostat design) when experimental conditions in the region of 20 K are required.

e. Air

Liquid air is a mixture of liquid nitrogen and liquid oxygen in the ratio of ~80 percent to ~20 percent. This is the principal product of the liquefaction process from which liquid nitrogen is distilled. It is not usually available commercially. Research facilities that produce their own liquid nitrogen will produce liquid air as the first step. For many applications the distillation of the liquid air to produce liquid nitrogen is not necessary as liquid air will suffice. The uses of nitrogen labeled (2) and (3) above are fulfilled by liquid air as well. Its temperature is between that of liquid nitrogen and liquid oxygen (~81 K). It has two major drawbacks.

1. The exact temperature is determined by the ratio of nitrogen to oxygen present. As the nitrogen has a lower BP, this tends to boil off. As a

result the temperature goes up as the liquid air gets older. It is therefore not of any use in cases where the temperature of the bath must be constant.

2. As the percentage of oxygen in the liquid air increases, for the reasons given above, the liquid becomes progressively more flammable. It is not dangerous to the extent that liquid hydrogen is, but care should be exercised when working with liquid air—particularly when working with old liquid air.

f. Oxygen

Liquid oxygen is generally not of any interest to the physicist interested in obtaining low temperatures except as a component of liquid air.

g. Neon

Neon allows access to a temperature range between those accessible using helium and nitrogen. The operating range of neon, as shown by its BP and TP is not very large. As neon is not readily available, helium is generally used instead in various of the cryostats described in the next section.

7.2 Cryostat design

Cryostats that you will see in the laboratory are for use with either liquid nitrogen or liquid helium. The designs of these two types are generally quite different and we will discuss them separately.

a. Liquid-nitrogen cryostats

Liquid nitrogen is inexpensive and not so cold as to condense and freeze most gases from the air. As a result, its transfer and use is quite simple and the cryostat needed to contain it can be quite simple. We discuss two common types of nitrogen cryostats: the Dewar type and the cold-finger type.

Dewar type. Dewar-type nitrogen cryostats consists merely of an insulated container into which the nitrogen is placed. This is essentially just a "Thermos bottle." It is generally constructed from glass or stainless steel, although inexpensive Dewars may be made of styrofoam. The glass and stainless steel types are double walled and use a vacuum space between the walls for insulation. Glass cryostats are typically silvered to reduce heat loss by radiation. The experiment is enclosed in some appropriate container and is placed in the bath. It is common to evacuate or to evacuate and fill with helium gas the region inside the experimental chamber in order to avoid

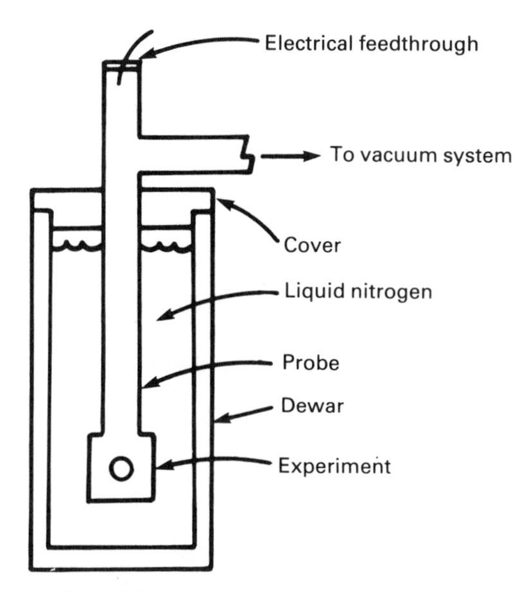

Figure 7.2. A simple liquid-nitrogen cryostat.

condensation of moisture on the experiment. The use of helium gas (referred to as exchange gas) improves the thermal coupling of the experiment to the nitrogen. Appropriate electrical feedthroughs are used if electrical connections are needed for the experiment. These are discussed at the end of this chapter.

A typical experimental arrangement is illustrated in Figure 7.2.

Cold-finger cryostat. The cold-finger design is somewhat more complex than the Dewar type but it has certain advantages over the simpler design. In this design the experiment is coupled to the nitrogen bath by a metallic thermal link (generally copper because of its exceptional thermal conductivity). Two examples are illustrated in Figure 7.3. The "Dewar" part of the cryostat in Figure 7.3*a* can be glass, stainless steel or even styrofoam. In Figure 7.3*b* stainless steel is used so that the copper cold finger may be linked to the bottom of the bath chamber. The advantages of this system over that shown in Figure 7.2 are as follows.

1. In cases where access to the experiment by light, x-rays, etc., is necessary, it is easier to place appropriate windows on this type of cryostat.
2. In cases where the experiment must be placed in a confined area (i.e., between the poles of a magnet) the cold finger arrangement is more convenient.
3. It is a simple matter to raise the temperature of the experiment above the temperature of the bath, by use of a heater attached to the

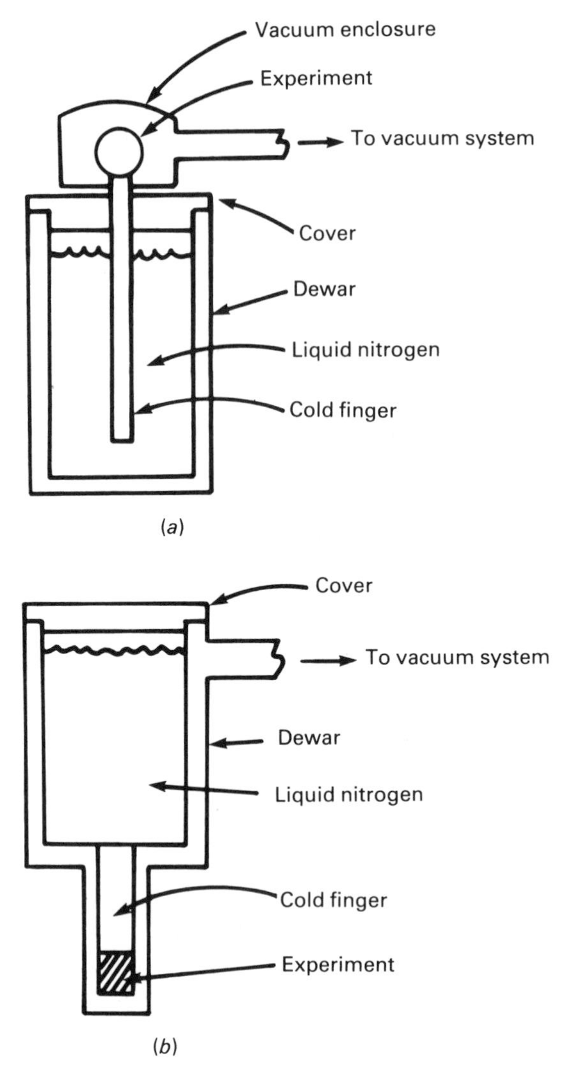

Figure 7.3. Two liquid-nitrogen cryostats using a cold-finger arrangement.

experiment. The degree of thermal coupling between the bath and the experiment can be controlled by placing a thermally resistive link in the cold finger. This can be done by replacing a section of the copper by a material with a lower thermal conductivity (e.g., stainless steel).

b. Liquid-helium cryostats (^{4}He)

Unlike liquid nitrogen, helium is expensive and cold enough to condense and freeze most gases present in air. As a result, both the transfer of liquid helium and the design of a helium cryostat are more complex than is the case for liquid nitrogen. The essential differences are that we must provide better insulation to prevent the helium from being lost, and we must avoid contact of the liquid helium with air. There are numerous designs for helium cryostats; here we discuss two of the most commonly encountered types. We conclude with a discussion of some of the considerations in the transfer of liquid helium.

Dewar-type liquid-helium cryostats. As illustrated in Figure 7.4 we see that a helium cryostat can be constructed that is similar to the Dewar type shown in Figure 7.2. Note that this is a double Dewar: the outer one is filled with liquid nitrogen as an intermediate-temperature bath. It is customary to make the Dewars out of glass to reduce heat transfer due to conduction.

The sample is enclosed in a probe. The probe is evacuated or filled with a small quantity of exchange gas. The amount of exchange gas controls the amount of thermal coupling between the helium-temperature bath and the experiment. If we want to increase the temperature of the sample, usually by electrical means (see section on temperature control), we should decrease the coupling between the experiment and the bath. Typically, helium gas is used as the exchange gas and pressures of a few Torr will generally provide sufficient coupling to keep the experiment at the same

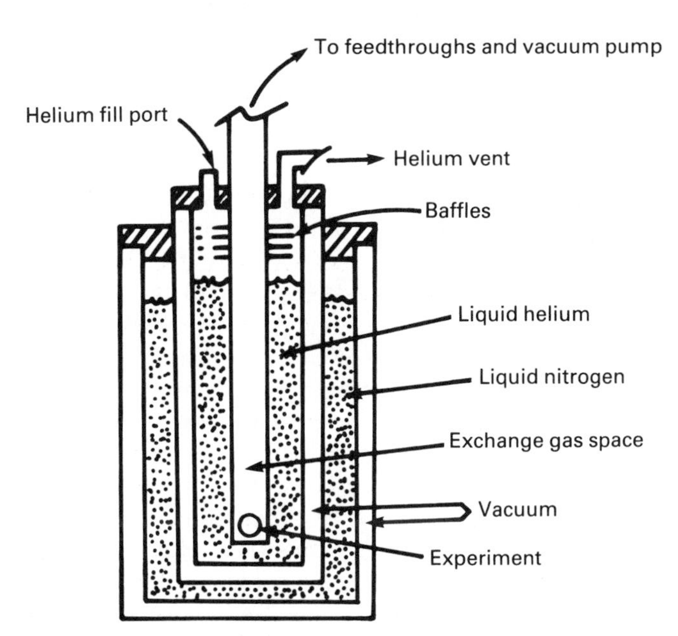

Figure 7.4. A liquid-helium cryostat of the Dewar type using an intermediate-temperature bath of liquid nitrogen.

temperature as the bath. Baffles attached to the top of the probe eliminate heat transfer to the helium bath by convection. It is important that the top of the helium Dewar be covered with an airtight cover. Two openings in the cover are necessary; a helium fill port, through which the helium is transferred into the Dewar and a helium vent for exhausting the gas as the helium evaporates. It is often convenient to include a helium level indicator as well; this is typically inserted through a third hole into the helium chamber.

This type of helium cryostat is among the simplest and most widely used. It has the advantage that it is relatively inexpensive and can be used for experiments at temperatures from 4.2 K up to room temperature. For ease of operation, the continuous-flow cryostat described next is often preferred.

Continuous-flow cryostats. Figure 7.5 shows two arrangements for continuous-flow helium cryostats. The difference in these two designs is as follows. The cryostat in Figure 7.5*a* uses a conventional intermediate-temperature nitrogen bath. The cryostat in Figure 7.5*b* uses *superinsulation* to reduce helium evaporation. The superinsulation consists of many layers of thin (~0.02 mm) aluminized mylar.

In these cryostats the sample is cooled by a flow of cold helium gas controlled by a throttle valve. The rate of helium gas flow is analogous to the amount of exchange gas in the Dewar type of cryostat and can be used in conjunction with an electrical heating system to control the temperature of the experiment from 4.2 K to about room temperature. In other respects the operation of the continuous flow cryostat is similar to the operation of the Dewar type. Because of the complexity of design of the continuous-flow cryostat, it is generally constructed largely of stainless steel, although various parts are sometimes made of aluminum or glass fiber.

Liquid-helium transfer techniques. A typical system for the transfer of liquid helium from a storage Dewar is illustrated in Figure 7.6. The helium space of the cryostat is often precooled with liquid nitrogen in order to reduce the amount of liquid helium needed for cooldown. It is necessary to make sure that all of the nitrogen has been purged from the cryostat before the helium is transferred. The purging should be done with helium gas, in order to avoid condensation of moisture and heavier gases present in compressed air. The exact details of the proper transfer procedure depend on the particular cryostat being used and some experience with a specific cryostat facilitates its use.

During the transfer of helium, it is common to use helium gas to pressurize the helium storage Dewar to force the liquid helium through the transfer tube. The transfer tube is generally surrounded by a vacuum jacket to minimize the transfer of heat to the liquid helium. A typical example is shown in Figure 7.7. In all cases the vacuum obtainable in jackets around

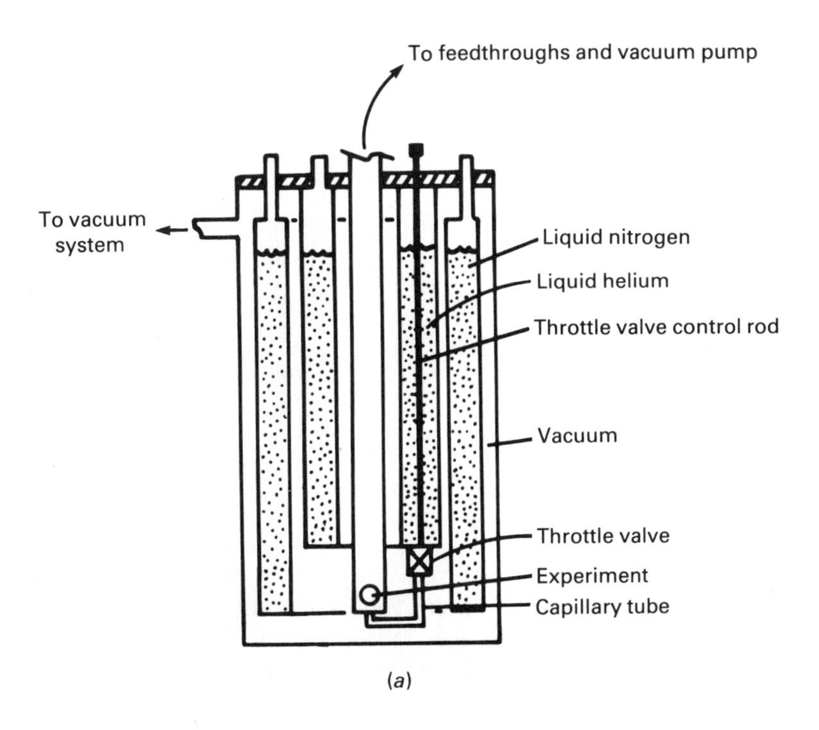

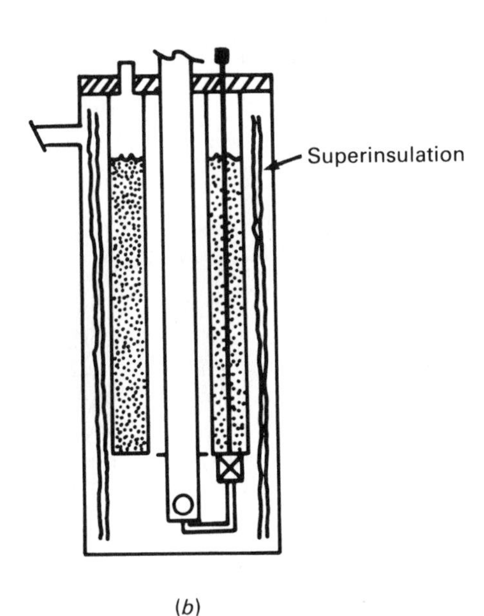

Figure 7.5. Continuous-flow liquid-helium cryostats using (*a*) an intermediate temperature bath of liquid nitrogen and (*b*) superinsulation.

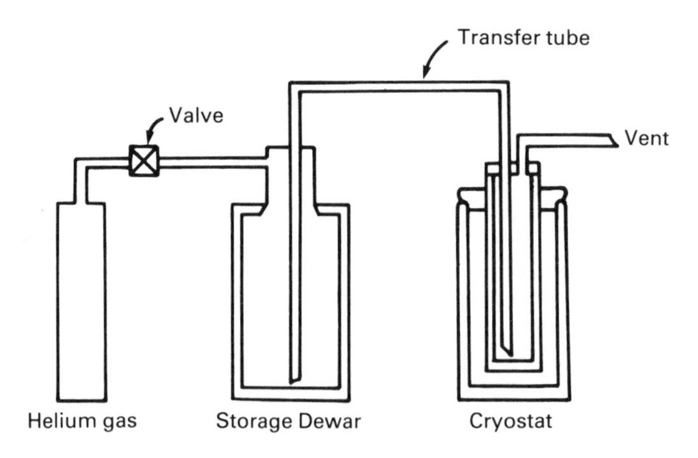

Figure 7.6. The arrangement for liquid-helium transfer.

liquid-helium-containing areas is improved as a result of cryopumping. The transfer tube is inserted prior to the transfer of helium and must be sealed where it enters the storage Dewar and the cryostat to avoid the condensation from air entering into regions at low temperature.

In many research or teaching facilities, helium gas that results from the evaporation of the liquid helium is recoverd to be sold or reliquefied. Recovery is accomplished by the connection of the vent to a gas storage container.

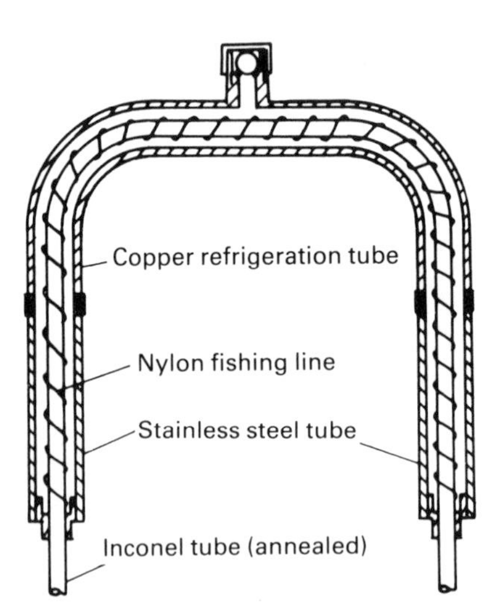

Figure 7.7. A stainless steel liquid-helium transfer tube. (From G. K. White *Experimental Techniques in Low Temperature Physics* 2nd ed. Oxford 1968).

7.3 Seals and feedthroughs

A few words concerning vacuum seals and methods of making electrical connections with the experiment are necessary to clarify certain aspects of cryostat design.

a. Permanent seals

Permanent seals between parts made of stainless steel, brass, or copper can be made using a variety of solders. in general, solders with a higher melting temperature are more difficult to use but provide greater strength. The choice of solder depends on the materials to be joined as well as the application; not all solders necessarily adhere well to all metals. An appropriate flux is necessary and must be chosen in accordance with the type of solder as well as the metals. The choice of torch depends primarily on the temperature needed to melt the solder as well as on the size of the metal pieces involved. Resulting seals are generally good over a wide range of temperatures. The process of soldering depends on the particular situation and in most cases it is an acquired art that is best learned from someone with experience as a machinist, technician, etc.

Aluminum parts may be welded (heliarc welding) to other aluminum parts, but this process requires special equipment and is best left to professionals.

Glass-to-metal seals are available commercially in a variety of standard sizes.

b. Room-temperature vacuum seals

Room-temperature vacuum seals that must be removed periodically are generally made with rubber O-rings. These are rubber toroids, generally with a circular cross section. A seal can be made between two flanges by compressing the O-ring between them. The O-ring is confined in a groove in one of the flanges, which is not quite as deep as the diameter of the O-ring. This allows for the compression of the O-ring and the formation of a seal. This is illustrated in Figure 7.8. The sealing properties of the O-ring are

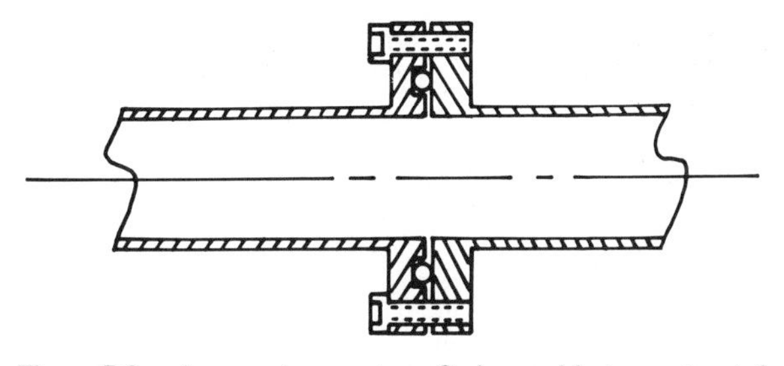

Figure 7.8. A room-temperature O-ring seal between two tubes.

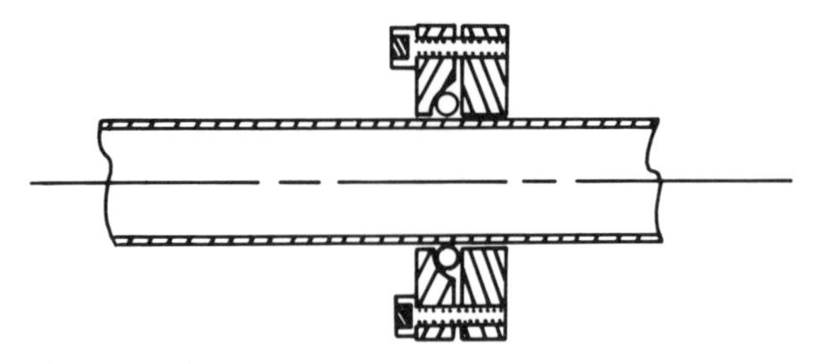

Figure 7.9. A room-temperature O-ring seal between a flange and the outside of a tube.

generally aided by the application of a *small* quantity of vacuum grease to the surface of the ring. Note the emphasis on small.

A similar seal between a continuous pipe and a flange can be formed as shown in Figure 7.9. A variation on this seal is shown in Figure 7.10 and enables a vacuum seal to be made around a rotating shaft.

Design criteria for these kinds of seals can generally be found in the technical notes of O-ring manufacturers' catalogues. O-rings can be reused many times.

c. *Low-temperature seals*

Indium seals. Seals of the O-ring type described above can be made that can be used at low temperatures, but since rubber becomes brittle some other O-ring material is necessary. Indium is the accepted material. This is readily available in the form of wire, and O-rings are made by forming a circle of the appropriate size and slightly overlapping the ends. Indium is a very soft metal and forms a seal as a result of its deformation when pressed between two surfaces. An indium seal can be cycled between room temperature and liquid-helium temperature many times without being damaged. However, once the seal is broken the indium O-ring must be

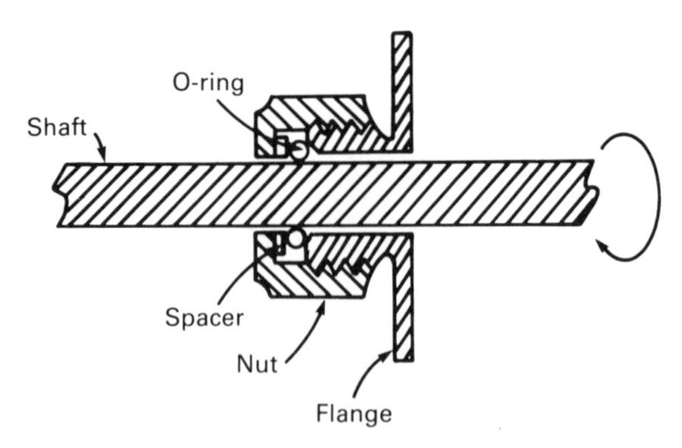

Figure 7.10. Rotating O-ring seal for use at room temperature.

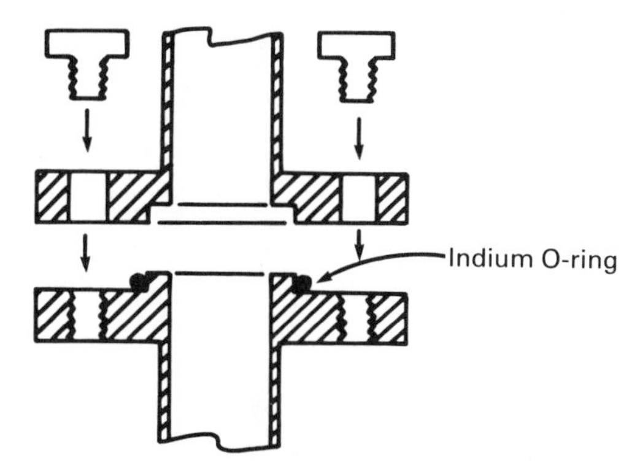

Figure 7.11. An indium O-ring seal for use at low temperatures, shown prior to assembly.

replaced. The indium itself should be saved as the metal is expensive and the production of new wire from indium scrap is not a difficult process. An example of an indium seal is shown in Figure 7.11. An O-ring groove of the sort shown in Figure 7.8 can also be used. It is important to ensure that the indium is clean before making the seal.

Wood's metal seals. Wood's metal (50wt.% Bi, 25wt.% Pb, 12.5wt.% Sn, 12.5wt.% Cd) has a very low melting temperature (~65°C). "Temporary" solder joints may be made and broken merely by heating Wood's metal with a heat gun. As the melting temperature is so low, no damage is done to the more permanent soldered joints. Several similar alloys are sometimes used, but Wood's metal is the most common. Various fluxes can be used with Wood's metal—zinc chloride solution is common. Further details are available in G. K. White, *Experimental Techniques in Low Temperature Physics.*

d. Electrical feedthroughs

Glass–metal feedthroughs. Two types of glass-to-metal feedthroughs are illustrated in Figure 7.12. The feedthrough consists of three parts: (1) a solid or hollow pin for electrical connection; (2) an insulating glass portion; and (3) a solderable metal outer part to be connected to the metal wall of the vacuum chamber. Although we have illustrated here feedthroughs with only one electrical connection, devices with up to a dozen or more electrical connections are commercially available. Feedthroughs with solid pins are used in most cases. A wire on the inside of the vacuum area is soldered to one end of the pin and a wire on the outside of the vacuum space to the other end. Hollow pins are used in cases where wires cannot have soldered joints. This is notably the case for thermocouple wires. In this case the wire is run through the hole in the pin and a vacuum seal is made by soldering

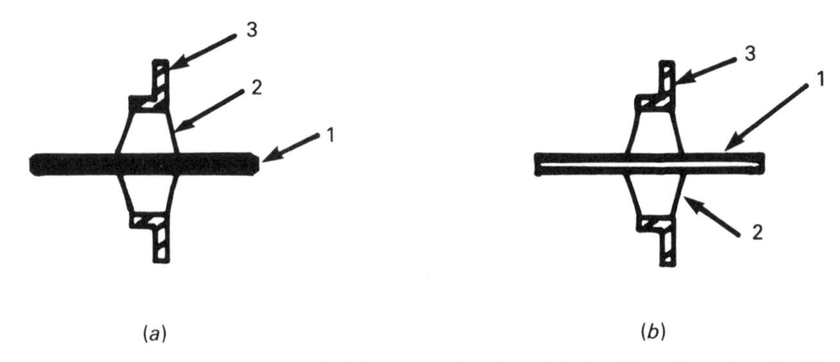

Figure 7.12. Glass-to-metal electrical feedthroughs: (*a*) solid pin; (*b*) hollow pin. The identity of the parts is described in the text.

around the wire. These are ideal for use at room temperature—that is, at the top of a cryostat. Although they can be used to some extent at cryogenic temperatures, they are not reliable for high vacuum.

Rubber feedthroughs. Figure 7.13 shows an arrangement for feeding wires through two rubber washers. Note that it is important to remove any plastic insulation from the wires where they pass between the rubber washers, otherwise air can leak into the system between the insulation and the wire. The rubber will provide good insulation between the wires and the

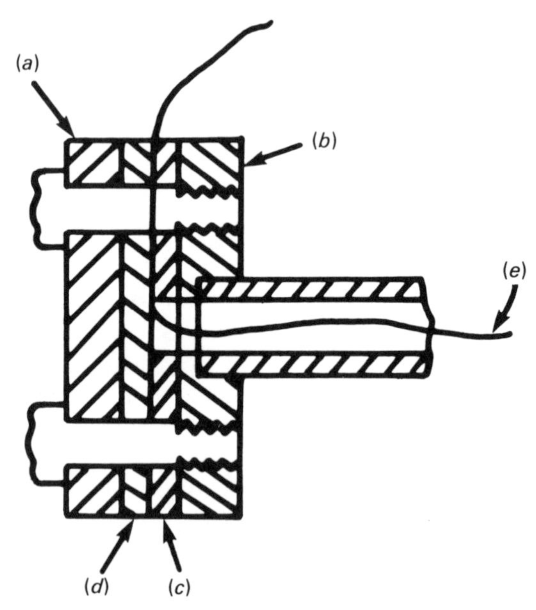

Figure 7.13. A rubber electrical feedthrough: (*a*) top flange; (*b*) bottom flange; (*c*) bottom rubber washer; (*d*) top rubber washer; (*e*) wires.

flange. This feedthrough is easy to construct in a small machine shop and is ideal for thermocouple wires as the wires themselves are continuous through the feedthrough. In order to ensure a good vacuum seal, the washers should be coated (lightly) on both sides with vacuum grease. These are obviously unsuitable at low temperatures.

7.4 Cryogenic level indicators

There is a variety of different types of level indicators in use. The discussion here is not comprehensive but will cover the more commonly used types.

a. Acoustic oscillation indicator

A simple device that is commonly used for periodically checking the level of liquid helium in storage Dewars and cryostats is illustrated in Figure 7.14. It consists of a long stainless steel tube (the length depends on the depth of the cryostat) about 4 mm in diameter. On the top end is a funnel-like device covered with a diaphragm made of rubber or thin metal. Oscillations are set up in the diaphragm when the open end of the tube is lowered below the level of the liquid helium. The frequency is in the order of 10 Hz. If the tube is raised so that the end of the tube is above the liquid helium level the frequency of the oscillations will increase considerably. The reason for the change in the frequency of the oscillators has to do with the resonant frequency of an open tube versus a closed tube. The reason why the oscillations occur in the first place is rather more complex and has been

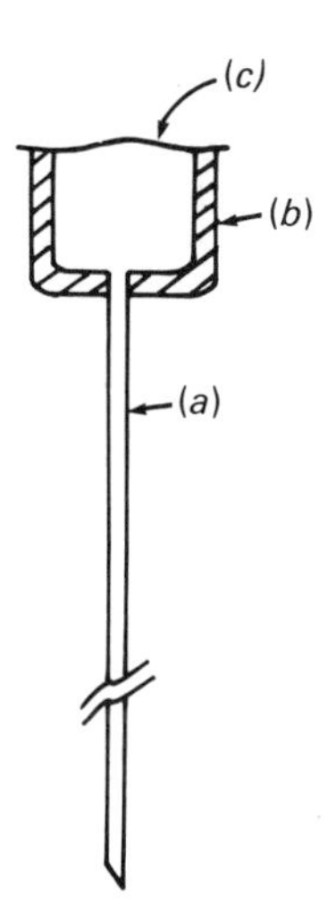

Figure 7.14. A liquid-helium level indicator: (*a*) stainless steel tube; (*b*) funnel; (*c*) diaphragm.

described by Hoare *et al. Experimental Cryophysics,* (London: Butterworths, 1961). By raising and lowering the tube the level of the liquid can be located to within about 0.5 cm. This method is effective only for helium, although it can be used with less ease for hydrogen and neon. This method is simple but can be used only for periodic checking of the helium level, for example, in measuring the level of a storage Dewar before and after a transfer. The three methods described next are suitable for more frequent monitoring of the liquid level.

b. Flotation devices

A straightforward method of measuring the level of a cryogenic fluid is by the use of a float. The simplest method is to attach to the float a light rod that protrudes from the top of the cryostat or Dewar. The level of the liquid is determined by the height of the rod above the top of the Dewar. This method is suitable for all types of liquids provided the float is made of a material with low enough density to float in the liquid of interest. This may not be a simple matter for liquid hydrogen, as its density is very low (0.07 g cm^{-3}). This method has largely been replaced for most liquid-helium and liquid-nitrogen cryostats by the more elegant methods described below.

c. Superconducting indicators

A piece of wire with a superconducting transition temperature above 4.2 K can be used as an effective liquid-helium level indicator. A typical design is shown in Figure 7.15. The wire is typically a niobium alloy (transition

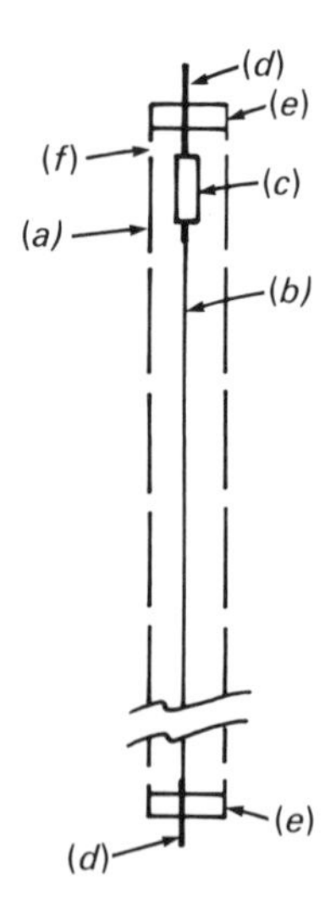

Figure 7.15. Superconducting liquid-helium level indicator: (*a*) outer housing; (*b*) superconducting wire; (*c*) heater resistor; (*d*) nonsuperconducting (normal) wires for electrical connections; (*e*) insulating end plugs; (*f*) holes for entry of liquid.

temperatures in the range of 10 K). The portion of the wire below the liquid helium is superconducting and has zero resistance. The portion of the wire above the liquid is normal. The total resistance is related to the height of the liquid and the proportion of normal wire. A four-point resistance measurement is necessary. The heater resistor is generally necessary to ensure that the wire above the liquid is warmed above the transition temperature. Given the known superconducting transition temperatures of metals, it is obvious and that this type of indicator is suitable only for helium. A control box that measures the resistance of the wire and is calibrated in inches or centimeters of helium is available for commercially made indicators.

d. Capacitance indicators

A long coaxial capacitor made of two lengths of stainless steel tubing, as shown in Figure 7.16, can be used as a level indicator. The capacitance is determined by the geometry of the cylinders and also by the dielectric constant of the medium between the cylinders. The dielectric constant of a liquid is different from the dielectric constant of the same substance in the gaseous state. This is obviously because of the difference in the density. Thus the total capacitance of the device depends on the level of the liquid. Commercially available gauges have "capacitance" meters that are calibrated in inches or centimeters of liquid. These devices are most commonly used for liquid nitrogen.

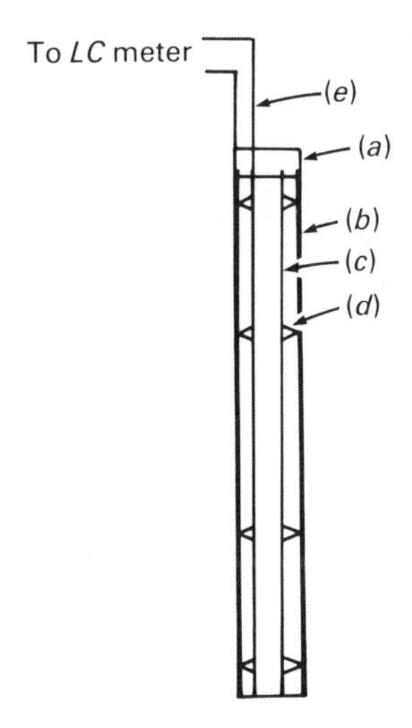

Figure 7.16. Cylindrical capacitor nitrogen depth sensor: (*a*) top plug; (*b*) outer stainless steel tube; (*c*) inner stainless steel tube; (*d*) spacers; (*e*) electrical connections.

7.5 Temperature control

We discuss two types of temperature control: (1) that in which the experiment is well linked (thermally) to the bath and the temperature is adjusted by changing the bath temperature, and (2) that in which the thermal link with the bath is weak and the temperature is regulated by supplying heat by means of an electrical heater. The first type of control results from a control of the fluid's vapor pressure and has been mentioned already. For most fluids this is usable only over a small temperature range and we discuss this only briefly. The second is more commonly used and is also applicable to the control of the temperatures of experiments well above room temperature. In this case the experiment is heated in a furnace (or oven), the heat output of which is sufficient to obtain an equilibrium temperature somewhat less than the desired experimental temperature. Fine temperature control is then accomplished by means of a small heater and a feedback circuit as described below.

a. Vapor pressure methods

The temperature of a cryogenic bath can be lowered by pumping. The temperature is regulated by controlling the vapor pressure over the liquid. The pressure is regulated by use of a manostat, the arrangement of which is shown in Figure 7.17. Details of two types of common manostats are shown in Figures 7.18 and 7.19. The pressure of the fluid is lowered by pumping with the bypass valve and valve G open until the desired vapor pressure (corresponding to the desired temperature) is reached. Note that since we are concerned with measuring P from 1 to 760 mm Hg with an accuracy of, say, ±1 mm Hg, a good-quality Bourdon gauge is genearally the best choice

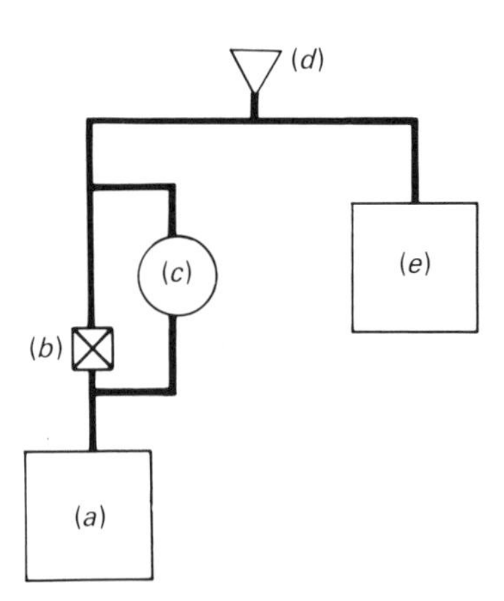

Figure 7.17. Pumping arrangement using a manostat: (*a*) vacuum pump; (*b*) valve; (*c*) manostat; (*d*) pressure gauge; (*e*) chamber.

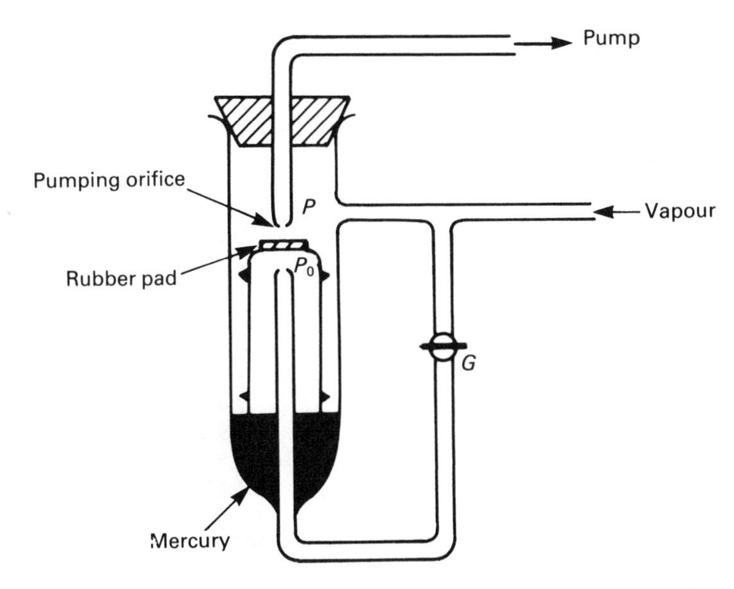

Figure 7.18. A cartesian-diver manostat. (From G. K. White, *Experimental Techniques in Low Temperature Physics,* 2nd ed. Copyright © 1965 by Oxford University Press.)

for the manometer, although a mercury manometer is suitable as well. At this point, pressure P_0 is equal to the desired pressure and valve G and the bypass valve are closed. The control valve formed by the pumping line and a rubber seal is cyled in order to keep P (and the vapor pressure) equal to P_0. Control of the pressure of liquid nitrogen to within 2–3 mm Hg is possible. This translates into temperature control of ~0.1 K (see Table 7.4).

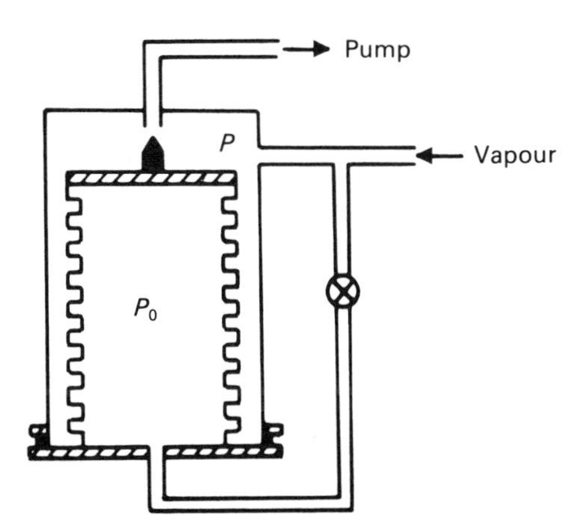

Figure 7.19. A bellows manostat. (From G. K. White, *Experimental Techniques in Low Temperature Physics,* 1st ed. Copyright © 1959 by Oxford University Press.)

Table 7.4. Vapor pressure of gases

	Temperature (K)					Temperature (K)			
P(mm Hg)	^4He	H_2	N_2	O_2	*P*(mm Hg)	^4He	H_2	N_2	O_2
800	4.265	20.44	–	90.69	90	2.575	14.77	62.94	73.80
780	4.238	20.37	77.58	90.43	80	2.514	14.56	62.39	73.09
760	4.210	20.27	77.36	90.18	70	2.447	14.30	61.77	72.29
740	4.182	20.18	77.14	89.93	60	2.373	14.04	61.06	71.39
720	4.153	20.09	76.91	89.67	50	2.290	13.73	60.24	70.39
700	4.124	20.00	76.67	89.41	45	2.244	13.57	—	69.81
680	4.094	19.90	76.43	89.14	40	2.194	13.40	59.28	69.15
660	4.064	19.80	76.18	88.86	35	2.140	13.21	—	68.46
640	4.033	19.70	75.93	88.58	30	2.081	12.99	58.08	67.67
620	4.001	19.60	75.68	88.29	25	2.016	12.74	—	66.74
600	3.969	19.50	75.41	87.99	20	1.942	12.46	56.48	65.67
580	3.936	19.40	75.13	87.69	18	1.908	12.32	—	65.17
560	3.902	19.30	74.85	87.38	16	1.872	12.17	55.65	64.59
540	3.867	19.18	74.56	87.06	14	1.833	12.02	—	64.00
520	3.832	19.06	74.27	86.72	12	1.789	11.83	—	63.32
500	3.795	18.94	73.97	86.37	10	1.739	11.62	—	62.50
480	3.757	18.82	73.66	86.01	9	1.712	11.50	—	62.07
460	3.719	18.69	73.34	85.65	8	1.682	11.36	—	61.58
440	3.679	18.57	73.00	85.29	7	1.649	11.23	—	61.01
420	3.638	18.43	72.66	84.89	6	1.613	11.08	—	60.39
400	3.595	18.29	72.30	84.48	5	1.571	10.88	—	59.68
380	3.551	18.15	71.92	84.06	4	1.522	10.65	—	58.80
360	3.506	18.00	71.54	83.62	3.5	1.495	10.53	—	58.29
340	3.459	17.83	71.13	83.16	3.0	1.464	10.38	—	57.73
320	3.410	17.67	70.70	82.68	2.5	1.428	10.23	—	57.06
300	3.358	17.50	70.25	82.18	2.0	1.387	10.02	—	56.28
290	3.332	17.41	70.01	81.92	1.5	1.336	—	—	55.28
280	3.305	17.32	69.77	81.65	1.0	1.270			
270	3.277	17.22	69.53	81.36	0.9	1.254			
260	3.248	17.13	69.28	81.07	0.8	1.236			
250	3.219	17.03	69.02	80.78	0.7	1.217			
240	3.189	16.93	68.75	80.48	0.6	1.195			
230	3.158	16.82	68.48	80.16	0.5	1.170			
220	3.126	16.71	68.18	79.83	0.4	1.140			
210	3.094	16.61	67.88	79.49	0.3	1.104			
200	3.060	16.49	67.57	79.14	0.2	1.056			
190	3.025	16.36	67.25	78.77	0.1	0.982			
180	2.988	16.23	66.91	78.39	0.08	0.960			
170	2.951	16.10	66.56	77.98	0.06	0.933			
160	2.911	15.97	66.20	77.57	0.04	0.897			
150	2.871	15.83	65.81	77.13	0.02	0.841			
140	2.828	15.68	65.42	76.65	0.01	0.791			
130	2.783	15.52	64.98	76.16					
120	2.735	15.35	64.52	75.64					
110	2.685	15.18	64.03	75.08					
100	2.632	15.00	63.50	74.47					

Source: From G. K. White, *Experimental Techniques in Low Temperature Physics* (New York: Oxford University Press, 1959).

b. Resistance heating

The temperature of an experiment can be raised above the bath temperature by supplying a current to an electric heater thermally coupled to the experiment. In this method the temperature is measured by some appropriate sensor and is compared to a set temperature for the purpose of controlling the temperature of the experiment. There is generally an additional sensor that is used to make an accurate temperature measurement of the experiment. If the experient is too cold, heat is supplied by the electric heater; if the experiment is too hot, the heater is turned off and the experiment is cooled by the bath of cryogenic fluid. In a continuous-flow liquid-helium cryostat, the equilibrium temperature of the experiment, determined by the setting of the throttle valve, is kept somewhat below the final desired temperature and the heater is used to regulate the temperature. Figure 7.20 shows a simple method of doing this. R_{set} is set to equal the resistance of the thermistor at the required temperature. If $R_{th} \neq R_{set}$, then $V_1 \neq V_2$ and there is an output from the amplifier. In cases where a thermistor is used, if heating is required, then $R_{th} > R_{set}$ and $V_1 < V_2$, so the output of the amplifier is negative. The diode ensures that this current will get to the heater. If the bridge unbalance were due to $R_{th} < R_{set}$, a condition resulting when the experiment is too hot, the amplifier output would be positive and would be blocked by the diode. Hence there would be no heating. When a platinum thermometer (or something with a positive temperature coefficient of resistivity) is used as the sensor, the direction of the diode must be reversed.

An arrangement similar to the above for use with a thermocouple (voltage-producing sensor) is shown in Figure 7.21. In this case the voltage of the measuring thermocouple is compared by a differential amplifier to a set voltage produced by a voltage-divider circuit.

In both of these cases, the amplifier is generally not a single op-amp but usually comprises several stages of amplification with adjustable gains and offsets at each stage.

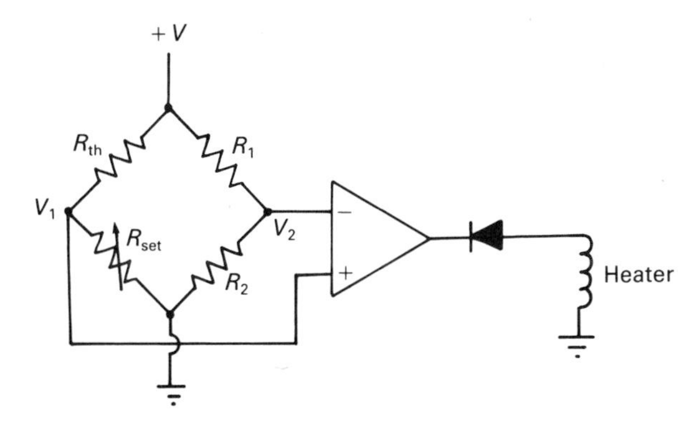

Figure 7.20. A temperature controller using a thermistor as a temperature sensor.

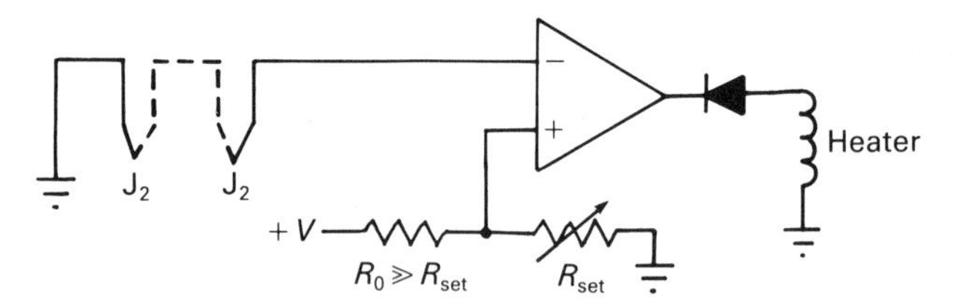

Figure 7.21. A temperature controller using a thermocouple as a temperature sensor.

Commercially available temperature controllers are often much more complicated than those discussed here and feedback to the heater consists not only of a signal proportional to the temperature error but also of signals proportional to its time derivative and its time integral. This is done in order to minimize the instabilities of the temperature (i.e., oscillations about the set point), and errors between the set and actual temperatures.

8
Light sources

8.1 Noncoherent light sources

We consider three types of light sources in this section: sources that produce line spectra; sources that produce band spectra; and continuum sources. Lasers and laser diodes (including LEDs) will be discussed in the next section.

a. Sources of line spectra

The de-excitation of atomic electrons results in the emision of photons. As the energy levels in an atom are quantized, the photons produced by this method are of discrete wavelengths. Let us look in detail at the quantization of energy levels in the simplest of atoms—hydrogen.

Although the problem of an electron in an atom is a quantum-mechanical one, we can obtain satisfactory results for the hydrogen atom without getting involved with too much theory. We begin with a classical picture of the electron orbiting the nucleus (a proton). We consider a circular orbit, so that the radial acceleration is given by v^2/r. From Newton's laws and the Coloumb force we write the equation of motion for the electron as

$$\frac{m_e v^2}{r} = \frac{e^2}{4\pi\epsilon_0 r^2} \tag{8.1}$$

where ϵ_0 is the permittivity of free space. The total energy (kinetic plus potential) is given by $\mathscr{E} = T + V$ and in this case is

$$\mathscr{E} = \tfrac{1}{2} m_e v^2 - \frac{e^2}{4\pi\epsilon_0 r} \tag{8.2}$$

These two equations can be combined to give

$$\mathscr{E} = -\frac{e^2}{8\pi\epsilon_0 r} \tag{8.3}$$

Now let us go back to the electron-in-a-box problem; recall Figure 1.3. The electron in a circular orbit is rather like this. The electron wavefunction must match up with itself when the electron travels all the way around the orbit. To put it another way; the circumference of the orbit must be an integral multiple of the electron wavefunction wavelength, otherwise the wavefunction would interfere distructively with itself. We therefore write

$$n\lambda = 2\pi r \qquad n = 1, 2, 3, \ldots \tag{8.4}$$

Using equation (1.4) we find $\lambda = 2\pi\hbar/p$ (where p is the momentum), so that equation (8.4) becomes

$$n\hbar = pr \tag{8.5}$$

We note here that pr is the angular momentum. Writing $p = m_e v$ we combine equation (8.5) with equation (8.1) to obtain

$$\frac{n^2\hbar^2}{r} = \frac{e^2 m_e}{4\pi\epsilon_0} \tag{8.6}$$

Using this for r in equation (8.3) we obtain the allowed energy levels as

$$\mathscr{E} = \frac{e^4 m_e}{2(4\pi\epsilon_0\hbar n)^2} \tag{8.7}$$

This is generally written as

$$\mathscr{E} = -\frac{2\pi\hbar cR}{n^2} \tag{8.8}$$

where R is the Rydberg constant. For very heavy atoms, R is found to be $1.09738 \times 10^7\ \mathrm{m}^{-1}$. For hydrogen it is about 0.055 percent smaller than this value because the motion of the electron perturbs the motion of the proton. Note that $1/0.00055 = 1820$; the ratio of the proton to the electron mass.

The integer n is referred to as the *principal quantum number* and designates the energy state of the electron. The energy is measured relative to the energy of two charges at infinite separation. $\mathscr{E}$ is negative and as n becomes large $\mathscr{E}$ approaches zero. The $n = 1$ state is the ground state and all other values of n represent excited states.

Figure 8.1 shows the energy states of atomic hydrogen. The various series represent transitions to the ground state or one of the excited states. The energy of the transition, that is, the energy of the photon that is emitted, is given by

$$\Delta\mathscr{E} = 2\pi\hbar cR\left(\frac{1}{n_1^2} - \frac{1}{n_2^2}\right) \tag{8.9}$$

The wavelengths associated with the different series are given in Table 8.1. We see that if we are interested in visible light sources we should consider the Balmer series. We can construct a light source that will emit photons at the wavelengths given by these transitions. The general design of the source is as shown in Figure 8.2. Hydrogen at a pressure of a few Torr is contained in a glass tube, which has a "U" shaped geometry, although straight tubes

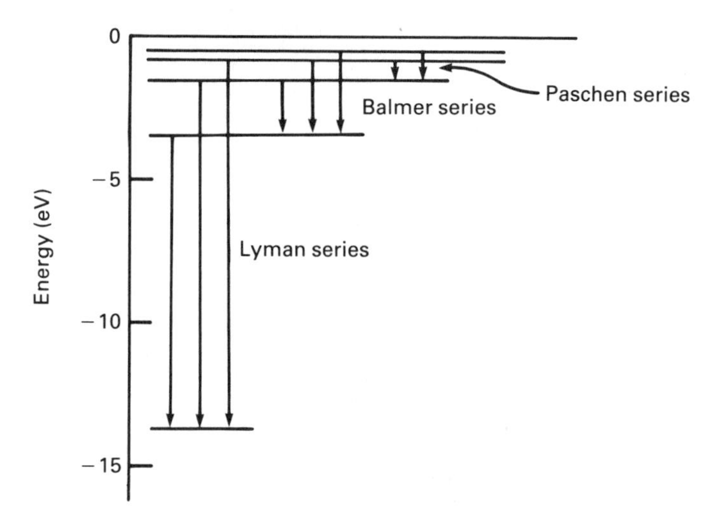

Figure 8.1. The energy levels of atomic hydrogen. Typical transitions from various series are illustrated.

also exist. A large potential (5–10 kV) is applied between the electrodes. The electrical discharge causes an excitation of the atomic electrons and the subsequent de-excitation results in the characteristic emission spectrum. The electrodes are commonly hollow nickel cylinders. The large surface area of this geometry reduces sputtering (deposition of the electrode metal on the inside of the Pyrex tube).

This type of source is the common discharge tube and this design may be used for producing characteristic line spectra of gaseous elements other than hydrogen by placing the appropriate gas in the tube. Spectra of metals with sufficiently high vapor pressure (e.g., mercury, cadmium, etc.) can be produced by placing a small quantity of the metal in the tube. A quantity of gas (e.g., argon) is placed in the tube and this gas begins the discharge that vaporizes the metal. Thus there are present in the spectrum emission lines due to the gas as well as to the metal. These are generally much weaker

Table 8.1. Atomic transitions in hydrogen

n_1	n_2	Designation	Wavelengths region[a]
1	2, 3, 4, . . .	Lyman series	Far UV
2	3, 4, 5, . . .	Balmer series	Near UV and visible
3	4, 5, 6, . . .	Paschen series	Near Ir
4	5, 6, 7, . . .	Bracket series	IR
5	6, 7, 8, . . .	Pfund series	IR

[a] UV = ultraviolet; IR = infrared.

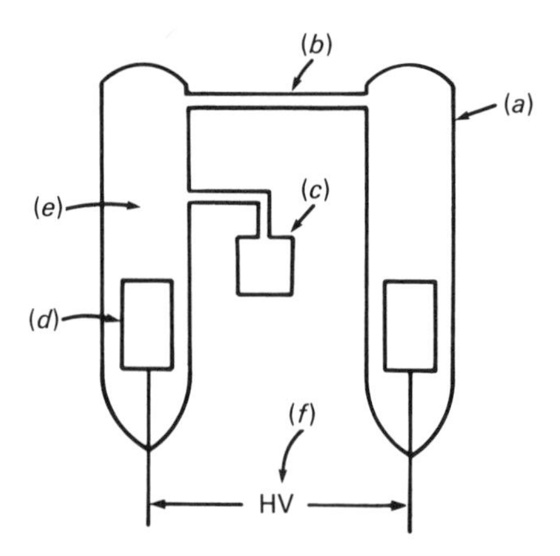

Figure 8.2. A gas discharge tube: (*a*) Pyrex tube; (*b*) area of maximum emission; (*c*) vapor trap; (*c*) electrode; (*e*) volume of gas; (*f*) high-voltage source.

than the metallic lines and we would normally choose a gas that did not have lines at wavelengths that would interfere with the metallic spectrum.

Some elements do not respond to excitation by this method. In these cases a small quantity of the element may be attached to the tip of the carbon electrode of an arc lamp. The resultant spectrum is a combination of that from the desired element and that from carbon. The relative intensity of the carbon lines may be reduced if a slit is placed in front of the source and centered on the middle of the arc. Characteristic line wavelengths in the visible region for various common line sources are given in Table 8.2. As a guide, the approximate range of wavelengths corresponding to each spectral color is given in Table 8.3.

b. Sources of band spectra

Molecules have discrete energy states and as a result the photons emitted during transitions between states are of specific frequencies. For a molecule there are three contributions to the energy:

$$\mathscr{E} = \mathscr{E}_{\text{vib}} + \mathscr{E}_{\text{rot}} + \mathscr{E}_{\text{el}} \tag{8.10}$$

where $\mathscr{E}_{\text{vib}}$ is the vibrational energy, and $\mathscr{E}_{\text{rot}}$ is a rotational energy. $\mathscr{E}_{\text{el}}$ is the electronic energy of the molecule, which is analogous to the principal quantum number of an atom.

The number of modes of motion (degrees of freedom) for a molecule depends on the number of atoms that make up the molecule. A molecule of N atoms has $3N$ modes of motion: of these, three are translational (along the x, y, and z axes) and three are rotational (about the x, y and z axes), except in the case of linear molecules, for which there are two rotational

Table 8.2. Wavelengths of some line sources (major lines)

Element	Wavelength (nm)					
	Red	Orange	Yellow	Green	Blue	Violet
H	656	—	—	—	486	410
						434
Na	—	615	589	498	466	449
		616	590	515	467	450
				568	475	
				569	475	
Hg	691	—	577	546	436	405
			579			408
K	759	—	587	556	—	445
	760			557		446
	763					450
	768					
	769					
Kr	—	—	—	—	—	427
						432
						436
						438
He	—	—	—	—	—	412
Ar	696	—	—	—	—	395
	707					404
	727					416
	738					418
	750					419
	751					420
						426
						427
						430
						434
Ne	627	598	585	533	—	—
	630	603	588	534		
	633	607	594	540		
	638	610				
	640	614				
	651	616				
	653	622				
	660					
	668					
	672					
	693					
	702					
	703					
	706					
	717					
	724					
	744					
	749					
	754					

Table 8.3. Color of photons of various wavelengths

Color	Wavelength range (nm)
Red	622–770
Orange	597–622
Yellow	577–597
Green	492–577
Blue	455–492
Violet	390–455

modes (the two directions perpendicular to the axis of the molecule). The remaining modes, $3N$-6 (or $3N$-5 for linear molecules) are vibrational.

The differences in the energies between rotational states are typically a few hundredths of an electron volt; the energy differences between vibrational states are tenths of an electron volt; while between electronic states the differences are a few electron volts, as they are in atoms. This gives us an idea of the wavelengths of photons emitted in various transitions. The energy is related to the wavelength as

$$\lambda(\text{nm}) = \frac{1.24 \times 10^3}{\mathscr{E}(\text{eV})} \tag{8.11}$$

Optical transitions correspond to energies of 1.6–3.2 eV.

A hypothetical energy-level diagram for a (nonlinear) molecule with three atoms is shown in Figure 8.3. Table 8.4 describes various transitions that can occur in this system—see the numbers 1–6 on the figure.

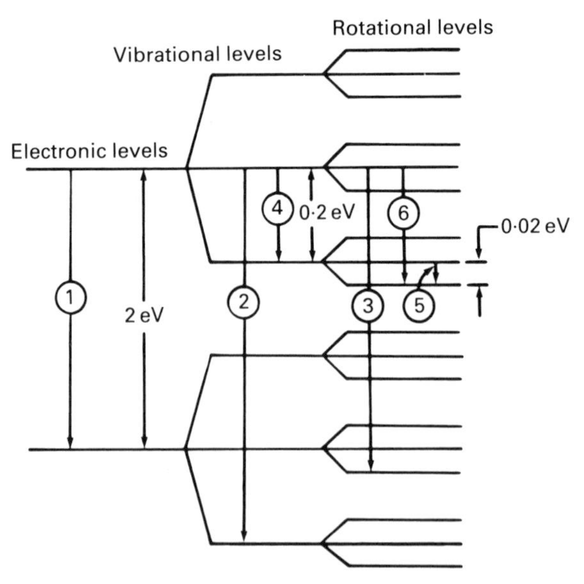

Figure 8.3. Energy-level diagram of a three-atom (nonlinear) molecule. The identity of the numbered transitions is given in Table 8.4.

Table 8.4. Identity of the transitions shown in Figure 8.3

No.[a]	Type	Energy (eV)	Wavelength (nm)
1	Electronic	2.0	620
2	Electronic + vibrational	2.2	563
3	Electronic + rotational	2.02	613
4	Vibrational	0.2	6.2×10^3
5	Rotational	0.02	6.2×10^4
6	Vibrational + rotational	0.22	5.64×10^3

[a] The numbers refer to the numbered transitions in Figure 8.3.

We see that photons from purely rotational transitions are in the far infrared or microwave region. Those from transitions that involve vibrational or vibrational and rotational states are in the near infrared and from those that involve electronic transitions, either alone or in conjunction with another transition, are in the visible.

We can see that the effect of the rotational state is to cause a small shift in the wavelength of the transition between electronic states. Thus the spectrum does not consist of single lines but rather of bands comprising of many lines that are separated by a few nanometers. Typically, the more complex the molecule, the more complex is the structure of the spectrum.

Sources of band spectra can often be prepared in the same manner as those of atomic line spectra.

c. *Continuum sources*

The most common continuum source is the tungsten incandescent bulb. The spectrum produced by a tungsten lamp is a "gray body" spectrum (see Figure 6.13). The location of the peak is, of course, a function of the temperature of the filament. You can see from the figure that most of the energy of this source is radiated in the infrared.

A standard fluorescent light bulb produces an energy spectrum that is roughly that of gray body. In this case the peak is at around 500 nm and essentially all of the photons are emitted in the visible. In addition to the continuum there are several emission bands. These result from electronic transitions in the gas filling in the tube and also in the phosphor coating on the inside of the tube. The particular structure of these bands depends on the specific compounds used in the tube.

An arc lamp frequently produces a continuum as well. In this case, however, the spectrum is quite complex and in addition to the continuum exhibits a large number of discrete lines due to transitions in the gas atoms near the arc.

8.2 Lasers

a. The theory of laser operation

Consider an atom with two electronic states—a ground state and an excited state—with a separation $\Delta\mathscr{E}$. If, at room temperature, $\Delta\mathscr{E} \gg k_B T$ we would expect that the probability of an atom being in the excited state would be very small. We can, however, induce transitions to the excited state by supplying just the right amount of energy. One way of doing this is with a photon of just the right frequency, given by $h\nu = \Delta\mathscr{E}$. An atom in the excited state will decay back to the ground state after, on the average, some time equal to the half-life of the state. This is because the excited state is unstable. Typical half-lives of lasing transitions are in the order of 10^{-4} s. This type of decay gives off a photon of energy $h\nu = \Delta\mathscr{E}$ and is referred to as spontaneous emission.

If an atom is in the excited state, we can *induce* a transition to the ground state after a time much shorter than the half-life by supplying energy. If a photon of energy $h\nu \approx \Delta\mathscr{E}$ (here the energy has to be only approximately correct) is incident upon the atom in the excited state, a transition to the ground state can occur. This is referred to as stimulated emission, and the result is two photons—the incident one with $h\nu \approx \Delta\mathscr{E}$ and the emitted one with $h\nu = \Delta\mathscr{E}$.

In a system at room temperature, a photon with $h\nu = \Delta\mathscr{E}$ is more likely to be absorbed and cause a transition from the ground state to the excited state than it is to cause stimulated emission. This is simply because at room temperature there are a lot more atoms in the ground state than in the excited state. We can, however, make stimulated emission more common than absorption by creating a situation in which there are more atoms in the excited state than in the ground state; such a situation is referred to as a *population inversion*. In this situation an incident photon with $h\nu \approx \Delta\mathscr{E}$ will be more likely to cause stimulated emission. The incident and emitted photons will, for quantum-mechanical reasons, emerge travelling in the same direction as the original incident photon and both will have the same phase. These two photons can then cause additional stimulated emissions, and in this way an avalanche process can occur and a large number of photons will be produced all traveling in the same direction and with the same phase. These photons are referred to as coherent. This is the principle on which the laser operates and is implied in its name: *l*ight *a*mplification by *s*timulated *e*mission of *r*adiation.

The light produced by a laser is, in a sense, monochromatic. We must, however, not take this terminology too literally. There is, of course, a width for $\Delta\mathscr{E}$ that results from the finite lifetime of the state and is given by Heisenberg's uncertainty principle as $\delta(\Delta\mathscr{E})\tau_{1/2} > h$. For a state of $\tau_{1/2} = 10^{-4}$ s, $\delta(\Delta\mathscr{E}) \sim 10^{-11}$ eV: this is in a transition of perhaps 2 eV, and can be neglected. However, there are mechanisms that result from the dynamics of atomic transitions in materials, the details of which are beyond the scope of this book, which do cause a spread of energies that can be in the range of 10^{-5} eV. This corresponds to a spread of wavelengths of about 0.003 nm. This is certainly beyond the resolution of any but the best spectrometers, but it is important, as we shall see shortly.

The amplification of light is enhanced by placing the emitting atoms inside a resonant cavity. This is generally a container with mirrors on the ends (the mirror on one end is not fully reflecting, in order to allow the beam to emerge). The frequencies at which this resonant enhancement of the light amplitude occurs are those for which the length of the cavity is an integral number of half-wavelengths. Hence there are a number of resonant modes of different wavelengths. Let us do a rough calculation to see how close these modes are together in wavelength. Consider a chamber of length $l = 50$ cm and light of approximately 600 nm. We have

$$\frac{n}{2}\lambda = l \qquad \text{or} \qquad \lambda = \frac{2l}{n} \tag{8.12}$$

We want to know how much λ changes between resonant modes n and $n+1$. From the above

$$\Delta\lambda = \Delta n \frac{\partial\lambda}{\partial n} = -\frac{2l\,\Delta n}{n^2} \tag{8.13}$$

For this example $n \approx 1.7 \times 10^6$, so $\Delta\lambda$ corresponding to $\Delta n = 1$ is $\Delta\lambda = -7 \times 10^{-4}$ nm. If we go back and look at the range of wavelengths produced by the atomic transition, we see that there may be five (or so) resonant cavity modes within the width of $\delta(\Delta\mathscr{E})$. This is illustrated in Figure 8.4.

Since our numbers were only rough estimates, the number of cavity modes in $\delta(\Delta\mathscr{E})$ might be several or it might be many. This has definite

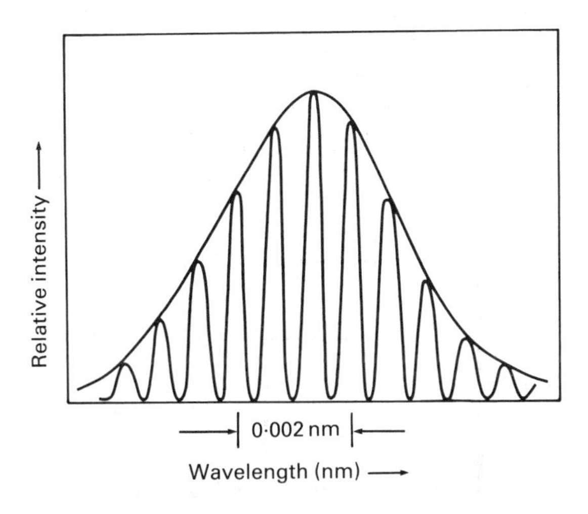

Figure 8.4. Resonant cavity modes in a laser output.

advantages and disadvantages. On the one hand, it means that the length of our resonant cavity is unimportant; there will always be several resonant modes for any length. Specifically, it means we do not have intentionally to make the cavity an integral number of half wavelengths long—it will always turn out that way anyway. On the other hand, it means that our "monochromatic" laser radiates not at a single wavelength but at several discrete wavelengths. This is not a concern for most applications.

In general, we see that the light produced by a laser has the following characteristics.

1. It is coherent (all beams are in phase).
2. It is intense: because it is coherent there is no destructive interference of the various components of the beam.
3. It is monochromatic, to the extent described.

The trick to building a laser is to be able to produce a population inversion. This requires supplying suitable energy to the system, but most of all it requires choosing atoms that have an appropriate energy-level diagram. Some specific examples are discussed below.

Population inversion and the three-level laser. The three-level laser is the simplest kind of laser that actually works. The energy-level diagram is shown in Figure 8.5. Atoms in the ground state are excited into the excited state. This can be done using an electrical discharge, as in a gas laser, or by optical excitation from a flash lamp, as in a solid-state laser. If we supply sufficient energy, then we can excite the majority of atoms into this state. However, this is a short-lived state (e.g., $\tau_{1/2} \sim 10^{-8}$ s) and atoms in this state will spontaneously decay very rapidly into a metastable state with $\tau_{1/2} \sim 10^{-3}$–10^{-4} s. Since $\tau_{1/2}$ is long, the atoms in this state do not decay so rapidly as those that are populating it. Thus atoms "collect" in this state and the result is a population inversion between the ground state and the metastable state. A few atoms in this sate then decay spontaneously and begin the avalanche of stimulated emission.

These lasers are typically *pulsed.* A flash-tube discharge, for example, causes the population inversion and is followed by a laser light pulse

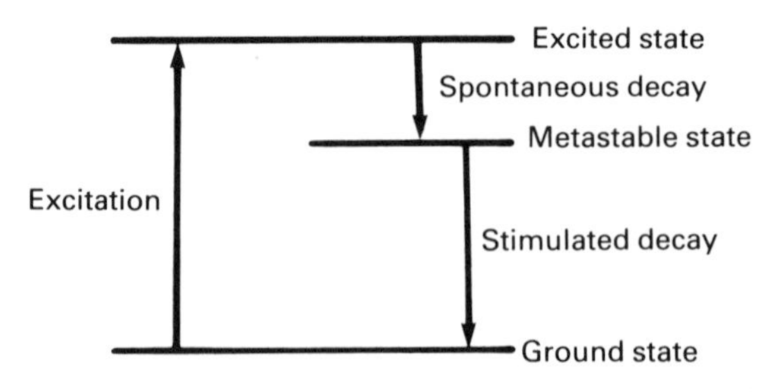

Figure 8.5. Energy-level diagram for a typical three-level laser.

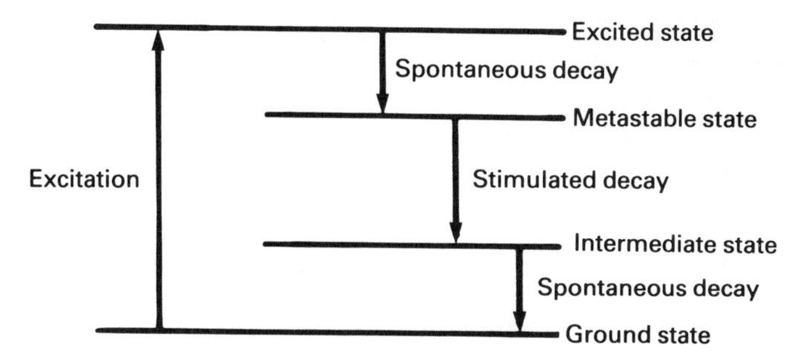

Figure 8.6. Energy-level diagram for a typical four-level laser.

associated with the stimulated emission. At this point the atoms have all returned to the ground state and the process can be repeated.

The four-level laser. A four-level energy diagram is shown in Figure 8.6. In this case the population inversion is between the metastable state and the intermediate (not the ground) state. This intermediate state is not normally populated except by atoms decaying from the metastable state. If the lifetime of this intermediate state is short we can keep the population of that state low and continuously add to the population of the metastable state by continuously exciting transitions from the ground to excited states. In this way we can, in some cases, use a four-level laser continuously because the population inversion will exist continuously.

We will divide lasers into four categories and discuss each separately: (1) solid-state lasers; (2) gas lasers; (3) dye lasers; and (4) semiconductor lasers.

b. Solid-state lasers

The active atoms in solid-state lasers are contained in a solid host material. This host material must, of course, be transparent to the light produced. The ruby laser is the best known in this group and we consider its operation.

The active ions in the ruby laser are Cr^{3+} ions that are impurities (0.05 wt.%) in an Al_2O_3 host. The ruby rod is typically cylindrical, 5–10 mm in diameter and 2–10 cm long. The ends are highly polished and are parallel to a high degree of accuracy. One end is provided with a fully reflecting surface; the other end (from which the beam emerges) is provided with a partially reflecting surface.

There are a number of energy levels in the Cr^{3+} ion, but for simplicity we discuss only those relevant to the laser action. A simplified energy-level diagram is shown in Figure 8.7. Lasing occurs at 694 nm, which is in the red. Excitation of the ground state atoms to one of two short-lived states is accomplished by optical pumping. This is a higher-energy transition than the lasing transition and the pumping radiation is at ~550 nm (green). A

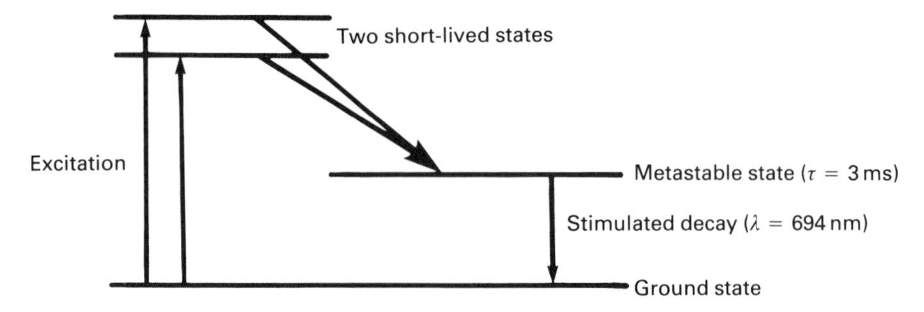

Figure 8.7. A simplified energy-level diagram for the Cr^{3+} ion.

convenient light source for pumping is the xenon flash tube. A variety of geometries for getting the pumping radiation into the ruby rod exist. Two are shown in Figures 8.8 and 8.9. In the geometry using the elliptical reflector the flash tube and the ruby rod are placed at the foci of the ellipse. The elliptical reflector has the property of focusing the light produced by the flash tube on the ruby rod. After each lasing action, atoms repopulate the ground state. However, before discharging the xenon flash tube a second time it is necessary to allow the ruby to cool down (literally). Typically, the xenon discharge tube is on for about 5 ms. The length of the lasing pulse must obviously be less than the half life of the metastable state (3 ms) and so is typically shorter than the xenon discharge.

A variety of other active ions (most of them rare earths) have been found to exhibit lasing action. Some of these are four-level systems and can therefore be used continuously. Tables 8.5 through 8.7 give some relevant information on these materials. You will note that most of the lasing occurs in the near infrared. As indicated in Table 8.6, some transitions are observed to provide continuous lasing action only when the lasing material is cooled significantly below room temperature. Most of the continuous laser materials can be used in the pulsed mode as well.

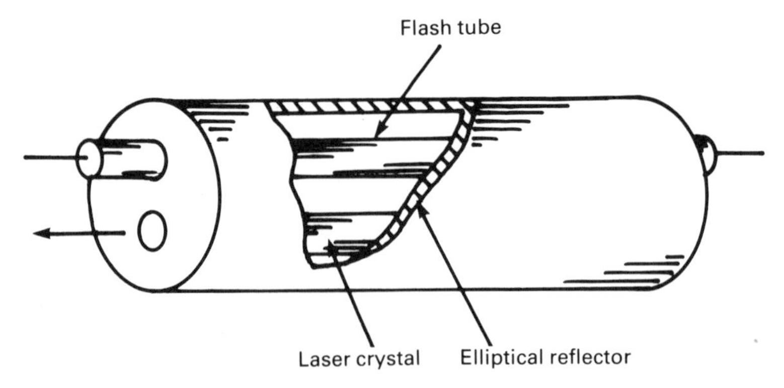

Figure 8.8. A solid-state laser using a linear flash tube and an elliptical reflector.

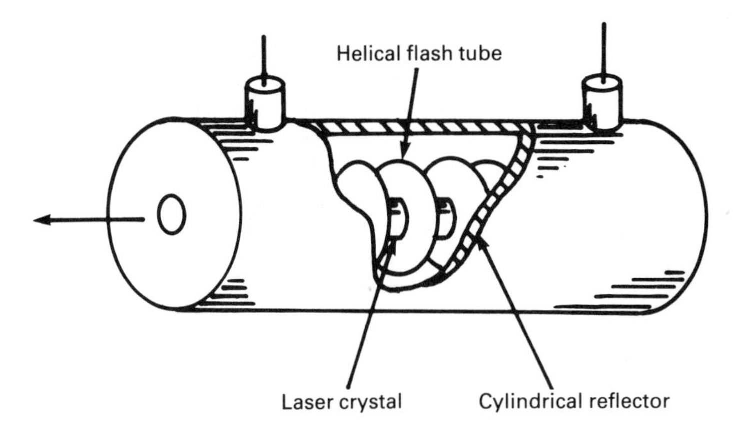

Figure 8.9. A solid-state laser using a helical flash lamp and a circular reflector.

Table 8.5. Solid-state lasers that operate continuously at room temperature (data from a compilation by R. B. Chesler and J. E. Guesic in *Laser Handbook,* F. T. Arecchi and E. O. Schultz-Dubois, eds. see bibliography).

Ion	Host material	Wavelength(s) (nm)
Nd^{3+}	$CaWO_4$	1058
Nd^{3+}	$CaMoO_4$	1067
Nd^{3+}	$Y_3Al_5O_{12}$(YAG)	1052, 1061, 1064, 1074, 1112, 1116, 1123, 1319
Nd^{3+}	$YAlO_3$	1064, 1079
Nd^{3+}	Glass	~1060
Nd^{3+}	La_2O_2S	1076
Nd^{3+}	$Ca_5(PO_4)_3F$(FAP)	1062

Table 8.6. Solid-state lasers that operate continuously at 77 K (data from a compilation by R. B. Chesler and J. E. Guesic in *Laser Handbook,* F. T. Arecchi and E. O. Schultz-Dubois, eds. see bibliography).

Ion	Host	Wavelength(s) (nm)
U^{3+}	CaF_2	2613
Dy^{2+}	CaF_2	2360
Ho^{3+}	$Y_3Al_5O_{12}$	2122
Tm^{3+}	$Y_3Al_5O_{12} + Cr^{3+}$	2097
Tm^{3+}	Er_2O_3	1934
Tm^{2+}	CaF_2	1116

Table 8.7. Solid-state pulsed lasers that operate at room temperature (data from a compilation by R. B. Chesler and J. E. Guesic in *Laser Handbook,* F. T. Arecchi and E. O. Schultz-Dubois, eds. see bibliography).

Ion	Host material	Wavelength(s) (nm)
Nd^{3+}	Glass	1047–1063
Nd^{3+}	$Y_3Al_5O_{12}$	1065, 1320
Nd^{3+}	LaF_3	1063
Nd^{3+}	$NaLa(MoO_4)_2$	1059
Nd^{3+}	$LiNbO_3$	~1090
Nd^{3+}	SrF_2	~1040
Nd^{3+}	YVO_4	1064, 1066
Nd^{3+}	$Ca(NbO_3)_2$	1060
Nd^{3+}	$SrMoO_4$	1064
Nd^{3+}	$CaMoO_4$	1067
Nd^{3+}	$PbMoO_4$	1059
Ho^{3+}	$Y_3Al_5O_{12}$	2097
Yb^{3+}	Glass	1015, 1060
Er^{3+}	Glass	~1540
Tm^{3+}	$Y_3Al_5O_{12}$	2013
Cr^{3+}	Al_5O_3	694

c. *Gas lasers*

The helium–neon laser is representative of gas lasers. This laser consists of a mixture of about 1 part neon to 10 parts helium. The neon atoms are active in the lasing process. The helium atoms, however, are important in obtaining the population inversion. There are three different decay processes in neon that can produce laser light. These are at 632 nm (red), 1152 nm (IR) and 3391 nm (IR). The 632 nm output is commonly used in

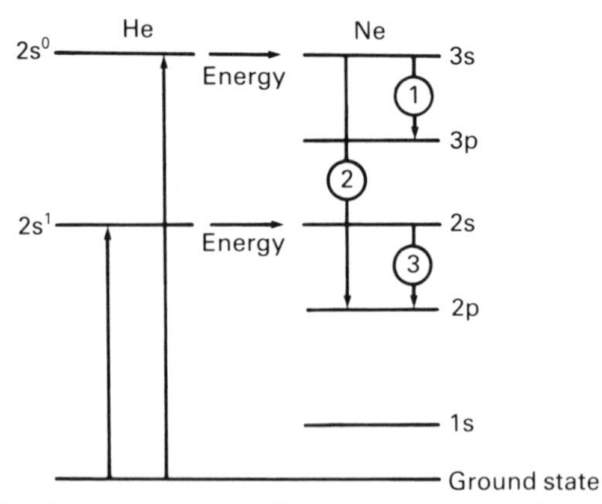

Figure 8.10. A simplified energy-level diagram for the He–He laser. The numbered transitions represent lasing at (1) 3391 nm (2) 1153 nm and (3) 632 nm.

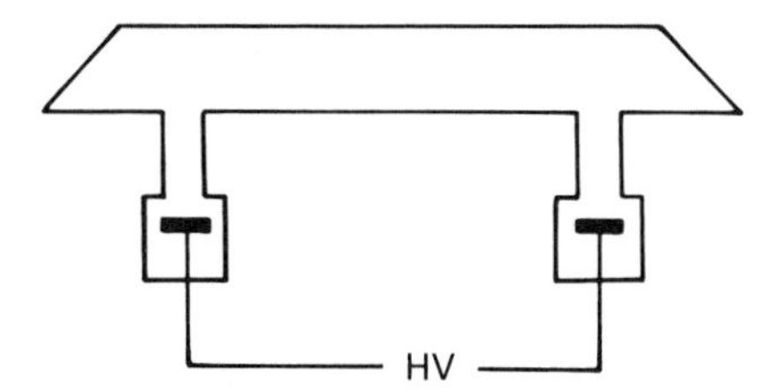

Figure 8.11. Arrangement of electrodes in a gas laser for excitation by DC electrical discharge.

laboratory optics experiments. A very simplified energy-level diagram is shown in Figure 8.10. The neon levels are actually split into a number of substates and there are a large number of possible transitions. Only those that are known to lase are shown in the diagram. The initial excitation causes a population of the He 2s levels. These are at nearly the same energies as two of the neon levels. Energy is transferred from the helium to neon atoms and the excited neon levels are populated, creating a population inversion.

Helium–neon and other gas lasers can be excited continuously using one of several methods. The most common are the following.

1. *DC electrical discharge.* This method uses electrodes inside the tube, usually mounted in extensions to the tube as shown in Figure 8.11 so that they do not interfere with the lasing channel.
2. *R.F. excitation.* Radiofrequency energy is applied to electrodes placed outside of the tube as shown in Figure 8.12. The tube itself behaves like a capacitor with the gas acting as the dielectric.

A variety of gases can be used in lasers. Most atomic or molecular lasers operate in the infrared. The 632 nm line in neon is about the lower limit for wavelength. Most molecular lasers, such as the CO_2 laser, have a number of closely spaced lines in their spectra, resulting from the large number of closely spaced energy states that exist in molecules because of their large number of degrees of freedom. Lasers that use ionized gas are more useful from the optical point of view as these have outputs in the wavelength range 200–700 nm. Table 8.8 gives the characteristics of some common gas lasers.

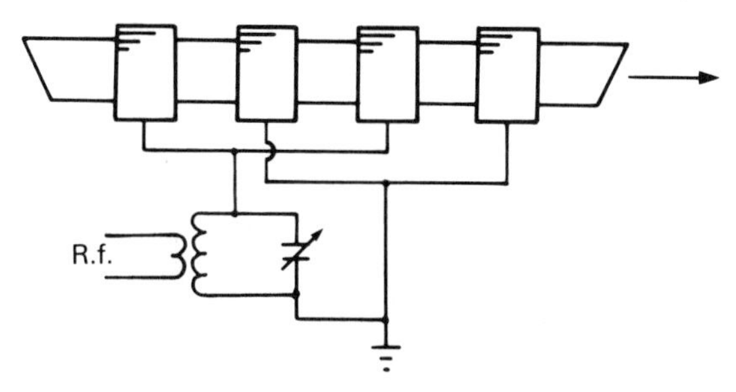

Figure 8.12. Radiofrequency excitation of a gas laser.

Table 8.8. Characteristics of some continuous-mode gas lasers

Gas	Type	Wavelength(s) (nm)
Xe He	Atomic	2026, 3507, 5575, 9007
He Ne	Atomic	632, 1152, 3391
CO	Molecular	4900 to 5700
CO_2	Molecular	9600, 10,600
H_2O	Molecular	~28,000
Xe	Ion	345, 378, 406
N	Ion	337
Ne	Ion	332
Ar	Ion	351, 364, 458, 466, 478, 476, 488, 496, 511, 514
Kr	Ion	521, 531, 568, 647, 676, 752, 793, 799

d. Dye lasers

In a dye laser the lasing atom is part of an organic molecule. Some of the organic compounds that have been used are commercial dyes; hence their name. These are dissolved in a suitable solvent. Because of the huge number of energy levels present in a complex organic molecule, there are numerous wavelengths at which lasing occurs. These lasers may be "tuned" to extract the specific wavelength of interest. Figure 8.13 shows a simple (although not very good) method of tuning a dye laser. Since the glass of the prism is dispersive, only a single wavelength will be refracted by the proper amount to be reflected back through the lasing cavity. Using this method, the wavelength of the laser beam can be narrowed to $\Delta\lambda \sim 20$ nm. A grating can be used in a similar manner (see Section 10.4) and $\Delta\lambda \sim 0.05$ nm can be achieved. The best device for tuning the dye laser is an etalon (see Section 10.3); with this we can attain $\Delta\lambda \sim 0.001$ nm, comparable with solid-state or gas lasers. The dye laser is excited by optical pumping. A flash lamp at an appropriate frequency is most commonly used to produce a pulsed output.

Table 8.9 lists some representative "dyes" and their common solutes along with the range of wavelengths at which they lase. There are many more and a more comprehensive list is given by Arecchi and Schultz-Dubois (see Bibliography).

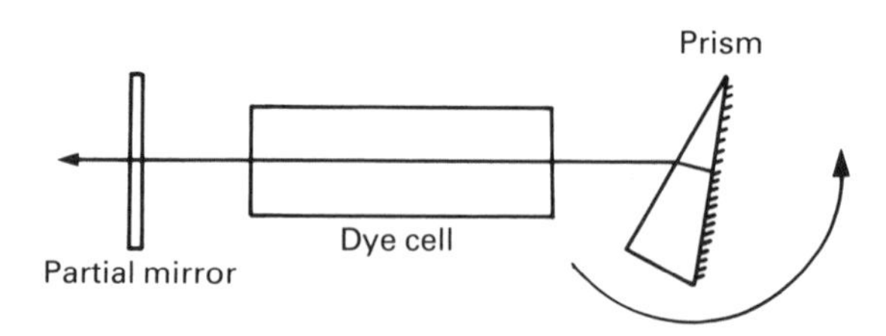

Figure 8.13. An arrangement for tuning a dye laser.

Table 8.9. Some "dye" laser materials (data from a compilation by F. P. Schafer in *Laser Handbook,* F. T. Arecchi and E. O. Schultz-Dubois, eds. see bibliography).

"Dye"	Typical solute	Wavelength range (nm)
p-Terphenyl	Ethanol	330–360
PPO-(2,5-diphenyloxazole)	Toluene	357–391
α-NPO (2-[naphthyl-(1′)]-5-phenyloxazole)	Ethanol	393–402
1,2-Di-4-biphenylethene	Toluene	~408
1,4-Distryrylbenzene	Toluene	401–402
Dimethyl-POPOP-NOPON (2,2′-*p*-phenylene-	Ethanol	423–431
bis-[5-(1-naphthyl)]-oxazole)	Toluene	430–445
7-Acetoxy-4-methylcoumarin	Ethanol	441–486
1,3-Diphenyl-*i*-benzofuran	Ethanol	484–518
Brilliant-sulfaflavine	Ethanol	508–574
Rhodamine 6G	Ethanol	560–640
Rhodamine B	Methanol	595–650
Cresyl violet	Ethanol	645–709
3-3′-Diethylthiadicarbocyanine iodide	Methanol	710–735
3,3′-Diethyl-2,2′-oxatricarbocyanine iodide	Acetone	~742
1,1′-Diethyl-2,2′-dicarbocyanine iodide	Glycerol	750–790
3,3′-Diethyltriatricarbocyanine bromide	Methanol	798–870

e. Semiconducting lasers

The transitions in semiconducting lasers are not transitions between well-defined atomic states such as we have discussed previously, but rather are "transitions" of electrons and holes across the energy gap. The result is then emission of a photon with energy of order $\mathscr{E}_g$. Obviously, the energy can be larger if an electron in a state not immediately at the edge of the conduction band but somewhat above it recombines with a hole that is somewhat below the top of the valence band. This means that the photons emitted have a fairly broad distribution of energies. There are four basic methods of exciting these types of lasers: (1) forward biasing (called *injection*); (2) reverse biasing (causing avalanche breakdown); (3) optical pumping (e.g., from another laser); and (4) electron-beam pumping. Different semiconducting materials respond to different methods of excitation. Avalanche breakdown works only in a small number of materials (e.g., GaAs; InP, and InSb), while the other methods work for many more. Because it is frequently successful and is particularly simple, forward biasing is by far the most commonly used method. Further discussion will be confined to this technique. Since the wavelength of the emitted photon is related to the energy of the transition by

$$\lambda \text{ (nm)} = \frac{1.24 \times 10^3}{\mathscr{E} \text{ (eV)}} \tag{8.14}$$

transitions have to be in the range of 1.8–3.0 eV in order to produce visible photons. This, as we know, is much larger than the energy gap of silicon or

Table 8.10. Room-temperature injection lasers (data from J. Guesic *et al.* Proc. IEEE *58*, 1419 (1970)).

Semiconductor	$\mathscr{E}_g$ (eV)	λ (nm)
GaAs	1.43	868
$GaAs_{1-x}P_x$	~1.4–2.0	~610–900
$Al_xGa_{1-x}As$	~1.4–2.0	~630–900

germanium. Fortunately, there are a number of semiconducting compounds with suitable gaps. Semiconductors have the convenient property that the energy gap can be adjusted by changing the composition. For example, gallium arsenide has an energy gap of 1.43 eV, corresponding to a photon wavelength of ~850 nm (in the near infrared). Gallium phosphide, on the other hand, has an energy gap of 2.26 eV, corresponding to a wavelength of ~550 nm. The energy gap, and hence the emitted photon wavelength, can be adjusted between these two values by using $GaAs_{1-x}P_x$. Table 8.10 lists the commonly used semiconductors that can be excited by forward biasing (injection lasers) and are usable at room temperature. There are many more, mostly with smaller energy gaps and hence with longer wavelengths, that can be used only at lower temperatures. This is because the thermally excited electron–hole pairs are much more prevalent in semiconductors with smaller energy gaps. As you can see, the common laser diodes are all based on gallium arsenide. Table 8.11 gives some other injection lasers that can only be used at low temperatures.

The simplest construction of a diode laser is in the form of a *p–n* junction. The active lasing region is confined to the region around the depletion layer and a lasing cavity is formed by cutting two of the ends of

Table 8.11. Low-temperature (only) injection lasers (data from J. Guesic *et al.* Proc. IEEE *58*, 1419 (1970)).

Semiconductor	$\mathscr{E}_g$ (eV)	λ (nm)
InP	1.36	910
$GaAs_{1-x}Sb_x$	~0.83–1.4	~900–1500
$InAs_{1-x}Px$	~0.39–1.4	~900–3200
GaSb	0.80	1550
$In_{1-x}Ga_xAs$	0.4–1.45	~850–3100
InAs	0.39	3100
$InAs_{1-x}Sb_x$	0.23–0.39	~3100–5400
InSb	0.236	~5200
PbTe	0.19	~6500
$PbS_{1-x}Se_x$	0.146–0.32	3900-8500
PbSe	0.146	8500
PbSnTe	0.045	28,000
PbSnSe	0.04–0.155	~8000–34,000

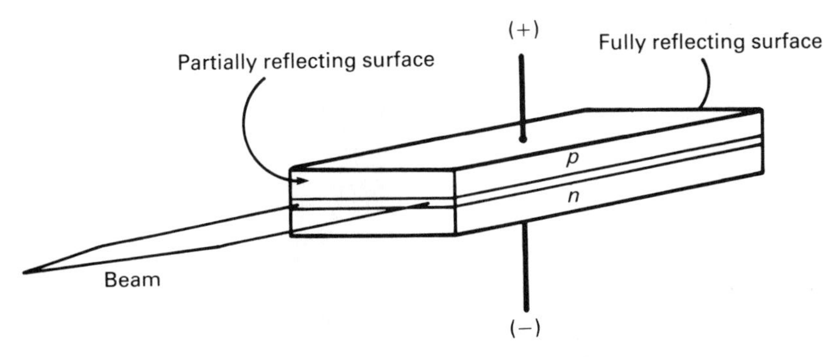

Figure 8.14. A simple junction-diode laser.

the semiconductor parallel and coating them (one fully and the other partially) with a reflective material. The beam emerges normal to the bias direction and from the end with the partially reflecting surface. This geometry is shown in Figure 8.14.

A certain minimum (threshold) current is necessary to obtain the lasing action. This is required in order to establish the equivalent of a population inversion. Figure 8.15 shows the light output of a gallium arsenide laser diode as a function of bias current. Below some threshold, the output is very low, and as you can see from the figure the light is incoherent as well.

Laser diodes, like other lasers, are subject to modes depending on cavity size. Although the range of output wavelengths can be fairly broad (~10

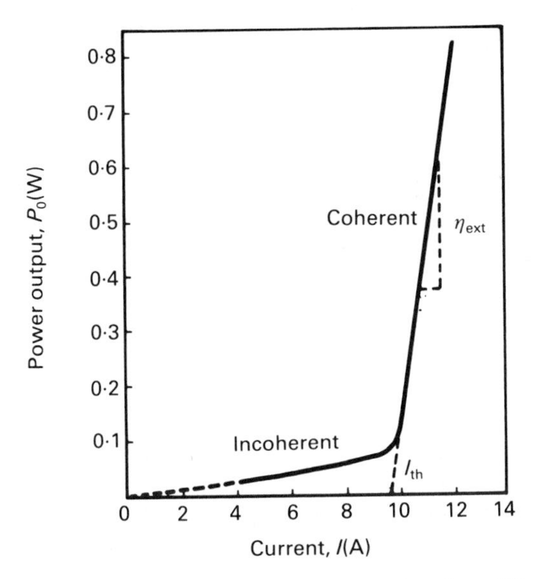

Figure 8.15. Power output of a gallium arsenide laser as a function of bias current. I_{th} is the lasing threshold current; η_{ext} is the change in efficiency. (Copyright © 1974, by RCA New Products Division, Lancaster, Pennsylvania. Used with permission.)

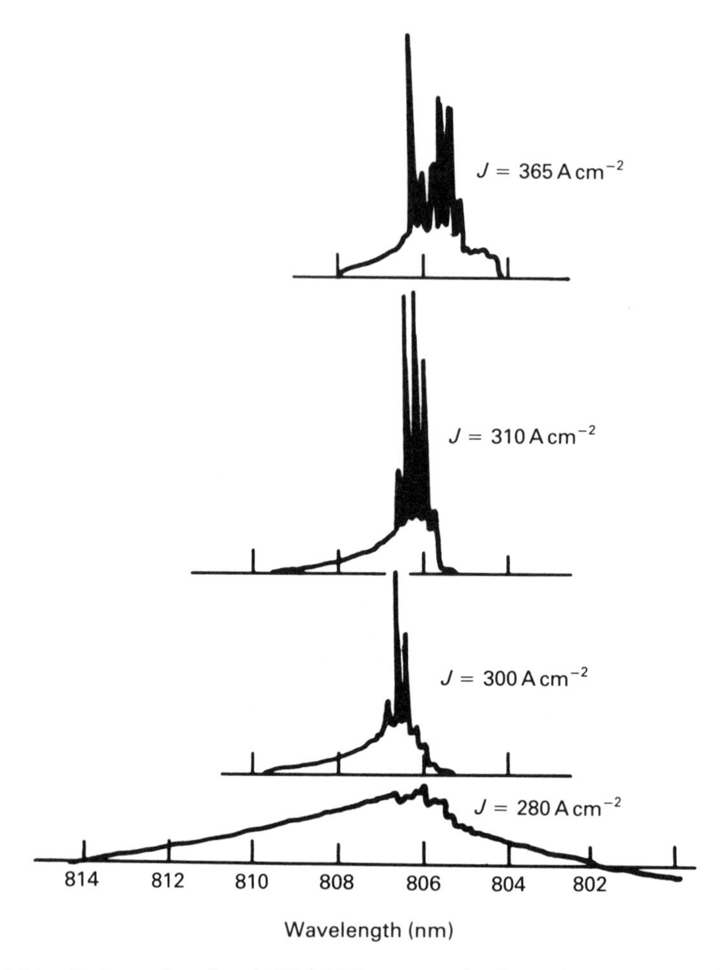

Figure 8.16. Cavity modes of an (AlGa)AS laser operating in continuous wave mode at 77 K for different current densities J. (From H. Kressel, "Semiconductor Lasers: Devices" in *Laser Handbook Volume 1,* F. T. Arecchi and E. O. Schulz-Dubois eds. Coptright © 1972 by North-Holland Publishing Co. Used with permission.)

nm), the cavity modes are fairly sharp. These are illustrated in Figure 8.16 for a typical laser diode. The various modes can be separated by the means discussed for dye lasers.

Most laser diodes are not simple p–n junctions, since these really do not work very well. Rather, a variety of more complex junction geometries are generally used. The common ones are shown in Figure 8.17. Popular commerical laser diodes are GaAs-based and are produced in the SH–CC and LOC configurations. Operation is typically pulsed with pulse periods of 0.1 to 1.0 μs and repetition rates of 10^4 to 10^5 Hz. Some devices can be used continuously.

A device that deserves mention at this point is the LED (light-emitting

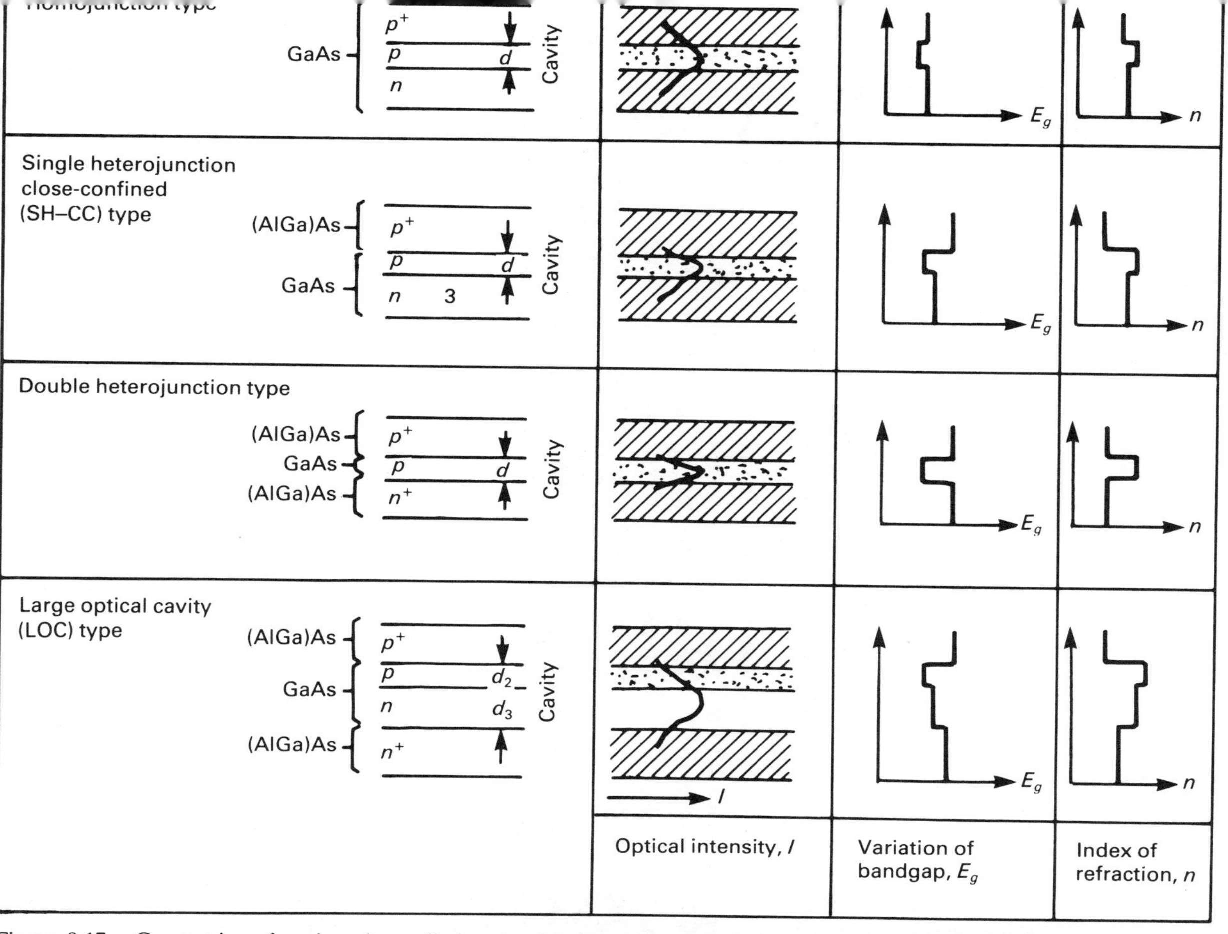

Figure 8.17. Geometries of various laser diodes showing intensity of optical output, variation in band gap and index of refraction. (Copyright © 1974 by RCA New Products Division, Lancaster, Pennsylvania. Used with permission.)

Table 8.12. LED materials

Semiconductor	$\sim\lambda$ (nm)	Colors
CdTe	855	Near IR
$Zn_{1-x}Cd_xTe$	530–830	Green, yellow, red
BP	640	Red
ZnSeTe	627	Red
ZnTe	620	Red
GaP	560–680	Yellow, red
SiC	456	Blue
GaN	340–700	Near UV, violet, blue, green, yellow, red

diode).These are not lasers, since their output is not coherent, but their operation is closely related to that of the laser diodes. The requirement for an LED is that the device emits light not that the light should be coherent or even particularly monochromatic. For these reasons the emission of light is not necessarily stimulated but can be the result of spontaneous transitions. This is equivalent to the incoherent region of Figure 8.15. There is no requirement for a resonant cavity and this simplifies production. A number of materials that make diode lasers that operate only at low temperature or do not lase at all are suitable for LEDs. Table 8.12 gives the characteristics of some commonly used LED materials. The range of different wavelengths available from a material results from the inclusion of different levels of impurities.

9
Optical components

9.1 Optical materials

We commonly think of lenses and prisms as being made of glass. Quartz, however, is often used as well. The situation is quite different for radiation of wavelengths that are not visible, as those materials which we think of as transparent can be opaque and vice versa. Figure 9.1 shows the wavelength ranges over which some optical materials can be used. It is interesting to note that glass can be used over the entire visible range and into the infrared but is useless for ultraviolet wavelengths. Here we discuss primarily the different kinds of glass that are used for optical purposes.

Different kinds of glasses have different indices of refraction μ. These indices of refraction are functions of the wavelength of the light. Figure 9.2 shows $\mu(\lambda)$ for some common glasses. The dependence of μ on λ is referred to as the *dispersion*.

Different glasses not only have, in general, different values of μ at a given wavelength but they also have different dispersions as well. The inverse of the dispersion of a particular glass is given by

$$V = \frac{1 - \mu_D}{\mu_F - \mu_C} \tag{9.1}$$

where μ_i are the indices of refraction at different wavelengths over the optical range, as given in Table 9.1. Thus the inverse of V is a measure of how curved the lines in Figure 9.2 are. V is sometimes referred to as the Abbe factor. Table 9.2 gives the optical properties of various kinds of glass. We see two clear distinctions here: crown glass and flint glass. These designations refer to glasses with low dispersion ($V > 55$) and high dispersion ($V < 55$) as indicated in Figure 9.3.

In general, a glass is a solid without a crystalline structure. Optical glasses are generally composed primarily of SiO_2. A variety of other oxides are added to the SiO_2 in order to obtain the desired optical properties. The name of the glass as given in Table 9.2 is an indication of the dispersion and in some cases also of the particular impurities that are added to the glass.

Lenses suffer from the problem of reflection of the light from the surfaces. Although the reflection may only amount to a few percent per

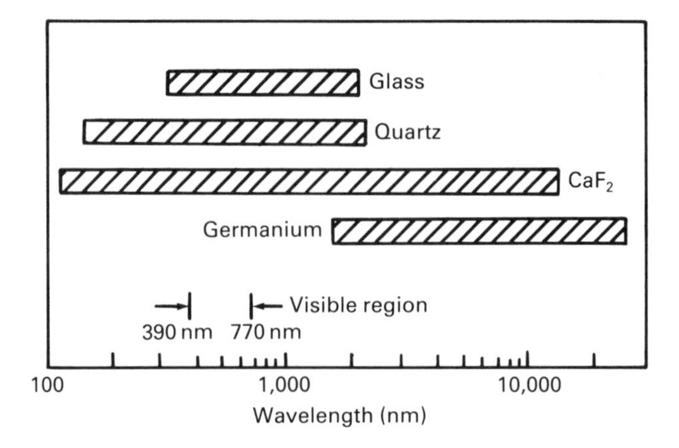

Figure 9.1. Regions over which various materials are transparent.

surface, an optical system may consist of many surfaces and the total loss may amount to one-third or more. Antireflective coatings can be used to eliminate this problem to a large extent. The coating consists of a thin transparent layer with an index of refraction that is different from that of the lens glass. Figure 9.4 shows the physical principles. An analysis of the reflections in the figure shows that the transmitted ray is a maximum when

$$\mu_c = (\mu_a \mu_g)^{1/2} \tag{9.2}$$

and the thickness of the coating is an odd number of quarter wavelengths.

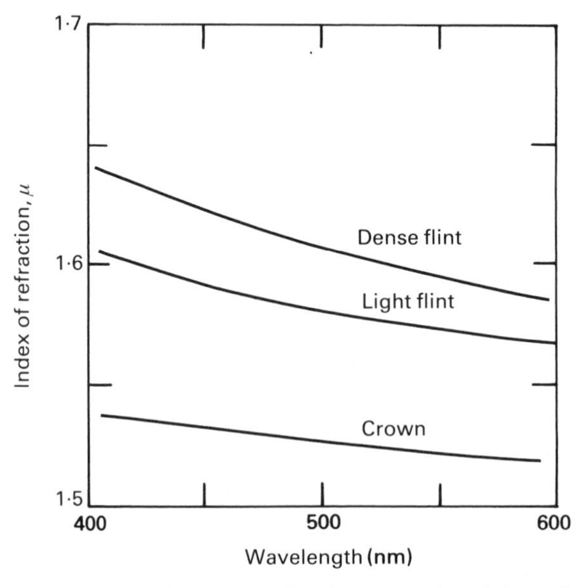

Figure 9.2. Indices of refraction as a function of wavelength for various glasses.

Table 9.1. Wavelength for the calculation of the dispersion

Wavelength	λ (nm)	Source	Color
D	589	Sodium	Yellow
F	486	Hydrogen	Blue
C	656	Hydrogen	Red

Table 9.2. Standard glass designations used in Figure 9.3

Glass	Designation
Fluor crown	FC
Borosilicate crown	BSC
Crown	C
Extra-light flint	ELF
Light flint	LF
Dense flint	DF
Extra-dense flint	EDF
Crown flint	CF
Light barium crown	LBC
Dense barium crown	DBC
Extra-dense barium crown	EDBC
Light barium flint	LBF
Barium flint	BF
Dense barium flint	DBF
Lanthanum crown	LaC
Lanthanum flint	LaF
Phosphate crown	PC
Dense phosphate crown	DPC

For single-layer coatings, magnesium fluoride is the most commonly used material; this has $\mu_c = 1.38$ and equation (9.2) is satisfied for many types of glass. Multilayer coatings can further eliminate reflection. These consist of two or more layers of coating material each with a different μ. Corresponding expressions for the indices of refraction similar to equation (9.2) can be obtained.

Plastic lenses are suitable for some applications. These generally have high dispersion (low V) and values of μ in the range of 1.0–1.3.

9.2 Lenses and mirrors

a. Basic properties of lenses and mirrors

The behavior of a lens is described by the use of *geometrical optics*. The basic principle on which all geometrical optics is based is the law of

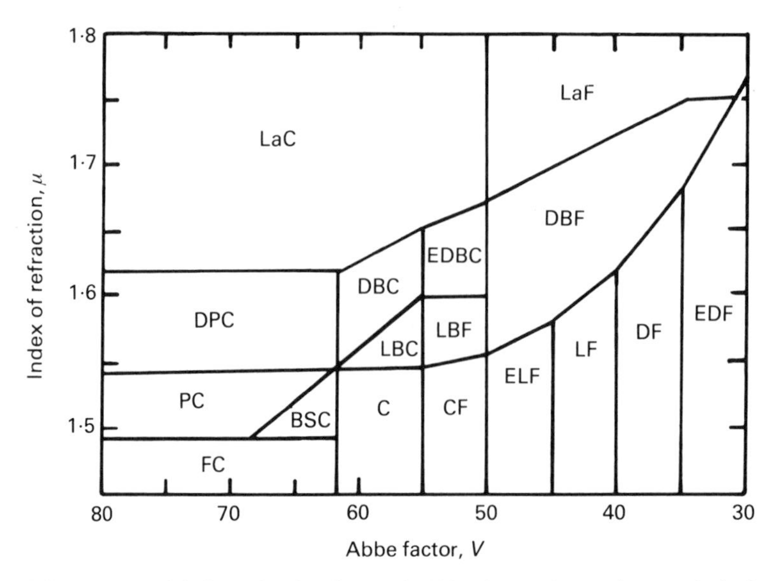

Figure 9.3. Range of index of refraction and Abbe factor for various optical glasses at 589 nm. The standard letter designations are given in Table 9.2.

refraction. Simply stated this is

$$\mu_1 \sin \alpha_1 = \mu_2 \sin \alpha_2 \tag{9.3}$$

where μ_i and α_i are the indices of refraction and angles of incidence on either side of a boundary (see Figure 9.5). Lenses work on the principle that light is refracted at a boundary between two materials of different μ.

We begin with something simple: a spherical refracting surface. We see this in Figure 9.6. The line RPC is the normal to the surface at the point of incidence of the ray QPB. Hence the radius r is equal to AC and the angle

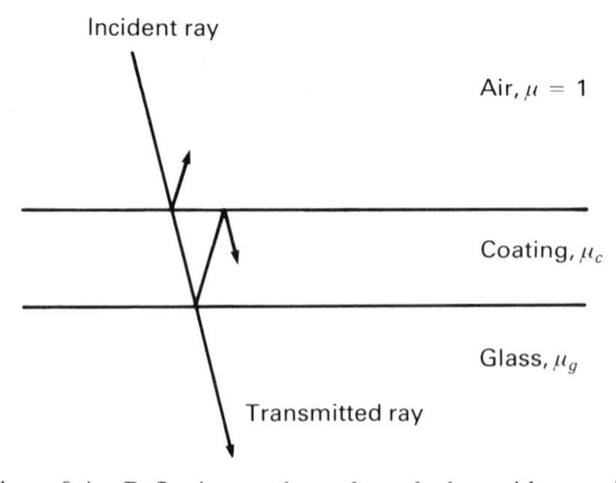

Figure 9.4. Reflections at the surface of a lens with a coating.

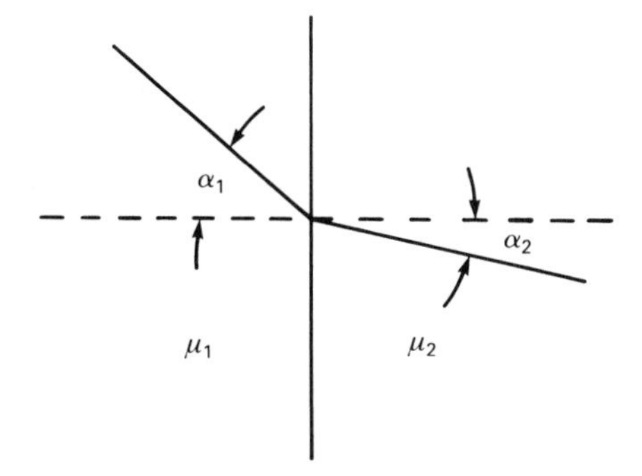

Figure 9.5. An illustration of the law of refraction.

of incidence is given in the figure by *I*. The refracted ray is *PB'* and the angle of refraction is *I'*. Angles *U* and *U'* are defined for convenience of the derivation. Look at the triangle formed by PCB. We write (see Figure 9.7*a*)

$$\sin U = \frac{x}{L - r} \qquad \text{and} \qquad \sin I = \frac{x}{r} \tag{9.4}$$

where we have defined *L* as the length from *A* to *B*. Now the above can be combined and α can be eliminated to give

$$\sin I = \frac{L - r}{r} \sin U \tag{9.5}$$

In the small angle limit $\sin \alpha = \alpha$ and

$$I = \frac{(L - r)U}{r} \tag{9.6}$$

Looking at Figure 9.7*b* we see that

$$\sin U' = \frac{x'}{L' - r} \qquad \text{and} \qquad \sin I' = \frac{x'}{r} \tag{9.7}$$

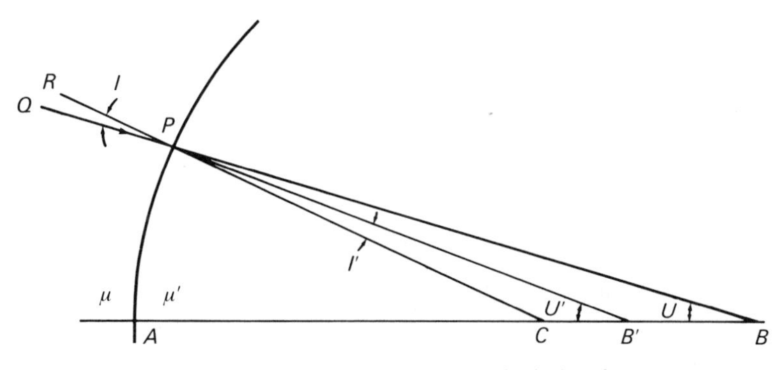

Figure 9.6. Refraction of light at a spherical surface.

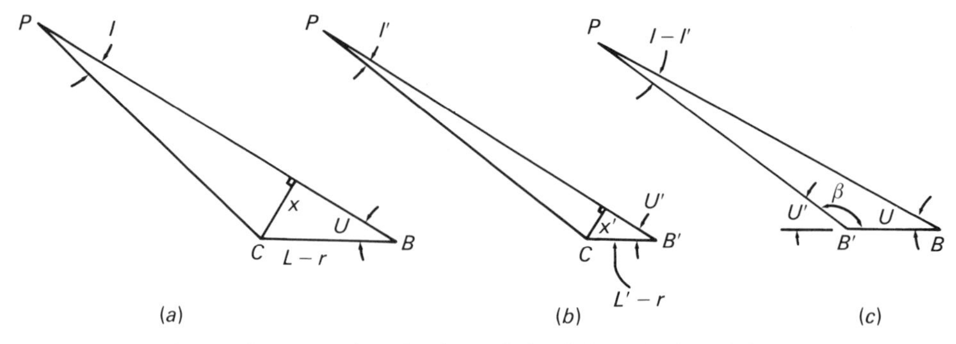

Figure 9.7. Geometric constructions for the analysis of the refraction of light at a spherical surface (see Figure 9.6).

Eliminating x' gives

$$\sin I' = \frac{L' - r}{r} \sin U' \tag{9.8}$$

where L' is defined to be the distance B' to A. In the small-angle limit this is

$$I' = \frac{(L' - r)U'}{r} \tag{9.9}$$

The law of refraction gives the angles in terms of the indices of refraction, so

$$\mu' \sin I' = \mu \sin I \tag{9.10}$$

or, in the small-angle limit,

$$\mu' I' = \mu I \tag{9.11}$$

Finally, we look at triangle BPB' (see Figure 9.7*c*). We know that

$$(I - I') + U + \beta = 180°$$

and

$$\beta + U' = 180° \tag{9.12}$$

Eliminating β gives

$$U' = U + I - I' \tag{9.13}$$

The equations we use to proceed are (9.6), (9.9), (9.11), and (9.13).

Eliminating I, I', U, and U' from these equations yields

$$\frac{\mu'}{L'} + \frac{\mu}{L} = \frac{\mu' - \mu}{r} \tag{9.14}$$

The above equation relates the object and image distances, L and L', respectively for a single spherical surface. This equation is the basis for the analysis of all lenses and lens systems. These may be considered as a series of interfaces that are, to a good approximation at least, spherical. Hence the system may be described by a repetition of equation (9.14) for all the surfaces. For a simple lens we write equation (9.14) for one surface as

$$\frac{\mu'}{L_1'} + \frac{\mu}{L_1} = \frac{\mu' - \mu}{r_1} \tag{9.15}$$

and for the second surface as

$$\frac{\mu'}{L_2'} + \frac{\mu}{L_2} = \frac{\mu' - \mu}{r_2} \tag{9.16}$$

For the ideal "thin" lens we assume $t \approx 0$ and the object of surface 2 is the same as the image of surface 1. This means that

$$-L_2 = L_1' \tag{9.17}$$

and equations (9.15) and (9.16) can be combined to eliminate L_1 as

$$\frac{1}{L_2'} - \frac{1}{L_1} = \frac{\mu' - \mu}{\mu r_1} - \frac{\mu' - \mu}{\mu r_2} \tag{9.18}$$

The L_1 on the left-hand side of equation (9.18) is the object distance of the first surface and the quantity L_2' is the image distance of the second surface. So we can write

$$\frac{1}{l'} - \frac{1}{l} = \frac{\mu' - \mu}{\mu}\left(\frac{1}{r_1} - \frac{1}{r_2}\right) \tag{9.19}$$

where l' and l are the image and object distances of the entire thin lens. The right-hand side of equation (9.19) is defined as the inverse focal length of the lens, that is, $1/f$:

$$\frac{1}{f} = \frac{\mu' - \mu}{\mu}\left(\frac{1}{r_1} - \frac{1}{r_2}\right) \tag{9.20}$$

Real lenses are not necessarily thin. For a thick lens we do not have $-L_2 = L_1'$ and the derivation above becomes somewhat more involved. The thin lens approximation provides a sufficiently accurate description of the real situation for the present purposes. Hence we arrive from the above discussion at the lens equation:

$$\frac{1}{l'} - \frac{1}{l} = \frac{1}{f} \tag{9.21}$$

Basic lens designs are illustrated in Figure 9.8. Lenses (*a*), (*c*), and (*e*) are referred to as *positive lenses* because they have positive focal lengths. These are sometimes called *convergent lenses*. Lenses (*b*), (*d*), and (*f*) are

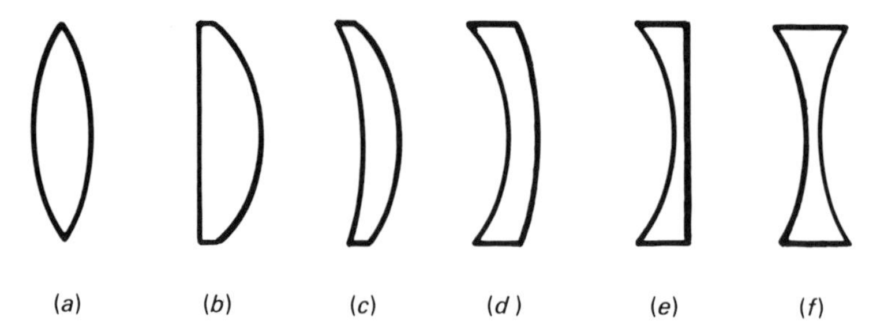

Figure 9.8. Various types of lens: (*a*) convexo-convex; (*b*) plano-convex; (*c*) positive meniscus; (*d*) negative meniscus; (*e*) plano-concave; (*f*) concavo-concave.

referred to as *negative lenses* as they have negative focal lengths. These are also sometimes called *divergent lenses.* The differences between these two kinds of lenses are illustrated in Figure 9.9. A convergent lens produces an image on the opposite side of the lens from the object. This image is a *real image.* A divergent lens produces an image on the same side of the lens as the object. This is a *virtual* image.

The power of a lens is defined as $1/f$. This is commonly measured in the unit *diopter.* This is given by the inverse of the focal length measured in meters. Hence a lens with $f = 1$ meter is 1 diopter and a lens with $f = \frac{1}{2}$ meter is 2 diopter, and so on. The larger the number of diopters the "stronger" the lens.

Let us look briefly at mirrors. The principles are similar to those of lenses except that we are dealing with reflection rather than refraction. This means

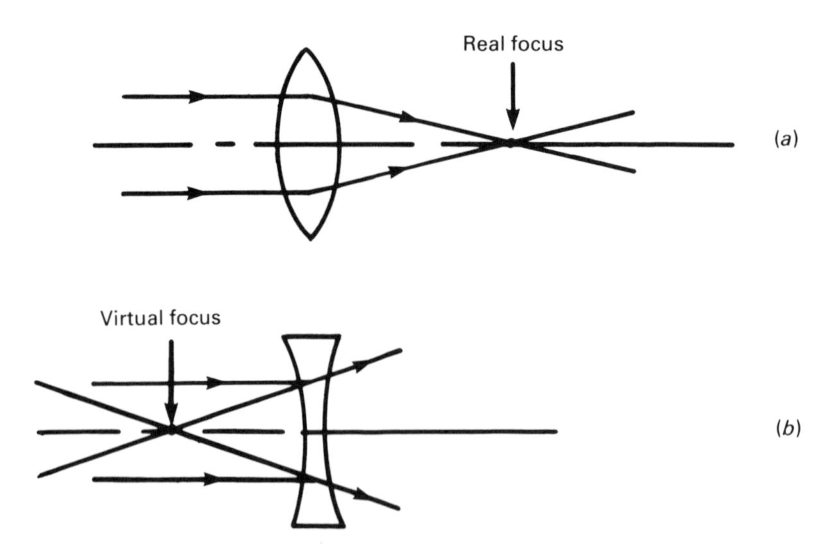

Figure 9.9. Image formation by (*a*) a positive (convergent) lens and (*b*) a negative (divergent) lens.

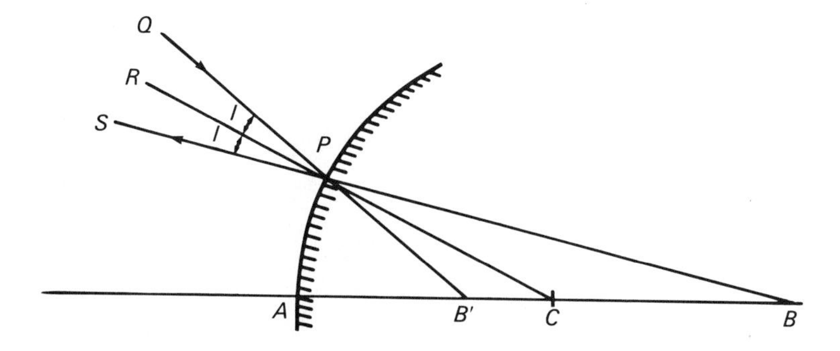

Figure 9.10. Reflection of light from the surface of a spherical mirror.

that the medium (and hence μ) for the incoming ray is the same as that for the outgoing ray. From equation (9.3) we see that this means that the angle of incidence is the same as the angle of reflection. Both angles, of course, are measured relative to the normal to the surface.

Consider as an example the convex mirror shown in Figure 9.10. The object distance is given by AB and the image distance by AB'. An analysis as before yields

$$\frac{1}{l} + \frac{1}{l'} = \frac{2}{r} \tag{9.22}$$

where r is the radius of curvature of the surface. The focal length in this case is $r/2$.

Now that we can deal with lenses and mirrors individually, we can, with a little practice, determine what a group of lenses and/or mirrors will do together. Using equations (9.21) and (9.22) we can determine the size and location of the image produced by an optical system given the size and location of the object. Rather than getting involved in an analytical analysis of a lens system, it is often convenient to understand the basic operation by means of ray diagrams. These are best illustrated by a couple of examples.

Optics example **1.** *The compound microscope*

The important rules to remember in making ray diagrams are that (1) rays going through the center of a lens (or mirror) are unaffected; (2) parallel rays pass through the focal point—and vice versa (rays passing through the focal point emerge parallel).

Consider the arrangement shown in Figure 9.11: f_o and f_e are the focal lengths of the objective lens and eyepiece, respectively. Note that the distance between the lenses is greater than the sum of the f and that $f_o < f_e$. Let us see what image is formed by the objective lens by an object to its left in the diagram. The ray diagram is shown in Figure 9.11. We note that the

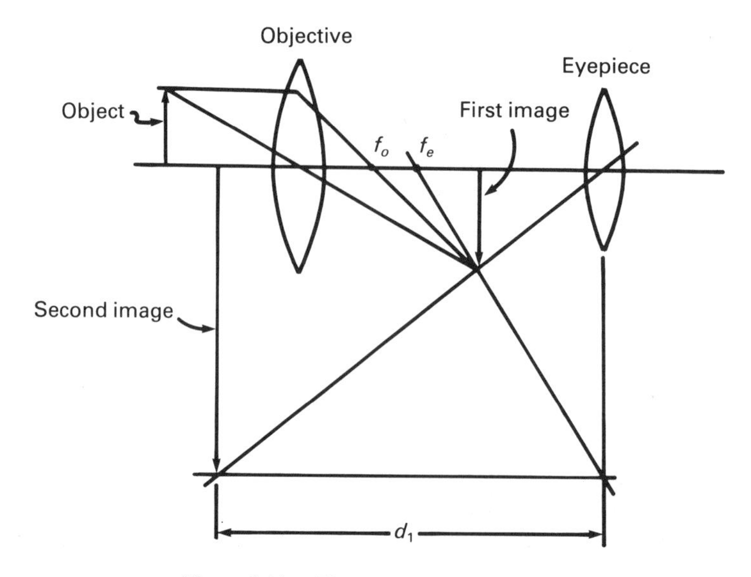

Figure 9.11. The compound microscope.

image is inverted and occurs between the eyepiece and the focal point of the eyepiece. The image due to the eyepiece remains inverted but becomes even larger. The image due to the objective is real, as it appears on the opposite side of lens from the object; and the image from the eye piece is virtual, as it appears on the same side of the lens as the object (that is the image of the objective).

In order for the resulting object to be in focus, the distance d_1 must be consistent with the operation of the eye, in this case about 20 cm or so.

Optics example 2. The Newtonian telescope

The essential optics of a Newtonian telescope are shown in Figure 9.12. The purpose of the diagonal mirror is to allow for convenient viewing. We can

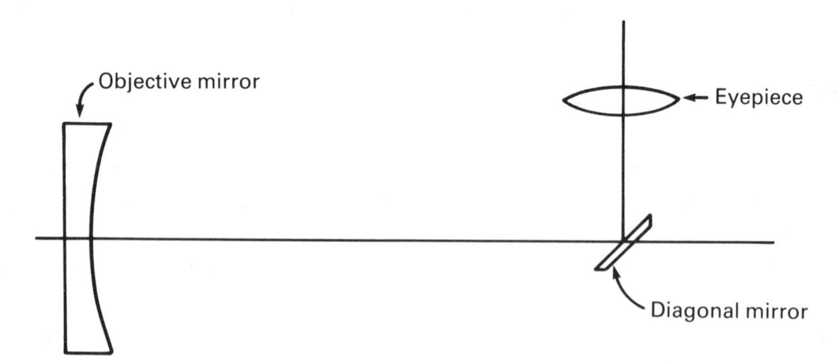

Figure 9.12. The Newtonian telescope.

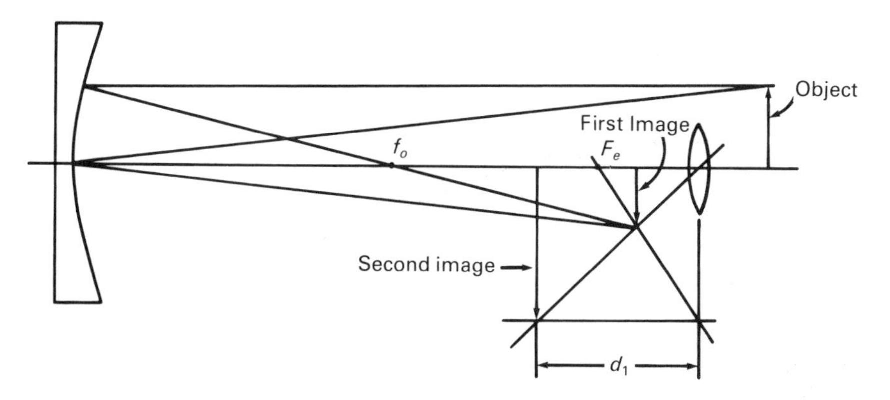

Figure 9.13. A ray diagram for the Newtonian telescope.

analyze the optics by removing the mirror and straightening at the optical path (see Figure 9.13). Here we note that $f_o > f_e$. The image of the mirror is inverted and occurs between f_e and the eyepiece. It is real, as it occurs on the same side of the mirror as the object (mirrors are the opposite of lenses in this respect). The second image, as for the microscope, is inverted and virtual. The location of the resulting image is such that $d_1 \approx 20$ cm and the image is clear to the viewer.

b. Lens aberrations

All of the lenses we have discussed so far have been ideal lenses. Real lenses are considerably more complex to analyze. The approximations given in the previous section are sufficient for most cases. However, the image that is produced is not perfect but suffers from a number of aberrations. Here we give a quantitative description of the important types of aberrations that can interfere with the operation of an optical system.

Chromatic aberration. Chromatic aberration occurs when more than a single wavelength of light is used. As the index of refraction of optical materials is different for different wavelengths of light, equation (9.20) indicates that the focal length is wavelength dependent. As a result, different wavelengths are focused at slightly different places (see Figure 9.14). This, of course, is not a problem for reflecting surfaces. This difficulty can be mostly eliminated by the use of *achromatic* lenses. An achromatic lens consists of two lenses of different indices of refraction. Figure 9.15 shows an example of an "achromat." The two lenses are in close proximity and in general are in contact. In some cases the lenses are actually cemented together. The precise values of the indices of refraction depend on the wavelengths of light that are of interest. In general, it can be shown that the glasses should be chosen in such a way that the inverses of their dispersions,

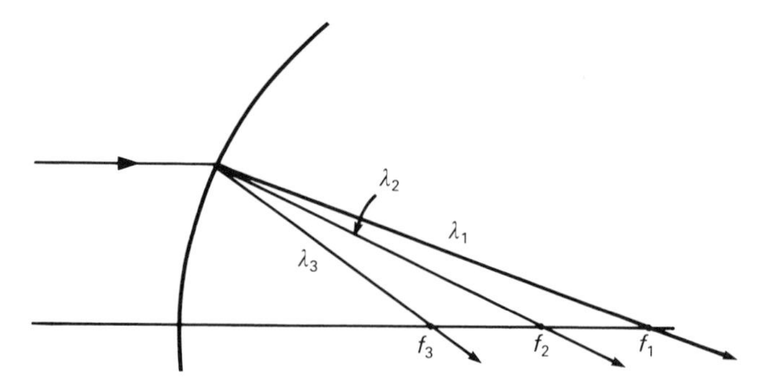

Figure 9.14. A demonstration of chromatic aberration from a spherically refracting surface. Note that $\lambda_1 > \lambda_2 > \lambda_3$.

V and V', are related to the focal lengths of the two components f and f', respectively, as follows:

$$\frac{f}{f'} = -\frac{V'}{V} \tag{9.23}$$

Note that V is always positive, but for a doublet as shown in Figure 9.15 one of the f's is negative.

Spherical aberration. For a spherically refracting surface, we can show that light rays that are parallel to each other but incident on the surface at different distances from the line of the center of curvature will be focused at different distances from the surface. This is shown in Figure 9.16. A surface that causes the rays that are farther from the axis to be focused closer to the surface is said to have a positive spherical aberration. This is true for convex surfaces. A concave surface focuses rays that are farther from the axis farther from the surface and is referred to as having a negative spherical aberration. The spherical aberration of an optical system is minimized by properly choosing optical elements such that the positive and negative

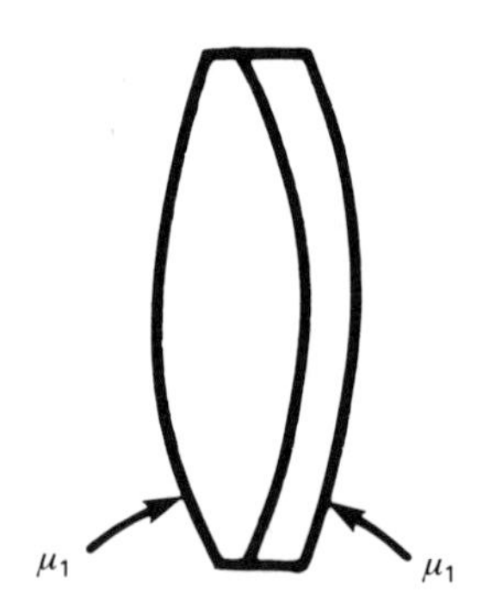

Figure 9.15. An achromatic lens.

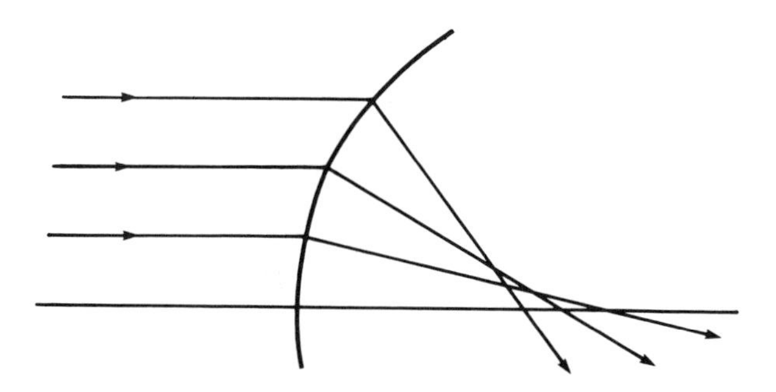

Figure 9.16. Demonstration of spherical aberration.

components of the spherical aberration cancel out. A crucial aspect of lens choice is the focal length. This in itself does not imply a particular spherical aberration. Let us consider a positive symmetric lens: see Figure 9.17*a*. The focal length of this particular lens is related to the differences in the radii of curvature of the two surfaces, $|(1/r_1 - 1/r_2)|$. This lens has some particular spherical aberration. We can make another positive lens with the same focal length by decreasing $1/r_1$ and increasing $1/r_2$ by the same amount. This is referred to as *bending* the lens. Progressive results are illustrated in Figure 9.17*b*, *c* and *d*. The four lenses shown here, although they have the same focal length, do not have the same spherical aberration.

Coma. The correction of an optical system for spherical aberration corrects the problem of longitudinal variations in the focal point on the axis. Even if this problem is corrected, it is not necessarily true that images that are focused off-axis will not suffer from a similar problem. This problem, when present, is called *coma*: the name originated from the "comet-like" appearance of a point image. The severity of this defect depends on how far the image is off-axis. Specifically, it is a function of the angle between the axis and the line drawn from the center of the lens to the image; this is the angle of *obliquity*. An optical system which is corrected for both spherical aberration and coma is called *aplanic*.

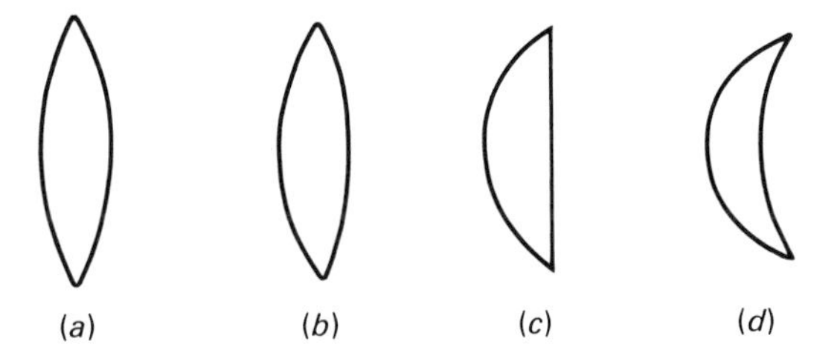

Figure 9.17. An illustration of bending: different lenses with the same focal length but different spherical aberrations.

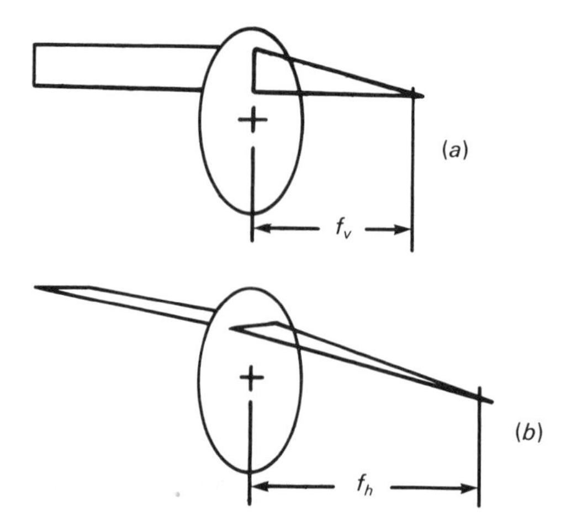

Figure 9.18. Astigmatism in a lens with spherical surfaces for (*a*) vertical image and (*b*) horizontal image. Note the different focal lengths for the two images.

Astigmatism. Astigmatism is a shortened form of the term "astigmatic difference of focus." It means the difference in the focal length for radial and tangential objects (see Figure 9.18). In the eye, for example, astigmatism can result from differences in the curvature of the lens in the horizontal and vertical directions. In a manufactured optical system we would not expect there to be any asymmetry in the lens that could result in astigmatism for images on the optic axis, and this is the case. However, as Figure 9.18 shows, even a symmetric lens can exhibit astigmatism for off-axis images. Uncorrected thin lenses generally exhibit a good deal of astigmatism. This is generally minimized in the design of optical systems by placing lenses so that there is some space between them.

Curvature of field. Curvature of field results because the surface onto which objects are focused by a spherical lens is not a plane but is curved—it is actually a parabola. This is illustrated for a positive lens in Figure 9.19.

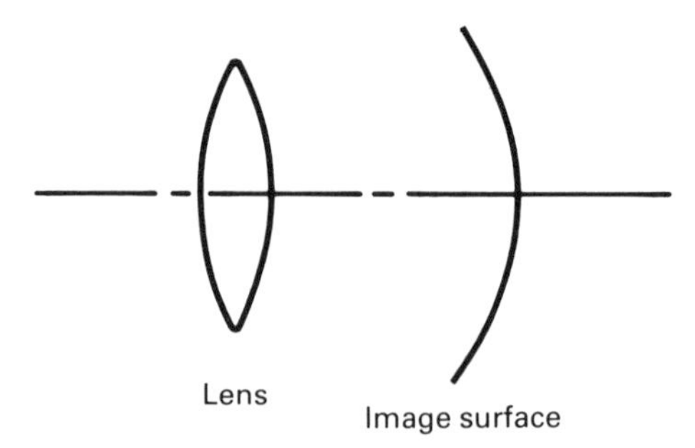

Figure 9.19. Curvature of field for a positive lens.

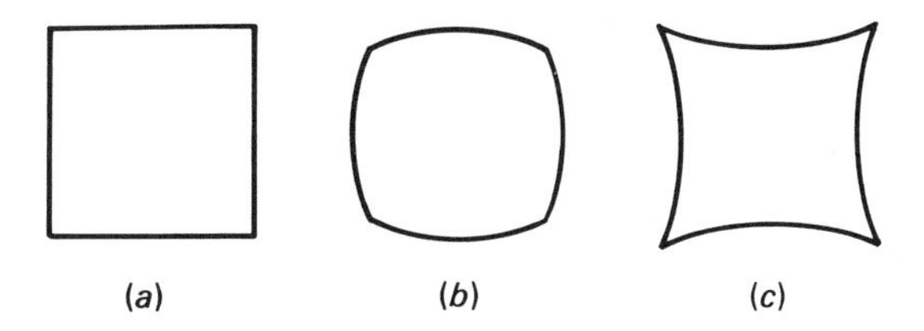

Figure 9.20. The two types of distortion for the image shown in (*a*): (*b*) barrel distortion and (*c*) pincushion distortion.

The only way of reducing this is by the use of optical systems that contain both positive and negative elements.

Distortion. Distortion is the deviation of the shape of the image from the shape of the object. It results from the situation in which the magnification of the image is a function of the distance from the optic axis. Two kinds of distortion can exist. These are pincushion distortion, and barrel distortion and they result when there is respectively a positive or negative proportionality between magnification and distance from axis. Figure 9.20 illustrates the differences. Distortion can be minimized by the proper choice of lens combinations and apertures. The aperture can be adjusted by placing diaphragms in the light path.

The six aforementioned aberrations are the six basic or Seidel aberrations. More complex ones exist as well. The Seidel aberrations are functions of three basic parameters of the optical system: (1) image height h; (2) aperture size s; and (3) angle of obliquity α. The aperture size is the diameter of the lens or the diameter of any diaphagm smaller than the lens. Table 9.3 shows the functional dependence of the six aberrations on these factors.

9.3 Optical fibers

The transmission of light along a fiber relies on the light in the fiber being totally internally reflected. Let us see what happens at an interface between

Table 9.3. Functional dependence of the six Seidel aberrations on h, s and α

Aberration	h	s	α
Chromatic	Independent	Independent	Independent
Spherical	Independent	s^2	Independent
Coma	h	s^2	α
Astigmatism	h^2	Independent	α^2
Curvature of field	h^2	Independent	α^2
Distortion	h^2	Independent	α^3

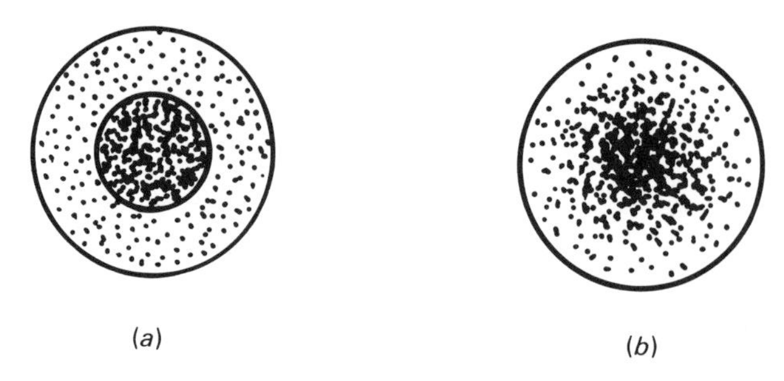

Figure 9.21. Structure of optical fibers: (*a*) singly clad fiber; (*b*) graded-index fiber. Note that the density of the pattern in the illustration is proportional to the index of refraction.

two materials. From Equation (9.3) we find that total internal reflection occurs for angles less than α_T:

$$\alpha_T = \sin^{-1}\left(\frac{\mu_2}{\mu_1}\right) \tag{9.24}$$

where μ_1 is the index of refraction of the fiber and μ_2 is the index of refraction of medium around the fiber. For a glass fiber in air we find $\alpha_T \approx 40°$, measured with respect to the normal. Thus if we shine light down a long cylindrical fiber we find that most of the light is totally internally reflected. If α is less than α_T by even a small amount, a significant proportion of the ray can be lost. In a single glass fiber, light can be lost if the fiber is in contact with any material or if there is any contamination of the outer surface by dust, grease, etc. For this reason it is important to use clad fibers or graded-index fibers. Examples of these fiber types are shown in Figure 9.21. Areas of different shading in the figure represent glasses with different indices of refraction. Let us consider some of the details of clad and graded-index fibers.

a. Clad fibers

First we consider the methods for preparing clad fibers. The simple method of preparing clad fibers is to first prepare a rod of material from two (or more) glasses, as shown in Figure 9.22. Initial rods might be several

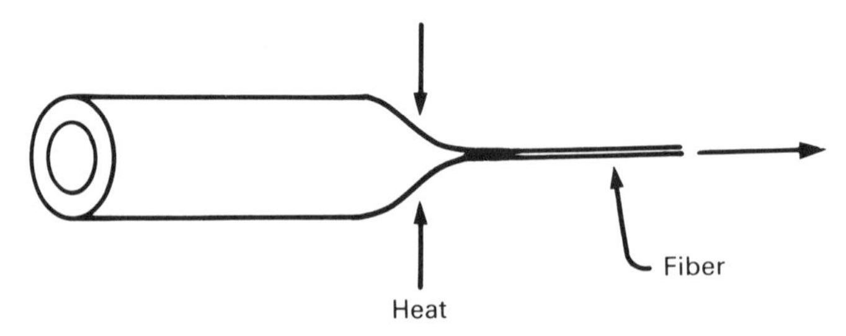

Figure 9.22. Drawing a clad optical fiber.

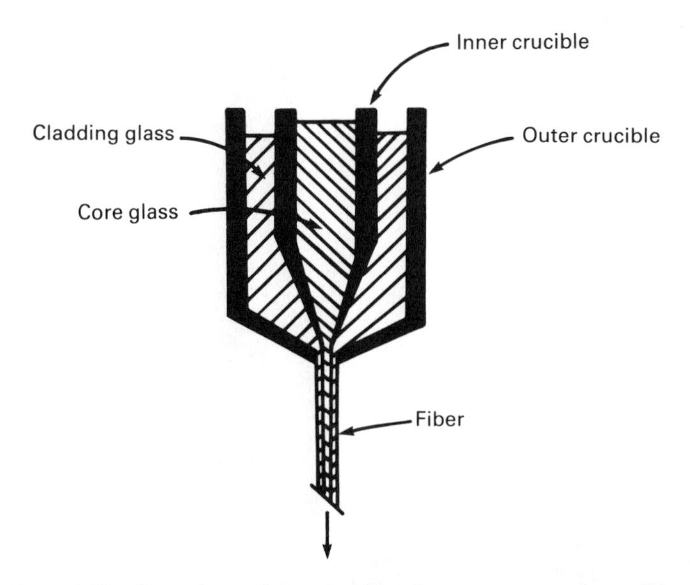

Figure 9.23. Preparing a clad optical fiber from two concentric crucibles.

centimeters in diameter. Heat is applied until the glass softens (~800–1200°C, depending on the type of glass) and a thin fiber (~150 μm diameter) is drawn from one end. The production of a singly clad fiber is illustrated but this method lends itself nicely to the production of multiply clad fiber as well. Because of the drastic reduction in diameter, a single centimeter of rod will yield several hundred meters of optical fiber. An alternate method is to draw the fiber from two melts contained in two concentric crucibles, as illustrated in Figure 9.23. A third method is to vapor-deposit layers of glass onto a cylindrical mandrel. Two or more layers can be deposited and then removed from the mandrel in the form of concentric tubes of glass. These can then be drawn out into a fiber.

In Figure 9.24 we consider the entry of a ray into a clad fiber. From

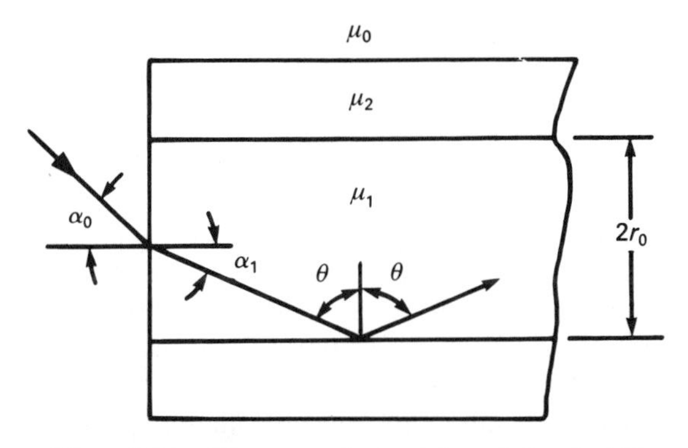

Figure 9.24. The entry of a ray into a clad optical fiber.

equation (9.24) we have the angle α_1 of the ray inside the fiber in terms of the angle of incidence α_0 at the end of the fiber as

$$\alpha_1 = \sin^{-1}\left(\frac{\mu_0}{\mu_1}\sin\alpha_0\right) \tag{9.25}$$

The diagram shows that

$$\theta = \tfrac{1}{2}\pi - \alpha_1 \tag{9.26}$$

For total internal reflection we have

$$\frac{\mu_2}{\mu_1} = \sin\theta \tag{9.27}$$

and we can solve for the incident angle:

$$\sin\alpha_{0T} = \frac{1}{\mu_0}(\mu_1^2 - \mu_2^2)^{1/2} \tag{9.28}$$

The numerical aperture (n.a.) of the fiber is defined as

$$\text{n.a.} = \mu_0 \sin\alpha_{0T} = (\mu_1^2 - \mu_2^2)^{1/2} \tag{9.29}$$

The angle of acceptance (i.e., the angle at which incident rays are totally reflected) is given from the n.a. as

$$\alpha_0 = \sin^{-1}\left(\frac{\text{n.a.}}{\mu_0}\right) \tag{9.30}$$

The greater the n.a. the greater the angle at which a ray may enter the fiber and become trapped.

As a particular ray will reflect back and forth many times within the fiber, it is important to know the actual path length of the ray in the fiber. From Figure 9.24 a straightforward derivation shows that the length a ray travels in a fiber of length L will be

$$l = L\sec\alpha_1 \tag{9.31}$$

The optical path length is thus given as

$$l_0 = L\mu_1\sec\alpha_1 \tag{9.32}$$

The maximum length will be traveled by those rays with an angle of incidence equal to α_T.

It is clear that some of the light will be lost upon bending the fiber. This is because rays incident at angles near α_T in the unbent fiber will, in the bent fiber be incident on the inside of the fiber at an angle that is too *small* and will be partially refracted into the cladding. The situation in a bent fiber is shown in Figure 9.25. We can see that all the light will be lost when we have bent the fiber to a point where rays initially traveling parallel to the fiber axis are incident upon the inside of the fiber at an angle less than α_T. Some

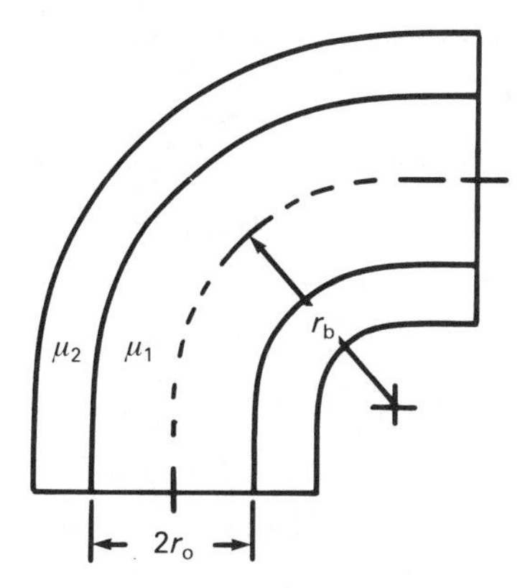

Figure 9.25. A cross section of a bent optical fiber: r_b is the radius of bending and r_0 is the radius of the light-carrying fiber.

straightforward (but messy) geometry leads us to the conclusion that the critical bending radius r_{bc} is given by

$$r_{bc} = r_0 \frac{\mu_1 + \mu_2}{\mu_1 - \mu_2} \tag{9.33}$$

To give you an idea of what this might be, let us consider a fiber with an inner radius of 100 μm, $\mu_1 = 1.8$, and $\mu_2 = 1$. We find that $r_{bc} = 1.1$ mm; so we see that a typical fiber can bend quite sharply before the light is totally lost.

b. Graded-index fibers

Two methods are commonly used to prepare graded-index fibers. The first is the double-crucible method shown in Figure 9.23. The two glasses are of the same type (e.g., borosilicate) but are doped with different impurities to give two different indices of refraction (e.g., thallium to give high μ and sodium to give low μ). If fiber is cooled slowly as it is drawn, the impurities will diffuse, eliminating the sharp boundary between the two different μ. With proper cooling, we can obtain whatever index-of-refraction profile we want.

A second method is to use the vapor-deposition technique. As layers of glass are deposited on the mandrel, the impurity composition of the glass and hence its index of refraction can be continuously varied. This index-of-refraction profile is then preserved in the drawn fiber.

In a graded index fiber the rays are not reflected at a sharp boundary but are bent smoothly back toward the fiber axis (see Figure 9.26). For a distance r from the fiber axis the angle θ that the ray makes with a line drawn parallel to the axis is given by

$$\mu(r) \cos(\theta(r)) = \text{constant} \tag{9.34}$$

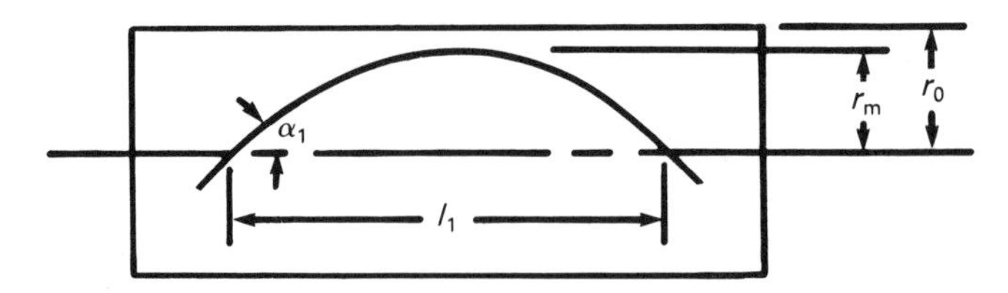

Figure 9.26. The path of a ray in a graded-index fiber.

The point r_m at which the ray is bent back is given from equation (9.34) as

$$\mu(r_m) = \mu(0) \cos \alpha_1 \tag{9.35}$$

Of course, for the fiber to be effective, we want r_m to be less than r_0. The ratio $\mu(r_0)/\mu(0)$ will define the angles α_1 for which the rays will be trapped in the fiber. From $\mu(r_0)$ and $\mu(0)$ we can define an effective numerical aperture of the form

$$\text{n.a.} = [\mu^2(0) - \mu^2(r_0)]^{1/2} \tag{9.36}$$

A graded-index fiber can be used to transmit images (not only light) in a rather ingenious way. If the fiber is an integral number of cycles long (l_1 in Figure 9.26) and this cycle length is independent of α_1, then, as Figure 9.27 shows, images can be transmitted. Let us see how we can meet this latter criterion.

The length per cycle is given by

$$l_1 = \int \mu(r)\, ds \tag{9.37}$$

where ds is the differential path length. We write

$$ds^2 = \left(1 + \left(\frac{dz}{dr}\right)^2\right) dr^2 \tag{9.38}$$

where r is the radius and z is the distance along the fiber axis. We find

$$\frac{dr}{dz} = \tan(\theta(r)) \tag{9.39}$$

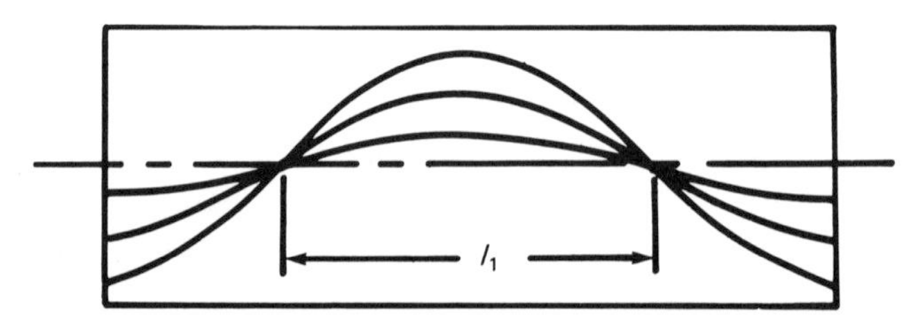

Figure 9.27. The principle of image transmission in a graded-index optical fiber.

and from equations (9.34) and (9.35),

$$\frac{dr}{dz} = \left(\frac{\mu^2(r)}{\mu^2(0)\cos^2\alpha_1} - 1\right)^{1/2} \tag{9.40}$$

Combining this with equation (9.38) we obtain

$$ds^2 = \left(\frac{\mu^2(r)}{\mu^2(r) - \mu^2(0)\cos^2\alpha_1}\right) dr^2 \tag{9.41}$$

and the integral in (9.37) becomes

$$l_1 = 4\int_0^{r_m} \frac{\mu^2(r)\,dr}{[\mu^2(r) - \mu^2(0)\cos^2\alpha_1]^{1/2}} \tag{9.42}$$

Obtaining a solution to this integral equation is not straightforward, so we will merely give the answer. We obtain a $\mu(r)$ for which this integral is equal to a constant length l_1 for all values of α_1. This is

$$\mu(r) = \mu(0)\,\mathrm{sech}\left(\frac{2\pi r}{l_1}\right) \tag{9.43}$$

In practice it is generally sufficient to approximate the sech in terms of a Taylor expansion to order r^2.

$$\mu(r) = \mu(0)\left(1 + \frac{2\pi^2 r^2}{l_1^2}\right) \tag{9.44}$$

By proper preparation procedures a radial dependence of the index of refraction can be obtained that follows equation (9.44). When cut to the proper length and polished, these fibers can be used to transmit images.

c. *Fiber bundles*

Fiber bundles may be coherent or incoherent. In a coherent bundle the arrangement of fibers at one end of the bundle bears a definite relationship to the arrangement of fibers at the other end of the bundle. In many cases the arrangement of fibers on the two ends is in fact the same. Each fiber transmits only light incident upon the end of that fiber to the other end of the same fiber. In this way the pattern of light at one end of the bundle (the image) is transmitted to the other end.

Incoherent fiber bundles have arrangements of fibres at one end that bear no relationship with the arrangement of fibers at the other end. Obviously, we cannot use these for transmitting images. We can use them for transmitting a number of independent signals or for transmitting large amounts of energy. Multiple fibers are of use in this application as the amount of energy a single fiber can carry without overheating is limited.

d. Applications

Single fibers with diameters as small as 10 μm can be used for optical scanning with extremely high resolution or in locations where space is very limited. The latter has numerous medical applications. Because of their small diameter, an optical fiber carrying even a few watts of energy will result in an output beam with extremely high energy density. This can be used for soldering of small electronic components or delicate (e.g., retinal) surgery. Care must be taken to avoid damage to the fiber, which can result from the dissipation of large amounts of heat.

Fiber bundles have widespread potential as a means of communication. A large number of signals in the form of modulated light can be carried in single bundle of fibers because each fiber can carry a signal independently. Bundles or cables for this purpose must be protected by an outer casing, usually of metal or plastic, or both, in order to prevent damage from abrasion, stress, bending, and corrosion.

10 Optical techniques

10.1 Light-intensity measurement

Various properties of light can be measured. In a simple experiment we could measure the intensity of the light. There is, however, often a good deal more information in a beam of light than merely its intensity. Other properties of the light that we might be interested in determining are its polarization, its phase relative to another beam of light, and its spectral content. The techniques of measurement of these three quantities are referred to respectively as polarimetry, interferometry, and spectroscopy. In this chapter we consider these techniques.

a. Photomultipliers

A photomultiplier tube (PMT) is a tube containing a photoemissive cell with a current amplifier. The photoemissive cell consists of a transparent *photocathode* that produces electrons when photons are incident upon it. The electrons are guided inside an evacuated tube by the use of electric fields. The electrons impinge upon the surface of plates called *dynodes,* which are coated with a material that has the property that it emits more electrons than strike it. The ratio δ of the number of electrons emitted by a dynode to the number of electrons impinging upon it is a measure of its *gain.* A tube would have something in the order of 10 dynodes, after which the electron current would be measured; the final dynode is referred to as the *anode.* The number of electrons observed at the anode for each electron produced by the photocathode is the *multiplication factor* M, and for n dynodes is given by

$$M = \delta^n \tag{10.1}$$

Three geometric configurations for photomultiplier tubes are in common use. These are illustrated in Figure 10.1. The focused arrangement is sometimes also referred to as a linear photomultiplier tube and the box and grid arrangement is sometimes known as a bucket type.

The electrons are directed by electric fields produced by applying increasingly more positive potentials to the dynodes, beginning with the

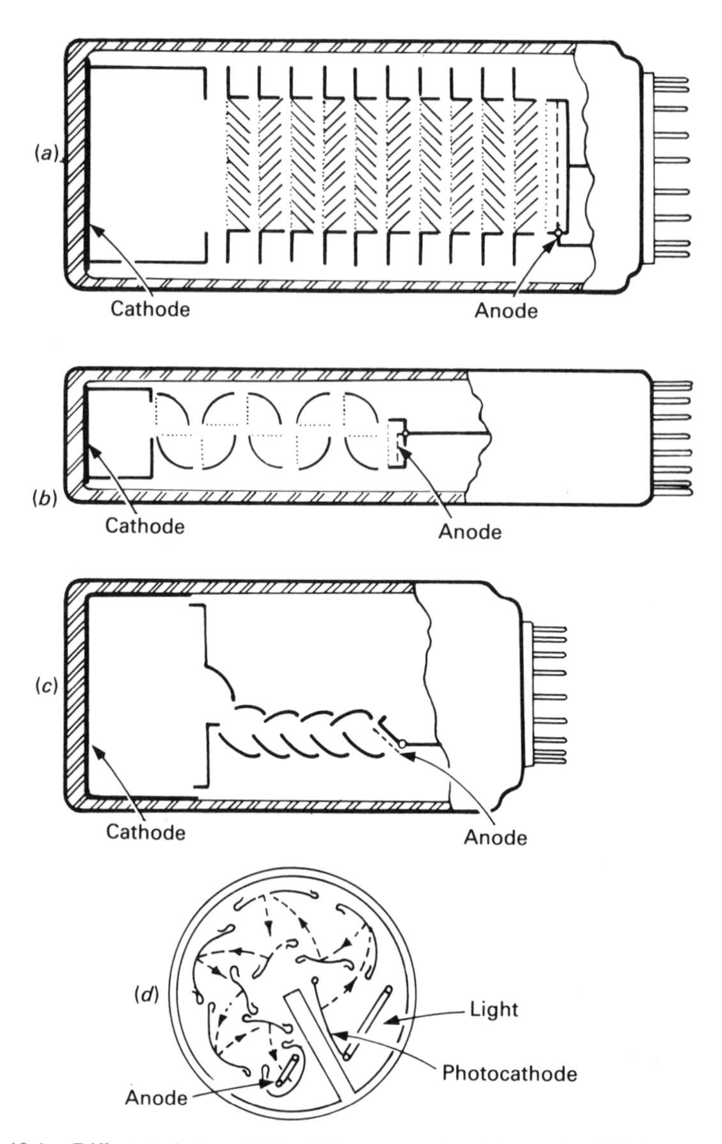

Figure 10.1. Different photomultiplier tube geometries: (*a*) venetian blind arrangement; (*b*) box and grid arrangement; (*c*) and (*d*) focused dynode arrangements. (Copyright © 1986 by Thorn EMI Electron Tubes Limited. Used with permission.)

photocathode and ending with the anode. A couple of possible arrangements for doing this are shown in Figure 10.2. The arrangement in Figure 10.2*a* is commonly used for light measurement and has a directly coupled output. That in Figure 10.2*b* is commonly used for scintillation counting (see Section 12.1) and has a capacitively coupled output. The voltage difference between dynodes is typically in the range of 100 V. The detail of

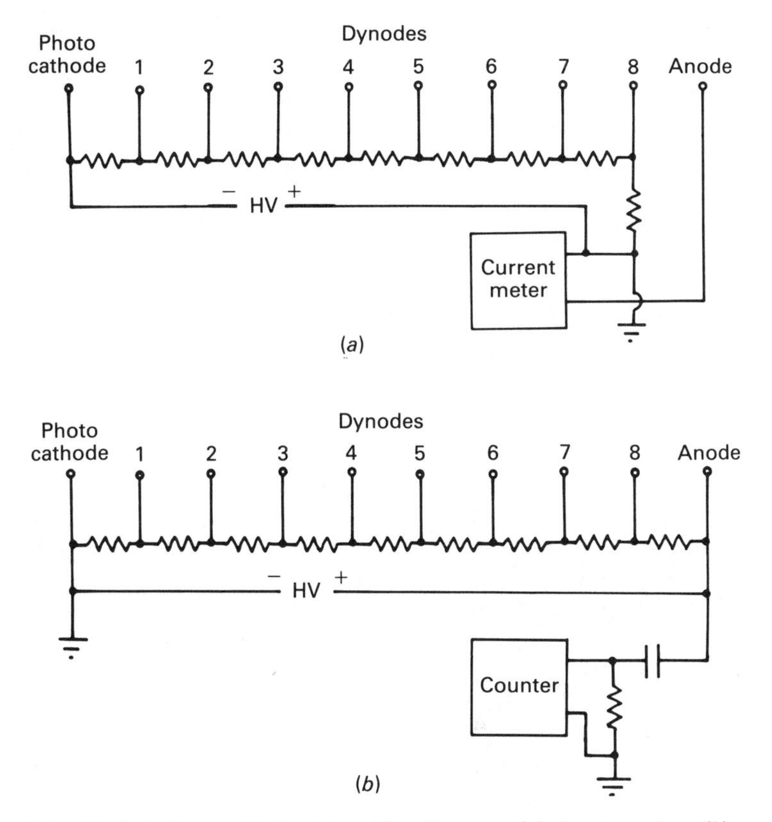

Figure 10.2. Typical photomultiplier tube wiring diagrams: (*a*) direct coupling; (*b*) capacitive coupling.

the electron paths inside the tube is complex and depends on the specific design of the tube.

Let us consider the choice of materials to be used for the construction of the photomultiplier tube. We begin with the photocathode. At low temperature a conduction electron in a metal cannot leave the metal because it does not have enough energy to overcome the potential barrier that exists at the surface. We can encourage an electron to leave the material by supplying energy. One way of supplying energy is with photons. The potential barrier at the surface has a magnitude ϕ and is referred to as the *work function.* If we supply energy in the form of photons with frequency ω, then electrons can be liberated from the surface when

$$\hbar\omega \geqslant \phi \tag{10.2}$$

When this is an inequality, the excess energy is given to the electron in the form of kinetic energy. Thus

$$\hbar\omega = \phi + \tfrac{1}{2}mv^2 \tag{10.3}$$

The *quantum efficiency* refers to the number of electrons given off for every photon with sufficient energy that is incident on the surface. Quantum efficiencies of 30–40 percent are the best that have been achieved.

However, photocathodes are most commonly used at room temperature, and here other effects are important. Some electrons on the high-energy tail of the Fermi–Dirac distribution have enough energy to escape from the surface. This is called *thermionic emission*—the higher the temperature, the more smeared out is the Fermi–Dirac distribution and the more thermionic emission occurs. In a photomultiplier tube the thermionic emission results in electrons being detected at the anode that do *not* result from a photon's incidence upon the photocathode. This is a type of noise—more specifically, it is referred to as *dark current.* PMTs are sometimes cooled to low temperatures to improve their performance for this reason.

Alkali metals are usually used for making photocathodes and different materials are denoted by a letter code, for example, S1, TU, etc. The photocathode consists of a semi-transparent layer coated on to a transparent window. The code refers to both the coating and the window material. Some common photocathodes are given in Table 10.1. The spectral response of the photocathode depends not only on the wavelength dependence of the quantum efficiency but also on the wavelength dependence of the transmittivity of the window. Two typical curves are shown in Figure 10.3. You see that the tube with the quartz window is much more sensitive to ultraviolet; this is because of the difference in the transmittivities of quartz and glass at these wavelengths (see Section 9.1). Different cathode materials produce different dark currents at room temperature. Some dark currents are listed in Table 10.2.

The minimum temperature T_{MIN} given in Table 10.2 is the temperature at which the dark current becomes too small to measure, about one electron per second per square centimeter. It is not worthwhile cooling the PMT to a temperature lower than this.

Table 10.1. Photocathode materials

Code	Substrate	Coating
S1	Glass	Cs–Ag–CsO_2
S5	Quartz	Sb–Cs
S8	Glass	Ag–Bi–Cs
S11	Glass	Sb–Cs–MnO_2
S13	Quartz	Sb–Cs–MnO_2
S20	Glass	Sb–Na–K–Cs
S20R	Glass	Sb–Na–K–Cs
T	Glass	Sb–Na–K–Cs
TU	Quartz	Sb–Na–K–Cs
D	Glass	Sb–K–Cs
DU	Quartz	Sb–K–Cs

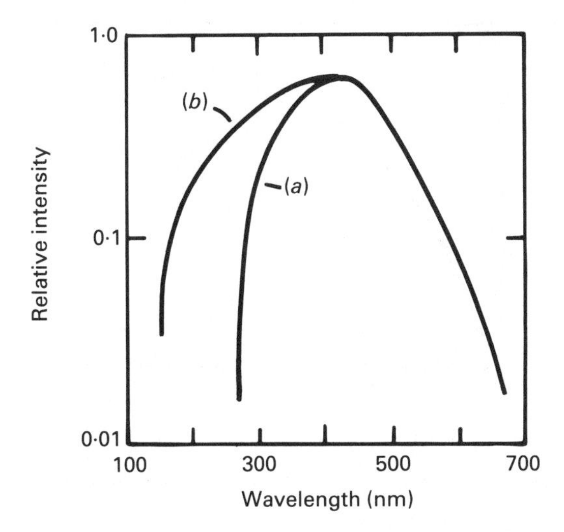

Figure 10.3. Spectral response of photomultiplier tubes: (*a*) typical response of a glass-window tube; (*b*) typical response of a quartz-window tube.

We turn now to the dynodes. Certain materials can emit one or more electrons when struck by an electron with sufficient velocity. The incident and emitted electrons are referred to as primary and secondary electrons, respectively. The number of secondary electrons per primary electron depends on

1. The nature of the surface
2. The energy of the primary electron
3. The incident angle of the primary electron

If we make the assumption that the primary electrons are incident perpendicular to the dynode surface, then we can look at δ as a function of primary electron energy. This is shown in Figure 10.4 for a silver–magnesium dynode. We see that this has a maximum at about 440 eV. This behavior is typical of most materials used for PMT dynodes. For low voltages between dynodes ($V < 300$ V), δ is roughly linear in voltage. So if

Table 10.2. Dark currents I_{DARK} of different cathodes

Material	I_{DARK} (295 K) (electrons/(s cm^2))	T_{MIN} (°C)
S1	5×10^6	−100
S20R	10^3	−40
S20	300	−40
S11	100	−20
D	10	0

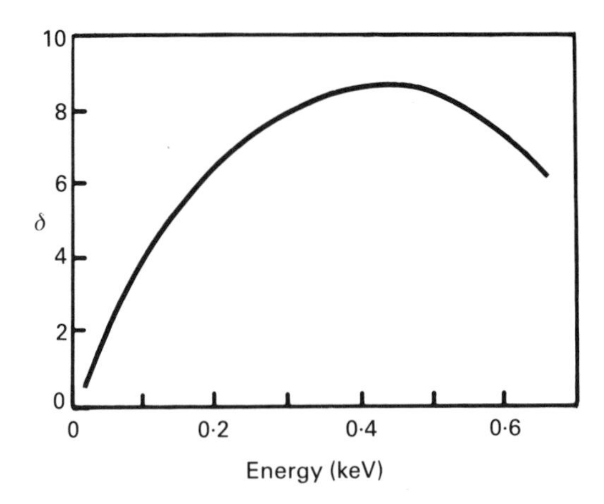

Figure 10.4. Secondary emission ratio for a typical silver–magnesium dynode as a function of primary electron energy.

we increase V by, say, 10 percent then we increase δ by 10 percent, and if there are 10 dynodes then we increase M as given by equation (10.1) by

$$(1.1)^{10} = 2.6 \tag{10.4}$$

Figure 10.5 shows a typical relationship between M and the voltage between dynodes. We see that M is a very sensitive function of V. This means that the supply voltage for the dynodes must be extremely stable in order to keep the gain of the PMT stable.

We can get a rough idea of the time response of one of these devices. Let us first consider how long it takes the pulse of electrons to reach the anode after a photon was incident on the cathode. We will make the following

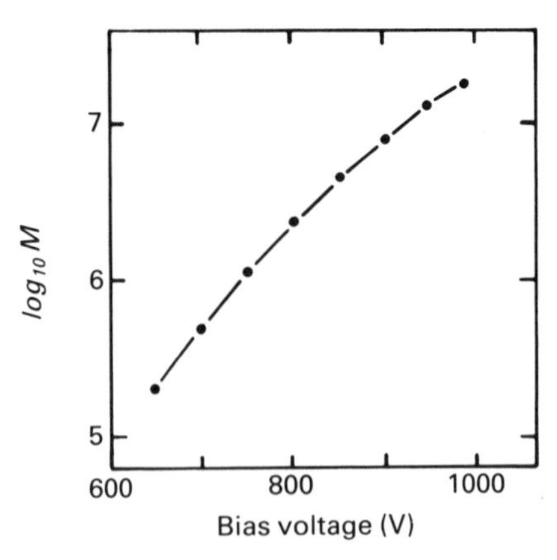

Figure 10.5. Total multiplication factor for a typical photomultiplier tube as a function of total bias voltage.

assumptions:

1. 10 dynodes are spaced evenly at 1 cm apart.
2. There is 100 volts between dynodes.
3. The electric field between dynodes is constant.
4. Secondary electrons are emitted with no kinetic energy.

We have 10 gaps for the electrons to transverse, the total time is 10 times the time to transverse one gap. We begin with the Lorentz force:

$$F = ma = -eE = \frac{eV}{d} \tag{10.5}$$

where $d = 1$ cm (0.01 m). Some simple kinetics (taking $a = \text{constant}$) gives

$$d = \frac{at^2}{2} \tag{10.6}$$

and with equation (10.5) we obtain

$$t = \left(\frac{2d^2m}{eV}\right)^{1/2} \tag{10.7}$$

Substituting in the appropriate values, we find $t = 3 \times 10^{-9}$ s so that the total time is 3×10^{-8} s (30 ns).

The situation in an actual photomultiplier tube is not so simple. The electric field between dynodes is not constant, therefore not all of the secondary electrons produced at one dynode by a single primary electron arrive at the next dynode at exactly the same time, because they may follow different trajectories. Thus the current pulse gets spread out in time as it progresses towards the anode. If there is a 1 percent spread caused by each dynode, then at the anode we will see about a 10 percent spread— about 3 ns. This spread in the arrival times for the electrons at the anode is the *rise time* τ of the output for the PMT. The focused dynode-type tube is designed to minimize the rise time and may have τ in the order of 1 ns. Other types of PMTs may have τ up to 10 ns or so. The time response is not generally important when the PMT is used in the directly coupled mode illustrated in Figure 10.2*a*.

A PMT is usually connected as shown in Figure 10.6. PMTs set up in this manner are very sensitive and can detect light that is much too faint to see with the unaided eye. Light levels in the range of a few photons per second can be measured if the tube is cooled to eliminate the dark current.

The advantages of the PMT as a light sensor are

1. High sensitivity
2. Fast response

However, the PMT is not suitable for all applications and has the following

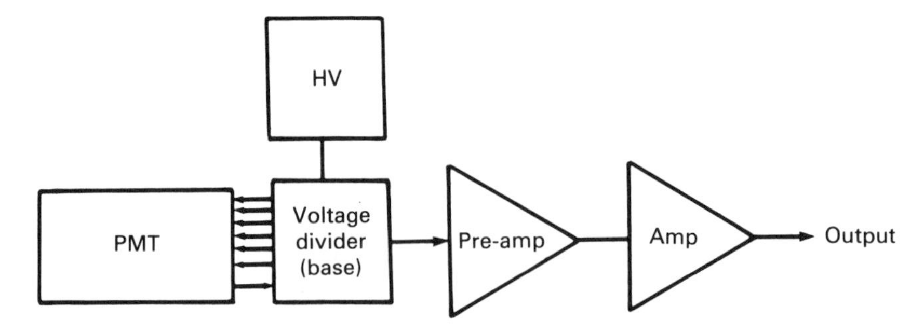

Figure 10.6. Electrical connections for a photomultiplier tube.

disadvantages:

1. Large size
2. Mechanical delicacy
3. Expense (great), not only of the PMT, but of a high-quality power supply and expensive amplifiers that are necessary as well

Concerning point 3, the set up shown in Figure 10.6 would cost approximately $2500.

Finally, a warning about PMTs. Because they are so sensitive, you have to be careful not to expose them to too much light. A PMT with the high-voltage connected would, if exposed to *normal room light,* produce such a large output current that it would permanently damage the anode. Obviously, this should not be done, and we have the most important rule for using PMT's:

Rule 1 Never expose a biased PMT to ambient light. An unbiased PMT that is exposed to high light level will not be permanently damaged, but it may take a day or so for the excitations in the photocathode and dynode materials to quiet down enough for the dark current to return to its usual level. So we have a second rule:

Rule 2 Never expose an *un*biased PMT to ambient light.

b. Photodiodes and phototransistors

Light, like other forms of energy, can cause atoms to become ionized. This can occur in a diode too, and the electron–hole pairs formed by ionization will contribute to the current and serve as a means for detecting the presence of light and measuring its intensity.

In most diodes a reverse bias of a few volts will produce a current that is very small and is essentially independent of bias. This is the reverse-saturation current I_s and it is proportional to the concentration of the minority carriers. It can be thought of as arising from the thermally excited

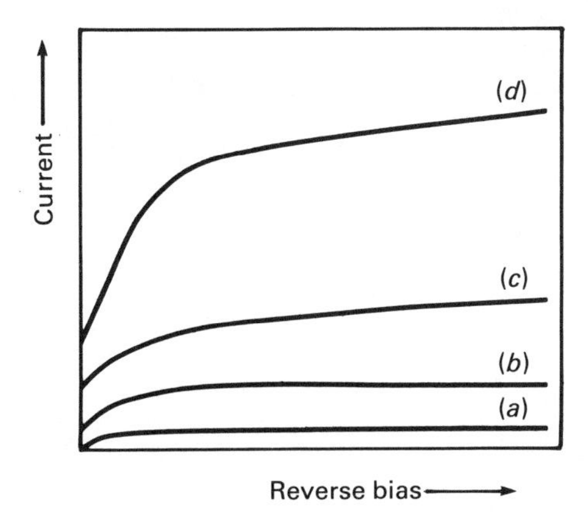

Figure 10.7. Current–voltage relationship for a reverse-biased photodiode with light intensity increasing from (*a*) through (*d*).

minority carriers. In a photodiode, as in a PMT, this thermally produced current is called the dark current and it can be decreased by lowering the temperature.

Figures 10.7 and 10.8 show the reverse current for different light levels for a typical photodiode. You can see that at a given reverse-bias voltage the current is fairly linear in the light intensity. This is, of course, if the energy of a single quantum of light is sufficiently energetic to cause ionization. This linear relationship results because the reverse current is linear in the number of ionization pairs, which is, in turn, linearly related to the light intensity. This linear relationship extends even to zero bias, so we can in fact use the photodiode unbiased. A simple unbiased setup for the

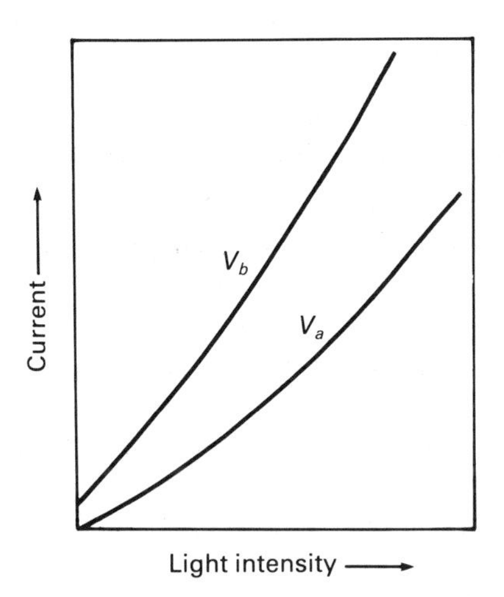

Figure 10.8. Current as a function of light intensity for a typical photodiode. The two curves represent different reverse-bias voltages where $V_a < V_b$.

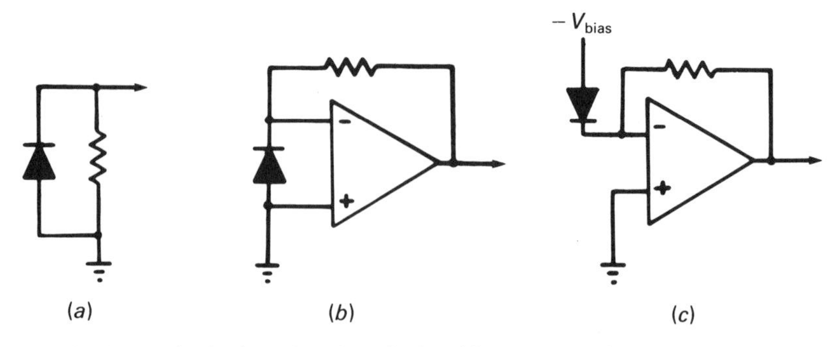

Figure 10.9. Some circuits for using photodiodes: (*a*) unbiased; (*b*) unbiased with op-amp; (*c*) biased with op-amp.

photodiode is shown in Figure 10.9*a*. Since we prefer to measure voltage rather than current, this circuit is a simple current-to-voltage converter; that is $V = IR$. Figure 10.9*b* shows a more practical current-to-voltage converter, and Figure 10.9*c* shows a similar setup in which the photodiode can be reverse biased. The advantage of reverse biasing is that the electric field carrying the minority charges across the depletion layer is larger and hence the charges move faster. This means faster response time for the photodiode.

The photodiode is sensitive to photons that strike the depletion region, but it is much less sensitive to photons that are incident in the p or n regions outside of the depletion layer. Generally, photodiodes have an area of a millimeter squared or so around the junction that is sensitive to light.

Photodiodes are often *pin* diodes. These are diodes that consist of three types of semiconducting material: p-type, intrinsic, and n-type—hence the name. The intrinsic material in the middle acts like a giant depletion region. This has some advantages:

1. It increases the size of the sensitive area.
2. It decreases leakage current—there is less chance of a majority carrier getting in the way.

These diodes can have response times of around 1 ns.

The spectral response of a photodiode is illustrated in Figure 10.10. You can see that there is a relatively sharp cutoff in the sensitivity at larger wavelengths; this occurs at about 1100 nm for silicon and at about 1700 nm for germanium. In order to cause an ionization event, the energy per photon must be greater than the energy of the energy gap. So

$$\hbar\omega = 2\pi\hbar\frac{c}{\lambda} > \mathscr{E}_g \tag{10.8}$$

For silicon, $\mathscr{E}_g = 1.1$ eV and for germanium, $\mathscr{E}_g = 0.72$ eV: these values explain the cutoff in Figure 10.10.

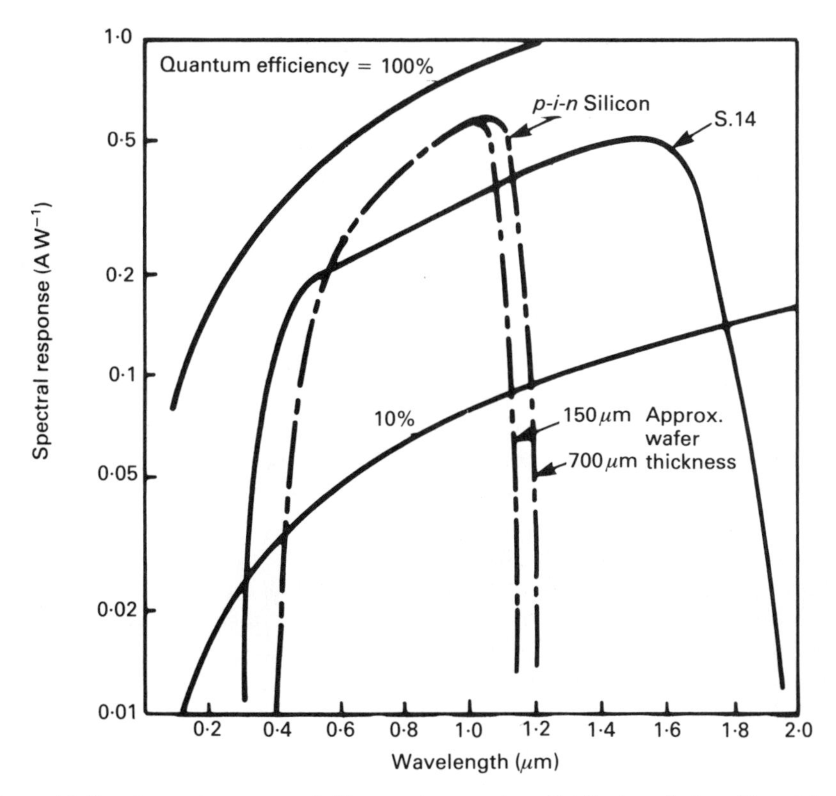

Figure 10.10. Spectral response of silicon and germanium (S-14) photodiodes. (Copyright © 1974 by RCA New Products Division, Lancaster, Pennsylvania. Used with permission).

Unfortunately, the sensitivity of photodiodes to low light levels is not especially good. The human eye can easily see light that the photodiode cannot. A typical output in current might be 1 μA per μW of incident light energy. Compare this with the sensitivity of the PMT—about 1 A per μW of incident light energy, 10^6 times greater. Photodiodes do have the advantage of being inexpensive—many cost under $1—and even a good quality *pin* diode may cost a few dollars. You can see from Figure 10.9 that the associated electronics are simple and as a result is generally inexpensive.

Since light must be able to reach the semiconductor, the package must be at least partially transparent. Two types of packages are common: in one the semiconductor is encapsulated in transparent plastic—often all faces except one are coated black. The other package is a metal can with a flat or lens-shaped plastic or glass window.

The phototransistor is an improvement on the photodiode. The collector–base junction is reverse-biased and the emitter–base junction is forward-biased. The light is detected in the region of the reverse-biased collector–base junction. The current for the phototransistor is in the order of 100 times greater than for the photodiode. Hence the phototransistor has a sensitivity of about 100–1000 μA per μW incident radiation energy. Unfortunately, this does nothing to improve the minimum light that can be detected. The limiting factor for the photodiode is the dark current; the transistor amplifies the dark current along with the signal of interest.

10.2 Polarimetry

a. The theory of polarization

Light possesses a dual nature. In one sense, it is a particle (the photon) and in certain circumstances it is advantageous to think of light in this way. In another sense, it is a wave, and for the purpose of discussing polarization we use this description. Figure 10.11 shows the relationship of the electric and magnetic oscillations that make up electromagnetic radiation. We note that these are in phase and orthogonal to each other and to the direction of propagation. This is not true for waves in anisotropic or absorbing media, but for the present discussion we adopt this simplified picture.

The direction of polarization can be taken to be either the direction of the **E** vector or the direction of the **H** vector. The convention presently seems to be to take the direction of **E** as the polarization direction.

There are three types of polarization commonly distinguished: linear, circular, and elliptical. Linear polarization is sometimes referred to as plane polarization. We discuss these three conditions.

Linear polarization. The case of linear polarization is illustrated in Figure 10.11. For this type of polarization the spatial and time variations of the

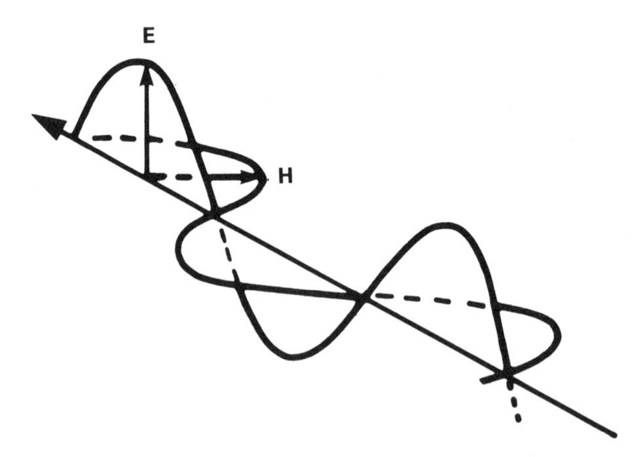

Figure 10.11. The relationship of the electric and magnetic field vectors in electromagnetic radiation.

orthogonal **E** and **H** fields can be written as

$$\begin{aligned} \mathbf{E} &= \mathbf{E}_0 \exp(i\mathbf{k} \cdot \mathbf{r} - i\omega t) \\ \mathbf{H} &= \mathbf{H}_0 \exp(i\mathbf{k} \cdot \mathbf{r} - i\omega t) \end{aligned} \tag{10.9}$$

where the $\mathbf{E}_0$ and $\mathbf{H}_0$ are real, time-independent orthogonal vectors. Normal, unpolarized light consists of a variety of wavelengths (unless, of course, it is monochromatic) and a random distribution of polarization directions. This unpolarized light may be linearly polarized by passing it through a polarizer. (Later we discuss the different types of polarizers available.) The polarizer allows only the component of light with its electric field along a particular direction to be transmitted and blocks the orthogonal component. At any instant, the net polarization of the light may be represented by a vector in a particular direction. This vector may be resolved into components parallel to and normal to the transmission axis of the polarizer. According to Figure 10.12, the magnitude of the transmitted field is given by

$$E_T = |\mathbf{E}| \cos \theta \tag{10.10}$$

E_B in the figure is the magnitude of the blocked field. Although we are dealing with the electric field here, we more commonly might measure the intensity of the light. From Poynting's theorem we have

$$I = \pi \, |\mathbf{E} \times \mathbf{H}|$$

or

$$I = \frac{\pi}{\mu_0 \omega} |\mathbf{E}|^2 \tag{10.11}$$

where μ_0 is the permeability of free space and ω is the angular frequency of the radiation. Hence from equation (10.10) we see that the transmitted

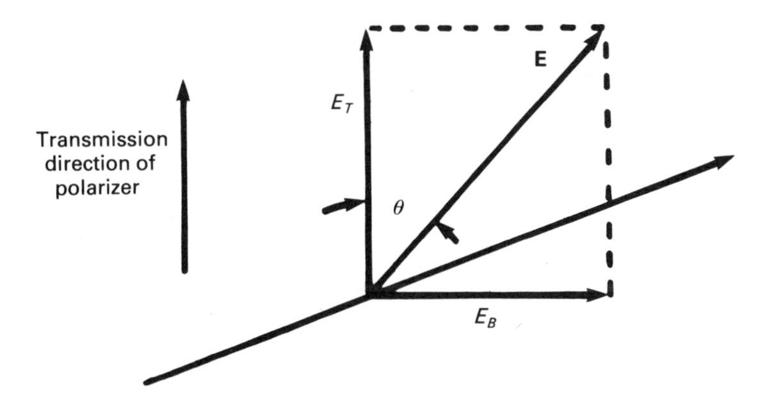

Figure 10.12. The linear polarization of light. The electric field vector of the incident wave is **E** and the transmitted and blocked components are E_T and E_B respectively.

intensity is

$$I = I_0 \cos^2 \theta \tag{10.12}$$

This gives the angular dependence of the light intensity.

Circular polarization. Circular polarization refers to the case where the electric field vector at a particular location in space is constant in magnitude but rotates in direction at a constant frequency ω about the axis of propagation. Figure 10.13 shows the relationship of the **E** and **H** vectors in a plane perpendicular to the direction of propagation. In this case the **E** and **H** vectors remain orthogonal but change their orientation with respect to a fixed x–y coordinate system at a constant frequency. If, when we look along the direction of propagation, the rotation is clockwise (as in Figure 10.13), the light is said to be right circularly polarized. If the rotation is counterclockwise, the light is left circularly polarized.

To justify the physical origins of circularly polarized light, let us go back to the linear case. Let us consider light that is polarized in the x direction. Its **E** vector is defined by,

$$\mathbf{E}_1 = |\mathbf{E}|\, \hat{\mathbf{i}} \cos(kz - \omega t) \tag{10.13}$$

Consider now another linearly polarized light ray of the same amplitude $|\mathbf{E}|$ polarized in the y direction and out of phase by 90° with the ray previously described. We can write the expression for this as

$$\mathbf{E}_2 = |\mathbf{E}|\, \hat{\mathbf{j}} \sin(kz - \omega t) \tag{10.14}$$

The total electric field is given by the vector sum of expressions (10.13) and

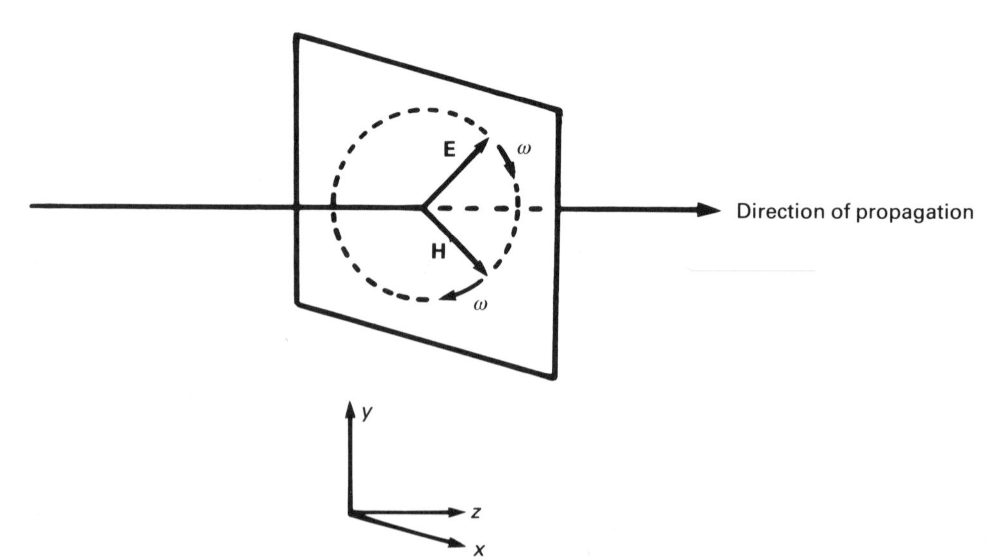

Figure 10.13. Electric and magnetic field vectors of right circularly polarized light.

(10.14) and is written as

$$\mathbf{E} = |\mathbf{E}|\,[\hat{\mathbf{i}}\cos(kz - \omega t) + \hat{\mathbf{j}}\sin(kz - \omega t)] \tag{10.15}$$

If we look at this expression for a fixed value of z, for example $z = 0$, then we see that this represents the expression for **E** which traces out a circle in the x–y plane. As t increases, the circle is traced out in a clockwise direction. If the second linearly polarized wave leads rather than lags the first by 90°, then the sign of expression (10.14) is negative and thus the sign of the second term in equation (10.15) is negative. This means that the circle is traced out counterclockwise with increasing t. These two cases correspond to right and left circularly polarized light, respectively. This means that circularly polarized light is the result of two orthogonal and out-of-phase linearly polarized rays.

Elliptically polarized light. If we return to the description of circularly polarized light as the superposition of two linearly polarized rays, but consider two rays of different amplitudes, we can write

$$\mathbf{E} = |\mathbf{E}_1|\,\hat{\mathbf{i}}\cos(kz - \omega t) + |\mathbf{E}_2|\,\hat{\mathbf{j}}\sin(kz - \omega t) \tag{10.16}$$

This electric field vector at some point in space (e.g., $z = 0$) traces out an ellipse; the degree of ellipticy is given by the ratio $|\mathbf{E}_1|/|\mathbf{E}_2|$.

We have mentioned the three basic types of polarization. It is possible, however, to have combinations of these types. In this sense, unpolarized light is a combination of linearly polarized rays with a random distribution of polarization directions and phases. It is thus possible to have partially polarized light that has an unpolarized component in addition to the polarized component.

b. Types of polarizers

Three basic classes of polarizers are in common use: reflection polarizers, birefringent polarizers, and dichroic polarizers. Below we also discuss quarter-waves plates and the production of circularly and elliptically polarized light.

Reflection polarizers. Figure 10.14 shows an arrangement for polarization of light by reflection. Reflected light is polarized because in a microscopic sense the incident beam causes atoms in the surface layer of the material to oscillate (in the simplest model these are harmonic oscillators). The reflected and refracted components of the beam result from the re-radiation of energy by the oscillators into (refracted) and out of (reflected) the surface of the material. Let us look down upon the plane formed by the incident, reflected, and refracted beams. The oscillations that give rise to the reflected beam are those that produce electric fields parallel to the surface of reflection. As light is a transverse wave, those electric fields must also be

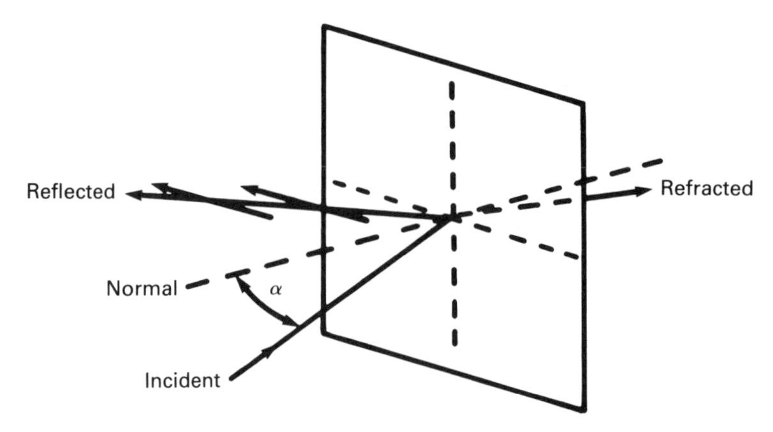

Figure 10.14. Polarization of light by reflection.

normal to the plane of incidence. The figure shows how horizontally polarized light is produced. This effect is maximized when the angle of incidence is equal to the Brewster angle α given by

$$\alpha = \tan^{-1}\left(\frac{\mu_2}{\mu_1}\right) \tag{10.17}$$

Here μ_2 is the index of refraction of the material into which the refracted beam penetrates and μ_1 is the index of refraction of the medium into which the reflected beam penetrates. Reflection at this angle ensures that the reflected and the refracted beams are normal to each other. For an interface between air ($\mu_1 \approx 1$) and glass ($\mu_2 = 1.5$) we find $\alpha = 56°$. For $\mu_1 = 1$, Figure 10.15 shows that α varies only slightly for values of μ_2 for available glasses.

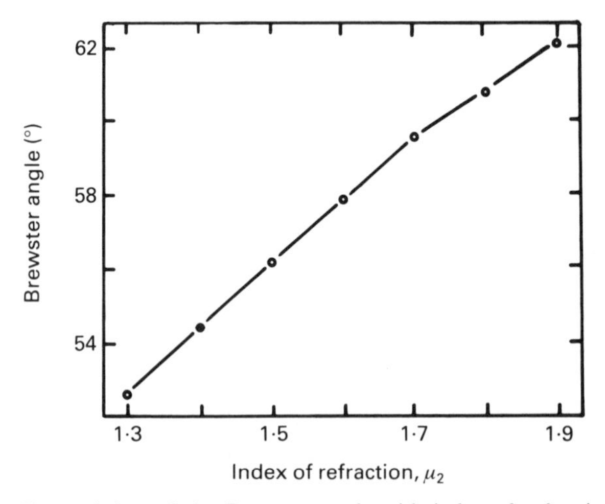

Figure 10.15. The variation of the Brewster angle with index of refraction. See equation 10.17.

Note that there is a small wavelength dependence of α because of the wavelength dependence of μ_1 (see, for example, Figure 9.2).

Birefringent polarizers. Certain types of crystalline materials have an anisotropic crystal structure. As a result, it is sometimes the case that the oscillations of atoms in the material that give rise to the refracted and reflected components of the beam do not occur with equal ease in all directions. The result is that the index of refraction for light polarized in one direction is not the same as the index of refraction for light polarized in a different direction. As a result, unpolarized light is split into two linearly polarized beams with orthogonal polarization axes. This effect is called *birefringence* or double refraction. A commonly known material that exhibits this property is $CaCO_3$ (known as Iceland spar). Figure 10.16 shows this effect. We can separate these components using a *Nicol prism.* In this the Iceland Spar is cut at a particular angle and a layer of Canada Balsam is placed between the two halves. The index of refraction of this layer is intermediate between that of the E and O rays in the Iceland spar. This enables the E ray to pass but reflects the O ray. Table 10.3 gives the relevant information and Figure 10.17 illustrates the principle.

Dichroic polarizers. Dichroic materials are crystalline materials that exhibit a very small amount of absorption for light polarized in one direction relative to a particular crystal axis and a very large amount of absorption for light polarized in an orthogonal direction. This selective absorption process is illustrated in Figure 10.18. Naturally occurring single

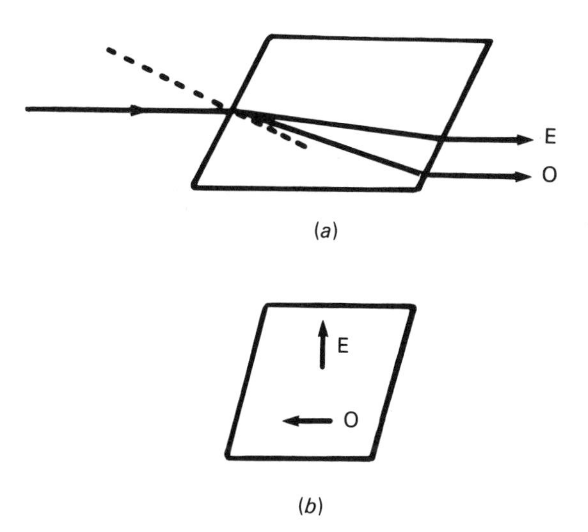

Figure 10.16. Birefringence of Iceland spar: (*a*) separation of the E (extraordinary) and O (ordinary) components of the beam; (*b*) end view of the crystal, showing the directions of polarization of the two beams.

Table 10.3. Indices of refraction for the Nicol prism

Ray	Index of refraction, μ
E	1.49
O	1.66
Canada Balsam	1.55

crystals of *tourmaline* are highly efficient polarizers and produce highly linearly polarized light. These are used for very precise measurements but are, unfortunately, rather expensive.

The conventional plastic polarizer that is commonly used consists of very small needle-like crystals of *herapathite* embedded in a plastic (cellulose acetate or polyvinal) matrix. The crystals are aligned so that their optic axes all point in the same direction. In an optical sense the polarizer behaves as a single crystal. These polarizers are inexpensive but are not suitable for some work requiring high precision.

Quarter-wave plates. If we cut a birefringent crystal parallel to its optic axis and consider only rays that are incident normal to the surface, we have produced what is called a *retarder*. Since the angle of incidence relative both to the surface and the optic axis is 0°, there is no birefringence. However, as the indices of refraction for the E and O components are different, their velocities in the crystal will be different. This means that a phase difference will be introduced between the two rays. The size of the phase difference will depend on the relative velocities for the E and O waves in the crystal, and on the thickness of the crystal. If the phase difference is $\pi/2$ (90°) the device is referred to as a quarter-wave plate or $\lambda/4$ plate (i.e., $2\pi/4 = \pi/2$). Similarly, half-wave and full-wave plates can be produced. If a beam of linearly polarized light is incident on a $\lambda/4$ plate when the optic axes of the plate are set at 45° ($\pi/4$) to the direction of the linear polarization, we will obtain circularly polarized light. Figure 10.19 illustrates how this works.

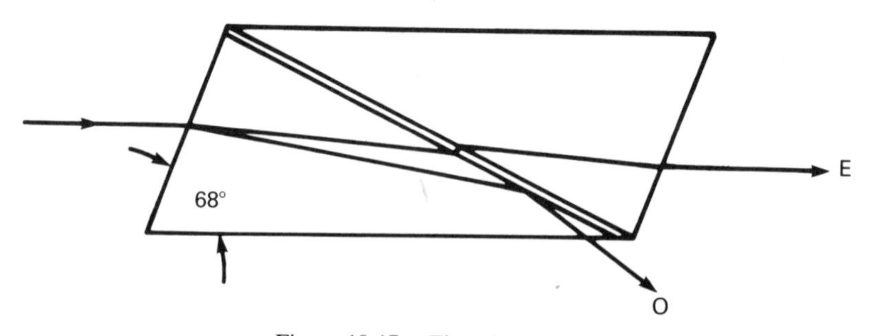

Figure 10.17. The Nicol prism.

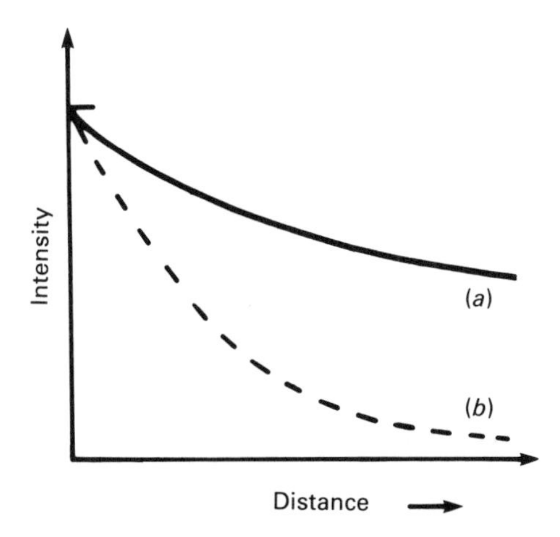

Figure 10.18. Intensity of the two orthogonal components of polarization of the beam as a function of distance within a dichroic polarizing material, (*a*) along polarization axis and (*b*) normal to polarization axis.

Rotating the $\lambda/4$ plate relative to the direction of polarization of the incident beam will result in elliptically polarized light, the ellipticity being a function of the angle of rotation.

c. *Applications of polarimetry*

Light passing through a material will sometimes become polarized. The nature of the polarization is characteristic of the optical properties of the material. Thus an analysis of the nature of the polarized light is a useful method of studying certain properties of solids, liquids, or even gases. The details of the experimental apparatus are determined by what properties of what kinds of materials we are interested in studying. An important aspect of many experiments is to determine the state of the polarization of a beam. A method for accomplishing this by the use of a linear polarizer (referred to as "the analyzer") and a $\lambda/4$ plate is described in Figure 10.20.

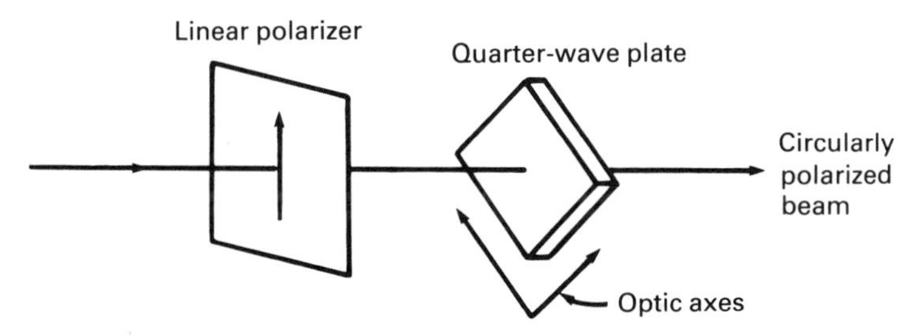

Figure 10.19. Use of a quarter-wave plate in conjunction with a linear polarizer to produce circularly polarized light.

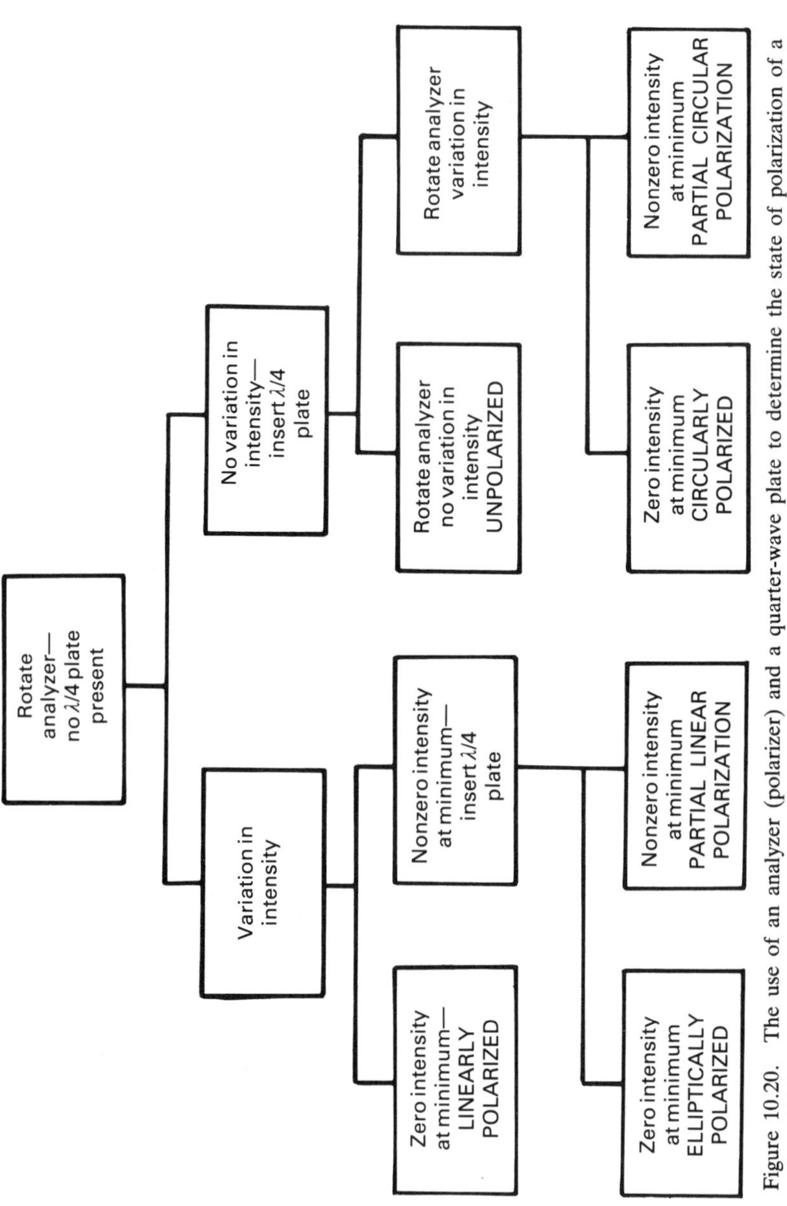

Figure 10.20. The use of an analyzer (polarizer) and a quarter-wave plate to determine the state of polarization of a beam of light.

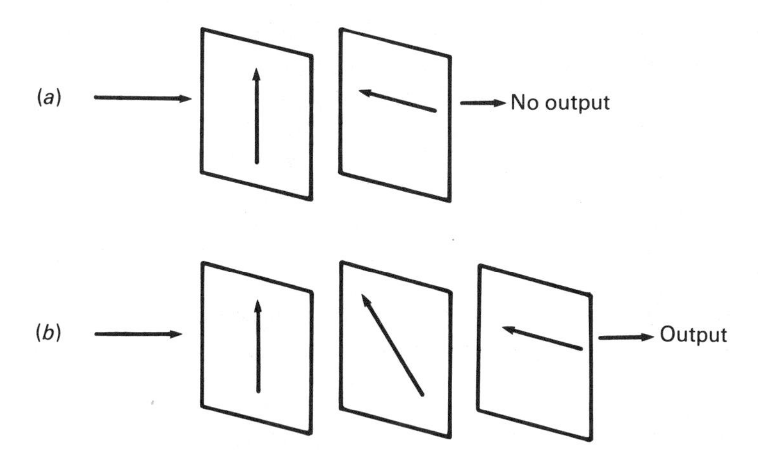

Figure 10.21. (*a*) Crossed polarizers giving zero output intensity: (*b*) Optical rotation caused by an additional polarizer.

A useful experimental technique is called optical rotation. Two polarizers set with their axes at right angles as illustrated in Figure 10.21*a* will not allow any light to pass. If certain types of materials are now placed between the polarizers (e.g., sugar solution, quartz, etc.), light emerges from the second polarizer. The substance has rotated the axis of polarization of the light from the first polarizer so that it is no longer normal to the polarization axis of the second polarizer. It is now necessary to rotate the second polarizer by some angle in order to prevent the light from being transmitted. The optical rotation can be either clockwise (positive) or counterclockwise (negative). The magnitude of the rotation depends on the nature of the substance and on its thickness.

The Kerr effect is another interesting phenomenon and refers to the ability of some materials to exhibit birefringence or, in the correct geometry, retardation effects when subjected to electric fields. A Kerr cell is shown in Figure 10.22. When the Kerr cell is not optically active (i.e., when

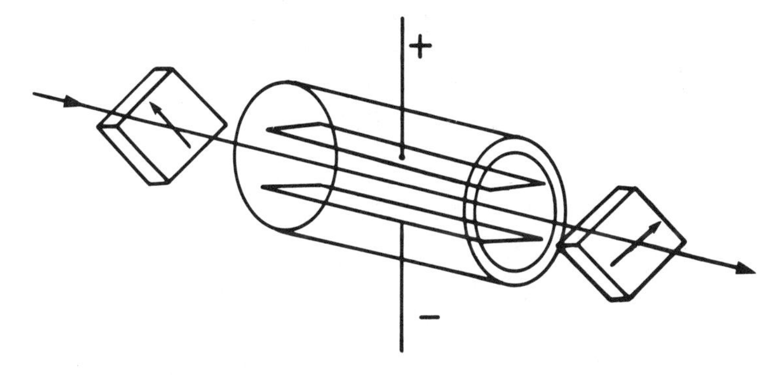

Figure 10.22. The use of a Kerr cell in conjunction with crossed linear polarizers.

it does not affect the polarization), no light is transmitted through the crossed polarizers. The cell contains some material (liquid) (e.g., nitrobenzene) that exhibits a large electro-optic effect when subjected to an electric field. Thus when voltage is applied to the plates of the Kerr cell, the cell will cause the linearly polarized light from the first polarizer to become elliptically polarized and *some* of this will be transmitted by the second polarizer. Thus when the voltage is *on* we can "see" through the cell; when the voltage if *off* we cannot. Applying a voltage pulse to the Kerr cell allows it to be used as an extremely fast shutter. Applying a square wave to the cell allows us to produce pulsed or chopped light. Applying other voltages to the cell enables us to modulate the intensity of the light however we want.

10.3. Interferometry

Interferometry refers to measurements that rely upon the *interference* of two (or more) light beams. There are numerous kinds of interferometers, some of which have very specific uses, others of which are more generally applicable to the study of optical phenomena. Here we discuss the construction, operation, and application of several of the more commonly encountered types of interferometers.

a. Michelson interferometer

The Michelson interferometer is the most widely known interferometer and is probably the most versatile. The general construction of the Michelson interferometer is illustrated in Figure 10.23. Light provided by the source S

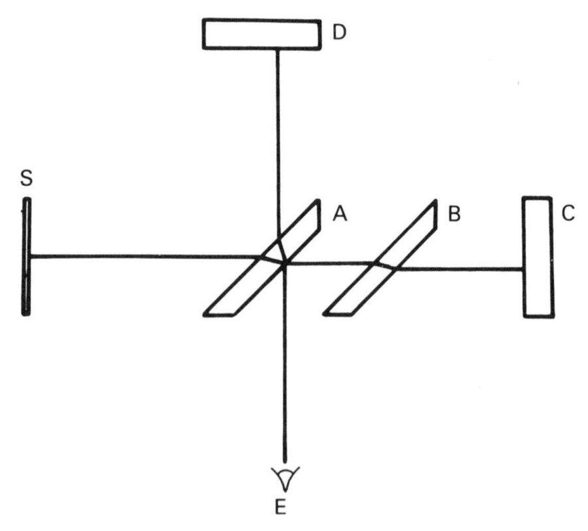

Figure 10.23. The Michelson interferometer: S, extended source; A, beam splitter; B, compensating plate; C, mirror; D, mirror; E, observer.

is split by the beam splitter A. The beam splitter consists of a partially reflecting mirror; Figure 10.23 illustrates the surface from which the reflection principally arises. Part of the beam travels to mirror D and back to the eyepiece E, where it is viewed. The mirror D (as well as mirror C) is a front-surface mirror, so reflections arise from the mirror without any refraction. The other portion of the beam passes through the compensating plate B to mirror C and back to the eyepiece by means of the beam splitter. Along the path from the beam splitter to the eyepiece, the beam is the superposition of beams from mirrors D and C. Thus what is viewed at the eyepiece is the spatially varying pattern caused by the interference of the two beams. Slight adjustments to the path-length of one beam can be made by rotating the compensating plate by a small angle about the vertical axis.

To begin with we assume that both mirrors are optically flat. When one of the mirrors is not exactly perpendicular to the other we can visualize two circumstances: one mirror rotated slightly about a vertical axis, or one mirror rotated slightly about a horizontal axis in the plane of the mirror. In these cases a spatial pattern of alternating light and dark bands will appear at the eyepiece. The bands will be vertical or horizontal, respectively, for the two situations described.

When the two mirrors are (exactly) perpendicular, the pattern observed at the eyepiece is one of alternating light and dark concentric rings. These rings occur only when the optical path-length of the two legs is different. To visualize this we straighten out the ray diagram as shown in Figure 10.24. If we look straight through the interferometer, the phase difference between the two beams is given in terms of the optical path-length difference x. At some angle θ, the path-length difference is x'. Rings occur at intervals of $x - x' = n\lambda/2$.

Here we discuss three of the many uses for the Michelson interferometer.

Measurement of the index of refraction of a solid. A plate of the material under consideration is placed in one leg of the interferometer. We begin by placing the plate of thickness d of the material normal to the beam. A fringe

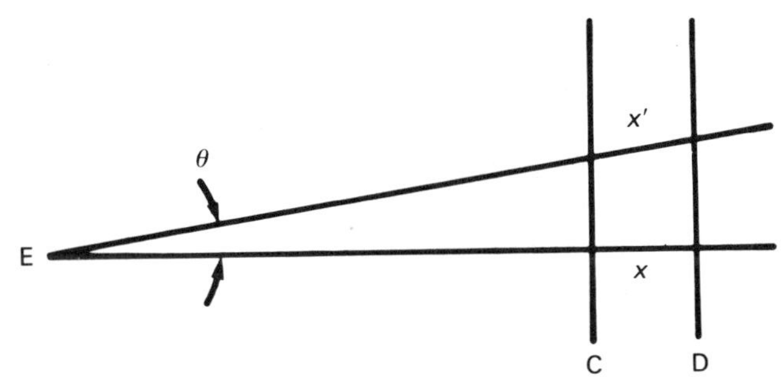

Figure 10.24. Formation of rings in the Michelson interferometer as a result of differences in the path length.

pattern is observed as a result of the difference in the optical path-length. We can rotate the compensating plate to make the number of fringes small—this makes the measurements easier. Now we rotate the material by an angle θ. This changes the optical path-length, that is the length measured in terms of the wavelength of the light, by an amount Δx:

$$\Delta x = 2d\mu\left(\frac{1}{\cos\theta} - 1\right) \tag{10.18}$$

The change in the number of circular fringes observed is given by

$$\Delta N = \frac{2\,\Delta x}{\lambda} = \frac{4d\mu}{\lambda}\left(\frac{1}{\cos\theta} - 1\right) \tag{10.19}$$

It is now an easy matter to find the index of refraction μ.

Index of refraction of a gas. An evacuated transparent container can be placed in one leg of the interferometer. Gas is let into the chamber and the index of refraction is given by the change in the number of fringes as

$$\Delta N = \frac{4d}{\lambda}\Delta\mu = \frac{4d}{\lambda}(\mu(P) - 1) \tag{10.20}$$

$\mu(P)$ is the pressure-dependent index of refraction and 1 is the index of refraction of a vacuum.

Examination of optical surfaces. The quality of a reflecting surface can be investigated by replacing one of the mirrors by the surface to be investigated. The result is a distortion of the interference pattern that is proportional to the departure from flatness of the test surface.

b. Twyman–Green interferometer

A modification of the Michelson interferometer known as the Twyman–Green interferometer is illustrated in Figure 10.25. The modifications are use of a point source rather than the extended source of the Michelson interferometer and insertion of collimating lenses in front of the eyepiece and the source. In the instrument shown in Figure 10.25, if the mirrors are perpendicular then the field as viewed in the eyepiece is uniformly lit, regardless of differences in the path-length. The differences in the path-length that give rise to the rings in the Michelson interferometer are compensated for by the addition of the lenses. When they are observed, the fringes in the Twyman–Green interferometer result from differences in the optical path-length over the field of view and are characteristic of imperfections in the optical components. As a result this is a highly sensitive method of testing optical components.

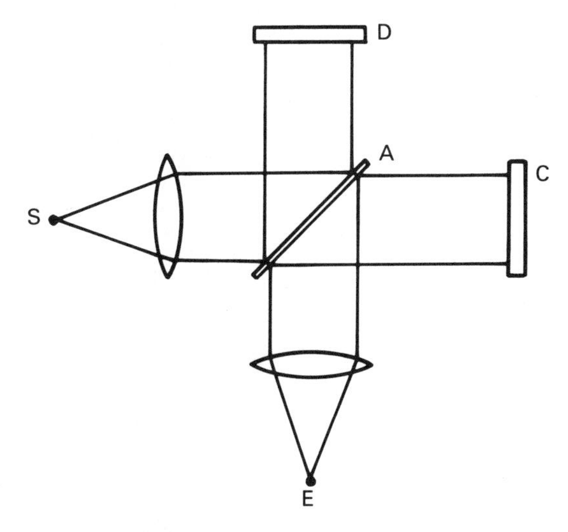

Figure 10.25. The Twyman–Green interferometer. The identity of the components is the same as in Figure 10.23 except that S is a point source.

c. *Fabry–Perot interferometer*

Fabry–Perot devices belong to a class of interferometers sometimes referred to as multiple-reflection interferometers. Figures 10.26 and 10.27 show the basic design of this device. There are two variations: that shown in Figure 10.26 is called the Fabry–Perot etalon and has fixed plates; the measurement is made by observing the interference pattern at the focal plane. The device shown in Figure 10.27 is called the Fabry–Perot interferometer; here the optical distance between the plates is scanned and the measurement is made by observing the intensity (usually with a photodetector) at a single focal point. Optically speaking, the path-length can be changed by keeping the plates fixed and changing the index of refraction of the medium between them. This can be done easily by, for example, changing the pressure of the gas. Note also from the figures that the etalon uses an extended source while the interferometer uses a point source. In the construction of either

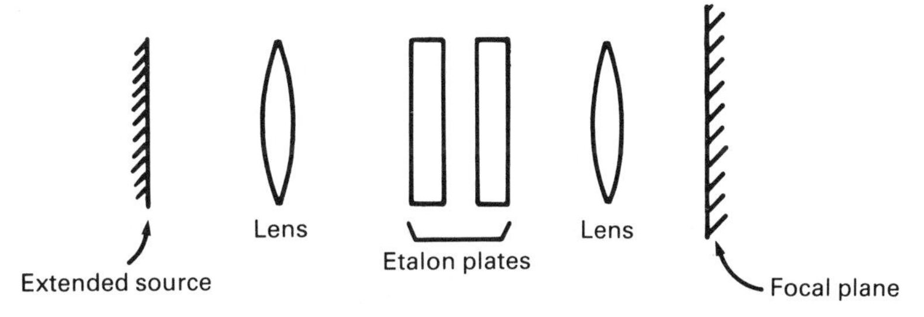

Figure 10.26. The Fabry–Perot etalon.

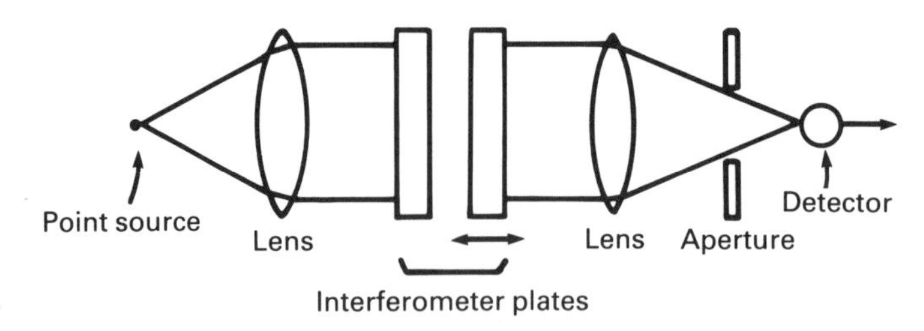

Figure 10.27. The Fabry–Perot interferometer.

device it is essential that the plates be held parallel. These plates are partially reflecting and must be flat to the order of 0.02λ. Note that a conventional optical flat is flat to about $\lambda/4$ or so. Thus Fabry–Perot devices require high-quality components.

Looking first at the etalon, for a monochromatic source, as we scan across the focal plane we will see alternating constructive and destructive interference resulting in a ring pattern. Since the pattern is the result of the interference of a large number of reflections, the rings are particularly sharp and allow for very precise measurements. The spacing of the rings will depend on wavelength and also on the etalon plate spacing.

In the interferometer, we see that alternating maxima and minima will be observed as the optical path-length between the plates is varied. The resulting pattern is related to the wavelength of the incident light. The derivation of this relationship is however, not straightforward (see, e.g., Fowles *Introduction to Modern Optics,* Chapter 3).

As a result of the multiple reflections involved, Fabry–Perot devices can be used for the precise measurement of fringe patterns. A common use for this device is to provide highly accurate measurements of the wavelength of monochromatic light. This method of measuring the wavelength is 10 to 100 times more accurate than the use of a conventional grating or prism spectrometer. This is the standard method for observing small splitting in optical spectra as a result from the Stark or Zeeman effects.

10.4. Optical spectroscopy

Spectroscopy is, in a general sense, the study of the distribution of the energy content of a signal. With respect to optics, spectroscopy relates to the measurement of the energies (or equivalently, the wavelengths) present in a beam of light.

There are innumerable different geometric configurations for optical spectrometers. We discuss here only the simple basic designs. In general, we divide the various designs into two types: grating spectrometers and prism spectrometers.

a. Grating spectrometers

To describe the operation of a diffraction grating we begin with the problem of diffraction from a single slit. Two basic types of diffraction can occur: Fraunhofer diffraction and Fresnel diffraction. Figure 10.28 shows the distinction between these two types. Fraunhofer diffraction deals with incident and diffracted waves that are plane waves; Fresnel diffraction deals with waves from point sources. In a sense, any diffraction can be thought of as Fraunhofer diffraction if the curvature of the wave fronts is sufficiently small. For a point source, the Fraunhofer approximation can be made if

$$2\lambda \gg \delta^2\left(\frac{1}{d'}+\frac{1}{d}\right) \tag{10.21}$$

where δ is the width of the slit, d and d' are the distances from the slit to the source and viewing points, respectively, and λ is the wavelength. Subsequent discussions will consider the Fraunhofer case only.

Consider the single slit, as shown in Figure 10.29. The amplitude of the electric field of light as a function of time is

$$|\mathbf{E}| = E_0 \sin(\omega t - kz) \tag{10.22}$$

where z is the direction of propagation. If we assume that ray A in the figure has zero phase shift, then ray B is given by

$$|\mathbf{E}| = E_0 \sin\left(\omega t - \frac{2\pi s}{\lambda}\sin\theta\right) \tag{10.23}$$

The total electric field vector for all rays energing from the slit at angle θ is

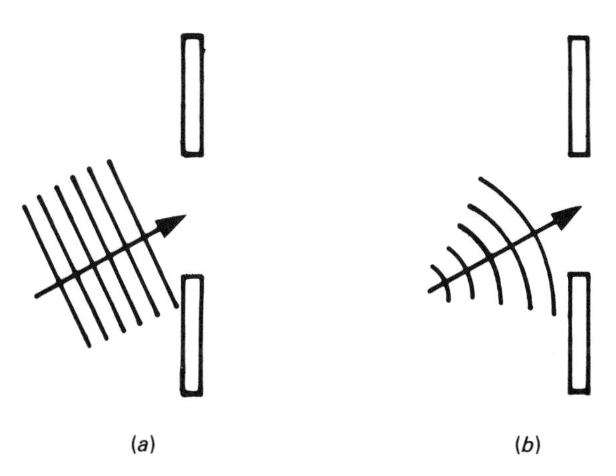

Figure 10.28. The diffraction of light wave by an aperture: (*a*) Fraunhofer diffraction; (*b*) Fresnel diffraction.

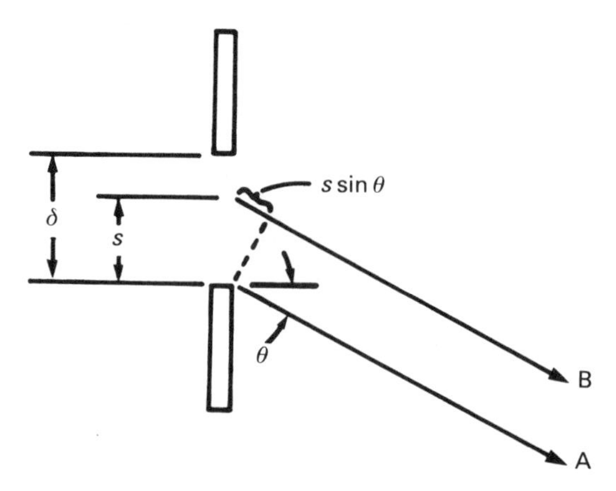

Figure 10.29. The geometry of a single slit and the diffracted beams A and B.

given by the integral of equation (10.23) over all ds:

$$|\mathbf{E}| = \int_0^\delta E_0 \sin\left(\omega t - \frac{2\pi s}{\lambda} \sin\theta\right) ds \tag{10.24}$$

The integration yields

$$|\mathbf{E}| = -E_0 \frac{\lambda}{\sin\theta} \sin\left(\frac{\pi\delta \sin\theta}{\lambda}\right) \sin\left(\omega t - \frac{\pi\delta}{\lambda} \sin\theta\right) \tag{10.25}$$

This is a wave of the form

$$|\mathbf{E}| = A \sin(\omega t - \phi) \tag{10.26}$$

where ϕ and A are given in equation (10.25) and represent the phase and the amplitude of the wave. The intensity is given by the square of the amplitude, so from the above we can write

$$I = I_0 \left(\frac{\sin\beta}{\beta}\right)^2 \tag{10.27}$$

where

$$\beta = \frac{\pi\delta \sin\theta}{\lambda} \tag{10.28}$$

Figure 10.30 is a plot of the single-slit diffraction pattern.

For multiple slits, the resultant diffraction pattern is the sum of the patterns from each slit individually. Figure 10.31 shows the geometry of a multiple-slit arrangement. A derivation along the lines of that above yields the expression

$$I = I_0 \left(\frac{\sin\beta}{\beta}\right)^2 \left(\frac{\sin\Delta}{\Delta}\right)^2 \tag{10.29}$$

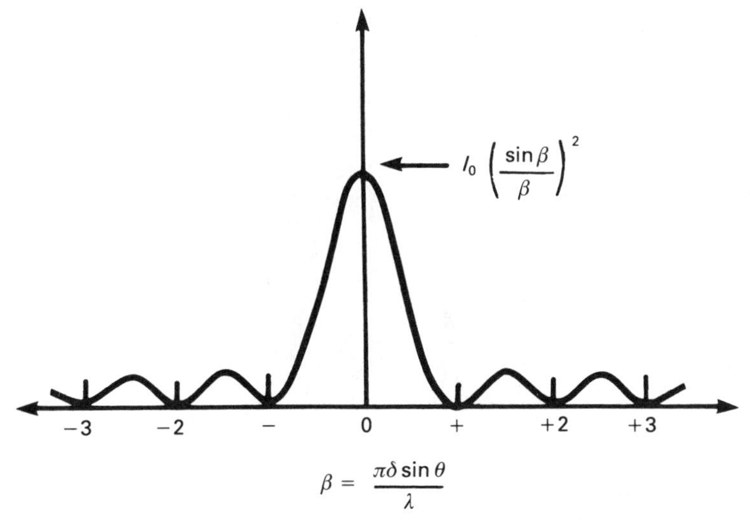

Figure 10.30. Fraunhofer diffraction pattern of a single slit.

Figure 10.31. Geometry of a multiple-slit diffraction grating.

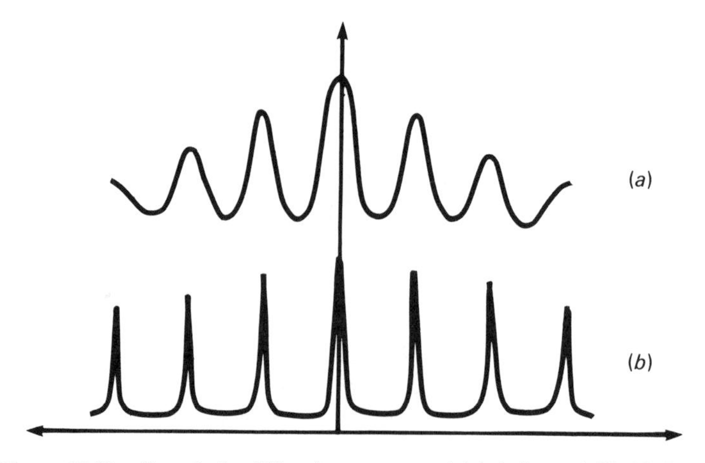

Figure 10.32. Fraunhofer diffraction patterns of (*a*) 4 slits and (*b*) 16 slits.

where β is defined as in equation (10.28) and

$$\Delta = \frac{\pi N d \sin \theta}{\lambda} \tag{10.30}$$

The total number of slits is N. Examples of diffraction patterns are shown in Figure 10.32. These are all patterns of monochromatic light and we see that the larger the number of slits the sharper the lines—this means the better the angular resolution. If we had two distinct wavelength components in the beam then we would see two distinct diffraction patterns on top of each other. The larger the number of slits the greater the wavelength resolution. We will see later how this relates to spectrometers.

The *unblazed* grating is a simple type of diffraction grating. This consists of a flat reflecting surface (e.g., glass with aluminum deposited on it). A number of closely spaced, parallel, rough grooves are scratched on to the surface. Figure 10.33 shows a cross-section of this type of grating. Light

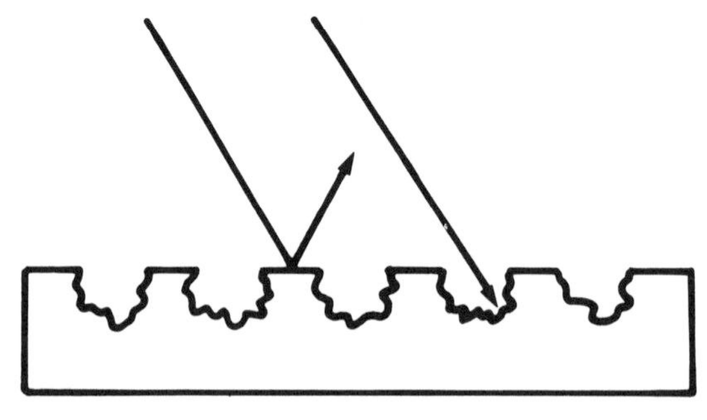

Figure 10.33. Cross section of an unblazed grating. The reflection of incident light is illustrated.

incident upon the remaining reflecting surfaces is reflected as from a plane mirror. Light incident upon the grooves is scattered incoherently and is greatly attenuated. Thus the reflected light behaves as the light transmitted through multiple slits. If monochromatic light were incident, a pattern such as that in Figure 10.32 would result. For light with different wavelength components, each line in the pattern will be split into the different wavelengths. A grating spectrometer looks at the details of the wavelength splitting of a particular order of diffraction fringe. Equation (10.29) applies and we see that the angular resolution and hence the wavelength resolution will improve for large N. Typical gratings would have dimensions of one to a few centimeters on a side. The number of "lines" per centimeter or per inch is generally specified by the manufacturer.

In the *blazed* grating the surface is cut in the form of a sawtooth, as shown in Figure 10.34. Light incident upon one face of the sawtooth pattern is reflected in one direction, while light incident upon the other face is reflected in a totally different direction. Thus from any particular direction the surface appears as regions of alternating high and low reflectivity. Gratings made by etching grooves on a glass plate using a diamond cutter are called master gratings; these are very expensive. There are various methods of producing a plastic copy of the master grating. These copies are called replica gratings and are much less expensive.

The simplest gratings are flat but some are curved. Curved gratings provide some focusing of the spectrum as well.

The gratings described above can be used in the reflection mode, but many can also be used in a refraction mode. A typical refraction spectrometer is illustrated in Figure 10.35 and a reflection type spectrometer is shown in Figure 10.36. For a particular wavelength, the location of a given line of order n for either geometry is

$$n\lambda = d \sin \theta \tag{10.31}$$

where d is given in Figure 10.31 and is the distance between the slits. The

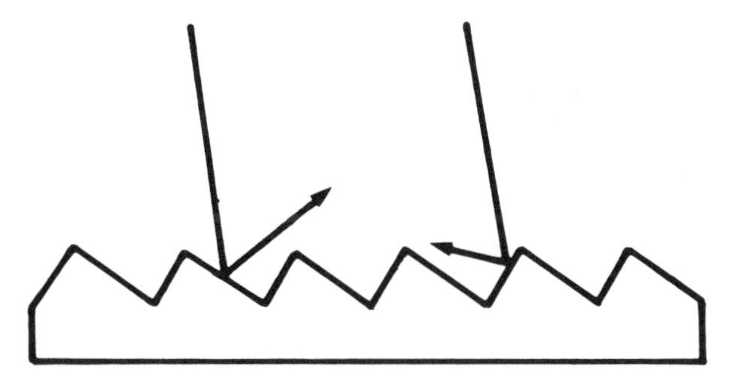

Figure 10.34. Cross section of a blazed grating.

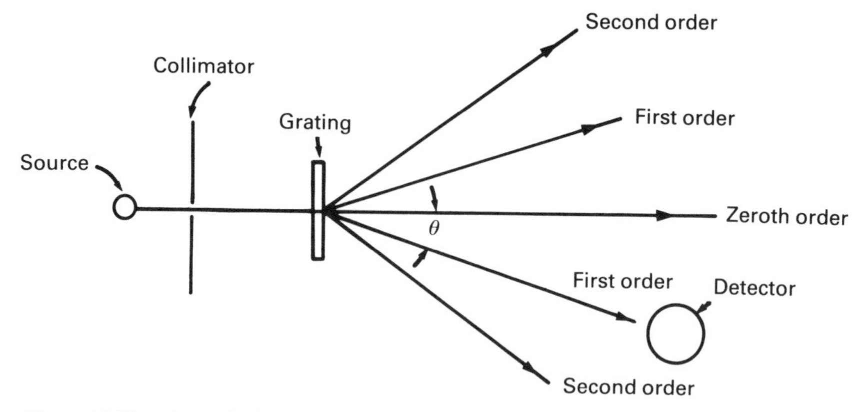

Figure 10.35. An optical spectrometer using a diffraction grating in the refraction mode.

wavelength resolution $\Delta\lambda$ is

$$\Delta\lambda = \frac{\lambda}{R} \tag{10.32}$$

where R is the resolving power. It can be shown that

$$R = Nn \tag{10.33}$$

This depends on the number of lines in the grating and the order of the fringe observed. Figure 10.32 shows that as the order increases the intensity decreases, so that there is a limit to the order we can use. In an ideal case we might look at $n = 2$ and we might have a high-quality diffraction

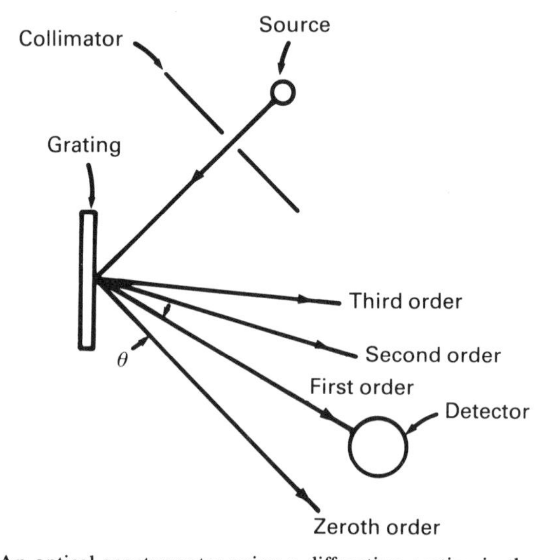

Figure 10.36. An optical spectrometer using a diffraction grating in the reflection mode.

grating with 10,000 lines cm and of a width of 2 cm. Thus we would find for $\lambda = 550$ nm:

$$\Delta\lambda = 0.014 \text{ nm} \tag{10.34}$$

The spectrum can be observed by rotating either an *eyepiece* or *photodetector* about the angle θ and observing the lines. The wavelength is thus related to the measured angle by equation (10.31). Alternatively, a *photographic film* may be placed at the viewing point and the entire spectrum may be photographed. The three different detection methods are incorporated into instruments referred to as spectroscopes, spectrometers, and spectrographs. However, the term spectrometer has acquired some degree of universal usage.

b. Prism spectrometers

Spectrometers may be constructed using prisms rather than gratings. We begin with a description of the optical properties of a prism.

The refraction of a light beam through a triangular prism is illustrated in Figure 10.37. The angle of incidence is I and Δ is the angle of deviation. Since the index of refraction of the prism glass is wavelength dependent, Δ is wavelength dependent and the prism will divide light up into its various wavelength components. If we rotate the prism we vary I and hence indirectly we vary Δ. Figure 10.38 is a plot of $\Delta(I)$ given by

$$\Delta(I) = (I - \theta) + \arcsin\left\{\mu \sin\left[\theta - \arcsin\left(\sin\frac{I}{\mu}\right)\right]\right\} \tag{10.35}$$

We see that a minimum in Δ occurs for a particular value of I. The corresponding value of Δ is called the angle of minimum deviation. This angle is, of course, a function of the wavelength of the light, the index of refraction of the glass, and the apex angle θ. In all cases, however, the minimum deviation occurs when the beam within the prism is parallel to the base. The measurement of $\Delta_{\min}$ is a convenient and accurate means of

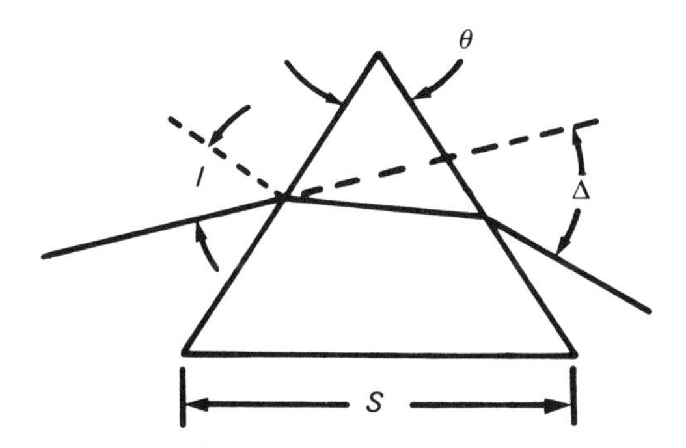

Figure 10.37. Refraction by a prism.

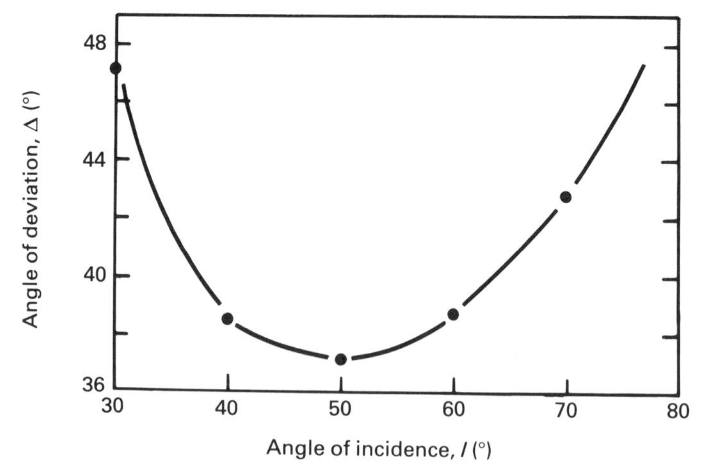

Figure 10.38. The deviation of a prism as a function of the angle of incidence for $\mu = 1.5$ and $\theta = 60°$.

determining the index of refraction of a prism. Application of equation (9.3) gives

$$\mu = \frac{\sin[\frac{1}{2}(\theta + \Delta_{\text{min}})]}{\sin(\theta/2)} \tag{10.36}$$

An arrangement for the prism spectrometer is illustrated in Figure 10.39. Unlike the case with grating spectrometer, only a single spectrim is formed in the prism spectrometer; that is, there are not spectra of different orders. The configuration shown in Figure 10.39 shows that the prism is oriented for the angle of minimum deviation. Although this orientation is not strictly necessary, it will improve the quality of the spectrum. This complication can be eliminated by use of one of the constant-deviation prisms shown in Figure 10.40. In prisms of these geometries, the angle of deviation is independent of the angle of incidence.

The resolving power of a prism spectrometer depends on the ability of the

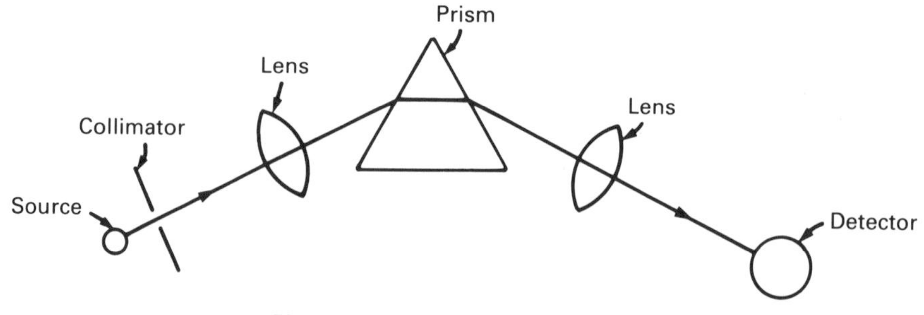

Figure 10.39. The prism spectrometer.

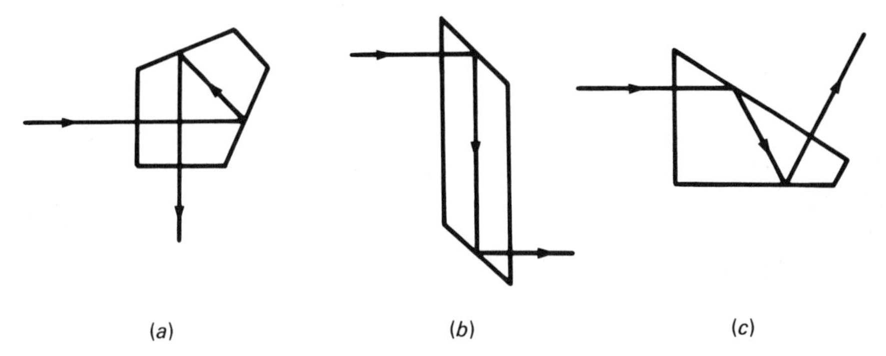

Figure 10.40. Some constant-deviation prisms: (*a*) pentaprism; (*b*) rhomboid; (*c*) un-named.

prism to separate light of different wavelengths. This ability is related to the dispersion of the prism. We write

$$\Delta\lambda = \frac{\lambda}{S}\left(\frac{\partial\mu}{\partial\lambda}\right)^{-1} \tag{10.37}$$

where S is defined in Figure 10.37. As an example, consider a prism with $S = 3$ cm for $\lambda = 550$ nm. From Figure 9.2, using light flint as an example, we find

$$\frac{\partial\mu}{\partial\lambda} = 1250 \text{ cm}^{-1} \tag{10.38}$$

and equation (10.38) gives

$$\Delta\lambda = 0.14 \text{ nm} \tag{10.39}$$

By comparison, we see that the grating spectrometer has a higher resolution.

11
Experimental aspects of nuclear physics

Experimentally, the physics of a nucleus is studied by the measurement of various properties of radiation that has either been produced by the decay of the nucleus or has some how interacted with the nucleus. Specifically, the properties of the radiation that are of particular interest are the time at which the radiation was produced, the direction in which the radiation propagates, and the energy distribution of the radiation. To understand the origin of various types of radiation relevant to nuclear physics it is important to understand the physics of nuclear decay processes. To understand the principles of radiation detection it is important first to review the processes by which the radiation interacts with matter. In this chapter we cover this background material in nuclear physics and also consider some of the practical aspects of preparing radiation sources. In Chapter 12 we present methods of detecting radiation and the relevance of these techniques to the measurement of some specific nuclear properties.

11.1 Nuclear decay and nuclear reactions

The nucleus of an atom consists of a number of nucleons, that is p protons and n neutrons. The number of protons determines the atomic number of the nucleus and tells us what element we are dealing with. The nuclear mass number A is given by the total number of protons and neutrons: $A = Z + N$. A *nuclide* is a nuclear species with a particular N and Z. This is frequently called an *isotope,* but the use of the word in this context is not strictly correct. A nuclide is designated as ${}^{A}_{Z}\mathrm{E}$ or just ${}^{A}\mathrm{E}$ where E is the symbol for the element. Some examples are ${}^{12}\mathrm{C}$ (or ${}^{12}_{6}\mathrm{C}$) and ${}^{86}\mathrm{Kr}$ (or ${}^{86}_{36}\mathrm{Kr}$). The second type of nomenclature is redundant, as the name of the element already tells us what Z is.

About 330 different nuclides occur naturally; of these about 25 are unstable but have very long half-lives. Another 35 unstable nuclides with shorter half-lives are found naturally because they are decay products of longer-lived naturally occurring unstable nuclides. Finally, another 1000 or so nuclides, all unstable, can be produced artificially. Some are long-lived but some have half-lives as short as $\sim 10^{-20}$ s.

Let us consider the basics of radioactive decay. If we have a large number of identical nuclides that all decay via the same process, then the number of decays per unit time will be proportional to the number of nuclides present. That is

$$-dN = \lambda N\, dt \tag{11.1}$$

where λ is a proportionality constant. This expression is easily integrated to give

$$N(t) = N_0 e^{-\lambda t} \tag{11.2}$$

where N_0 is the number of nuclides present at $t = 0$. What we usually measure is the number of decays per unit time—commonly called the *count rate*. This is given by the time derivative of equation (11.2) or

$$-\frac{dN}{dt} = \lambda N_0 e^{-\lambda t} = -\left(\frac{dN}{dt}\right)_0 e^{-\lambda t} \tag{11.3}$$

Taking the logarithm of this equation gives

$$\log_{10}\left(-\frac{dN}{dt}\right) = \log_{10}\left(-\frac{dN}{dt}\right)_0 - \lambda t \log_{10} e \tag{11.4}$$

This means that the count rate $(-dN/dt)$ versus t plotted on semi-log paper will result in a straight line with slope $-\lambda \log_{10} e$. See an example of this in Figure 11.1. The half-life is defined as the length of time required for half of

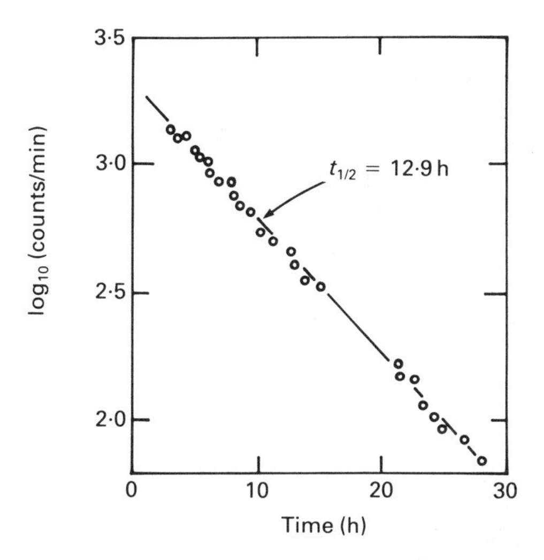

Figure 11.1. The count rate measured for the decay of a ^{64}Cu source as a function of time after source preparation.

the original number of nuclides to decay. Therefore,

$$\tfrac{1}{2}N_0 = N_0 e^{-\lambda t_{1/2}} \tag{11.5}$$

or $t_{1/2} = (\ln 2)/\lambda = 0.693/\lambda$. Using the value of λ obtained from the slope in Figure 11.1, the half-life can be determined. The mean life is given by

$$t_{\text{mean}} = -\frac{\int_0^\infty (dN/dt)t\,dt}{\int_0^\infty (dN/dt)\,dt} \tag{11.6}$$

We write this as

$$t_{\text{mean}} = -\frac{1}{N_0}\int_0^\infty t\,dN \tag{11.7}$$

and from equation (11.3) we have

$$t_{\text{mean}} = \int_0^\infty t\lambda e^{-\lambda t}\,dt \tag{11.8}$$

Integrating this gives

$$t_{\text{mean}} = \frac{1}{\lambda} = \frac{t_{1/2}}{0.693} \tag{11.9}$$

A nuclide can change to another state or into another nuclide for one of two reasons: (1) if it is unstable and decays naturally, or (2) if it is bombarded by something. Let us look at the first process.

a. Alpha decay

An α-particle is the same as a ^{4}He nucleus (two protons and two neutrons). It has a charge $+2e$. A nuclide that gives off an α-particle changes its identity as follows:

$$^{A}_{Z}\text{E1} \rightarrow {}^{A-4}_{Z-2}\text{E2} + \alpha \tag{11.10}$$

Generally there is some energy on the right-hand side in the form of a γ-ray as well. The right-hand side of equation (11.10) is only energetically favorable for nuclei with large A and Z. As a result, unstable light elements do not undergo α decay but rather decay by one of the β processes discussed below.

b. Spontaneous fission

A few heavy nuclei decay by spontaneous fission. This means that the nucleus breaks up into two or sometimes three large fragments. The result is often that there are a few neutrons left over and these are given off as well.

c. Beta decay

There are several similar process all collectively referred to as β decay.

β^- *decay* occurs when a neutron inside the nucleus changes into a proton. The process is

$$n \rightarrow p + e^- + \bar{\nu} \tag{11.11}$$

where the $\bar{\nu}$ is an antineutrino. The result is a change in the identity of the nuclide as follows:

$${}^{A}_{Z}\mathrm{E1} \rightarrow {}_{Z+1}{}^{A}\mathrm{E2} + e^- + \bar{\nu} \tag{11.12}$$

β^+ decay is similar but proceeds in the opposite direction;

$$p \rightarrow n + e^+ + \nu \tag{11.13}$$

So the change in the nuclide is

$${}^{A}_{Z}\mathrm{E1} \rightarrow {}_{Z-1}{}^{A}\mathrm{E2} + e^+ + \nu \tag{11.14}$$

Electron capture is equivalent to β^+ decay and can occur in nuclei that can undergo the β^+ decay process. It looks like

$$p + e^- \rightarrow n + \nu \tag{11.15}$$

A proton in the nucleus captures one of the orbital electrons and changes into a neutron and a neutrino. The net result is the same as equation (11.14).

d. Gamma decay

Gamma decay represents the decay of an excited nucleus into a lower-energy state. Nuclei, like atomic electrons, can exist only in specific quantized states and transitions between these states produce photons of a given energy. One process that can occur is the emission of a γ-ray photon. A second process is referred to as *internal conversion*. In this the energy $\mathscr{E}_\gamma$ that would have been given up to the γ-ray is used to free an atomic electron (usually an inner shell one). The energy given to the electron is

$$\mathscr{E} = \mathscr{E}_\gamma - \phi \tag{11.16}$$

where ϕ is the binding energy. The ratio of the probability for decay by internal conversion to the probability for γ-decay is called the internal conversion coefficient.

Reactions can be induced in stable nuclei (or in unstable nuclei for that matter) by the bombardment of the nucleus by particles. We consider five possible reactions (*e* to *i* below).

e. Scattering

In scattering, a particle interacts with the nucleus and the same particle, or another one identical to it, is expelled. An example is

$$^{26}\mathrm{Mg} + \mathrm{p} \rightarrow {}^{26}\mathrm{Mg} + \mathrm{p} \tag{11.17}$$

In words, we have written

$$\text{Target nucleus} + \text{incident particle} \rightarrow \text{residual nucleus} + \text{emitted particle} \tag{11.18}$$

A short-hand notation for the expression (11.17) is $^{26}\mathrm{Mg(p,p)}^{26}\mathrm{Mg}$. This particular reaction is called the (p,p) reaction, where the first p refers to the incident particle and the second p refers to the emitted particle. This scattering process may be either elastic or inelastic. In the example here an elastic process would not provide any energy to the magnesium nucleus so it would be left in its ground state. In an inelastic process, the incident proton would give some energy to the nucleus and would leave it in an excited state. This would then decay by γ-emission or something equivalent.

f. Transmutation

Transmutation is a process in which the emitted (heavy) particle is not the same as the incident (heavy) particle. An example is

$$^{26}\mathrm{Mg} + \mathrm{p} \rightarrow {}^{23}\mathrm{Na} + \alpha \tag{11.19}$$

that is, the $^{26}\mathrm{Mg}(\mathrm{p},\alpha)^{23}\mathrm{Na}$ reaction. Note that the number of protons and neutrons on each side of the equation must balance.

g. Radiative capture

Radiative capture is the capture of a low-energy neutron or proton by a nucleus. The residual nucleus is then left in an excited state and emits a γ-ray by de-excitation. An example is

$$^{26}\mathrm{Mg} + \mathrm{p} \rightarrow {}^{27}\mathrm{Al} + \gamma \tag{11.20}$$

or $^{26}\mathrm{Mg}(\mathrm{p},\gamma)^{27}\mathrm{Al}$. The (n,γ) process is another example of radiative capture and is the reaction that is used to produce radioactive sources in neutron reactors.

h. Photodisintegration

Photodisintegration is the inverse of radiative capture and represents the breaking up of a nucleus as a result of bombardment by a γ-ray. An

Table 11.1. Some metastable nuclear states

Nuclide	$t_{1/2}$	Energy (MeV)
^{184m}Re	169 d	0.188
^{181m}Ta	6.8 μs	0.0063
^{178m2}Hf	10 y	2.5
^{178m1}Hf	5 s	1.148
^{124m2}Sb	21 min	0.035
^{124m1}Sb	93 s	0.010

example is

$$^{27}\mathrm{Al} + \gamma \rightarrow {}^{26}\mathrm{Mg} + \mathrm{p} \tag{11.21}$$

This is, of course, called the (γ,p) reaction.

i. Induced fission

This process is similar to scattering or transmutation, in that we are dealing with the capture of a neutron or proton. However, in this case the nucleus breaks into two (or occasionally three) large fragments plus a few neutrons. An example is

$$^{235}_{92}\mathrm{U} + \mathrm{n} \rightarrow {}^{140}_{54}\mathrm{Xe} + {}^{93}_{38}\mathrm{Sr} + 3\mathrm{n} \tag{11.22}$$

j. Metastable states

Nuclides that have excited states that are reasonably long-lived are referred to as *metastable* and are designated with an "m". A few examples are shown in Table 11.1. If more than one long-lived excited state exists for a particular nuclide, these are designated m1 and m2.

11.2 Interactions of radiation with matter

Before we can begin to study the methods by which radiation (γ-rays, X-rays, α-particles etc.) can be detected, we have to understand the various ways in which this radiation can interact with matter. All radiation detectors, with the exception of the Cherenkov detector, rely upon the fact that some interaction will take place.

We divide "radiation" into four categories:

1. Heavy charged particles—α, p^+, . . .
2. Light charged particles—e^-, e^+, . . .
3. High-energy photons—X-rays, γ-rays
4. Heavy uncharged particles—neutrons

We have neglected the more exotic and difficult-to-detect types, such as neutrinos.

a. Heavy charged particles

Heavy charged particles can interact with matter in two ways: with nuclei or with atomic electrons. The first method is called *Rutherford scattering* and it is an elastic process that causes a deflection of the heavy charged particle but no change in its energy. It is the result of Coulombic repulsion between the heavy ion and the nuclear charge (heavy ions are nearly always positively charged.)

The second process is the one of primary interest here. The Coulombic interaction of the ion with the atomic electrons in the material causes the atoms to become ionized that is the (+) ion pulls electrons away from the atom. The amount of energy lost by the ion during each interaction is limited by the criterion that momentum must be conserved. A straightforward calculation shows the maximum energy that can be transferred to an electron in a collision with a particle of mass M as follows. Nonrelativistically the conservation of energy requires,

$$\tfrac{1}{2}Mv_i^2 - \tfrac{1}{2}Mv_f^2 = \tfrac{1}{2}mv_e^2 \tag{11.23}$$

where m is the mass of the electron, v_i and v_f are the velocities of the ion before and after the collision, and v_e is the velocity of the electron. Conservation of momentum gives

$$M\mathbf{v}_i = M\mathbf{v}_f + m\mathbf{v}_e \tag{11.24}$$

The maximum energy transfer occurs when v_i, v_f and v_e are all parallel so that we can replace vectors with scalar quantities in equation (11.24). Rearranging equation (11.23) gives

$$\tfrac{1}{2}m_e v_e^2 = \tfrac{1}{2}M(v_i - v_f)(v_i + v_f) \tag{11.25}$$

We can now find $(v_i - v_f)$ from equation (11.24). Since $M \gg m$, the velocity change for the ion when colliding with the electron is very small. Intuition tells us this and so would a rigorous derivation. So $v_i + v_f \approx 2v_i$ and equation (11.25) becomes

$$\mathscr{E}_e = \frac{4m\mathscr{E}_i}{M} \tag{11.26}$$

where $\mathscr{E}_e$ and $\mathscr{E}_i$ are the electron energy after collision and the ion energy before. For a typical example, consider an α-particle of energy 1 MeV colliding with an electron in an argon atom. We find

$$\mathscr{E}_e = 540 \text{ eV} \tag{11.27}$$

or only about 1/2000th of the ion energy. One thing to note here is that the ionization energy for the outer electron of argon is about 16 eV, so there is

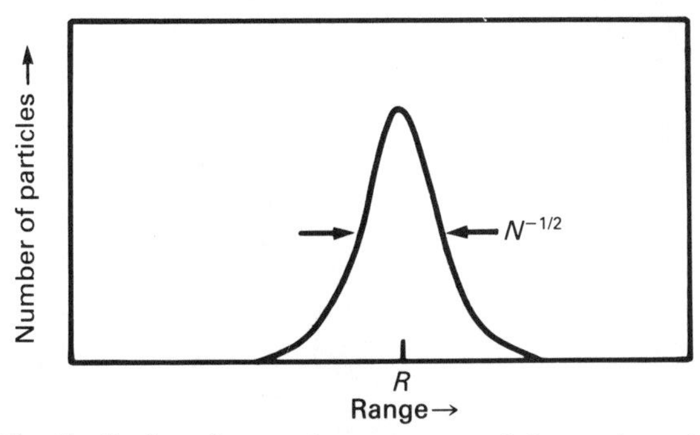

Figure 11.2. The distribution of ranges for monoenergetic heavy charged particles in matter.

more than enough energy transferred to cause ionization. On the average, then, this α-particle will have to undergo about $N = 2000$ interactions before being stopped. However, any particular α-particle will not undergo exactly 2000 interactions before stopping. We would expect a distribution of N of the order $\sqrt{N}$. Since the range of the particle is related to N, we expect an uncertainty in the range R as

$$\frac{\delta R}{R} = \frac{\delta N}{N} = N^{-1/2} \tag{11.28}$$

For $N = 2000$, $\delta R/R$ amounts to only about 2 percent, so we might see a distribution of ranges as shown in Figure 11.2.

The point is that the range of a heavy charged particle is very well defined. That is, many particles of the same energy will all have about the same range.

So the heavy charged particle travels through the material leaving a trail of ionization. The electron–ion pairs formed in this way can cause further ionization. In the end the total number of electron–ion pairs formed will be $\mathscr{E}_i/\mathscr{E}_0$, where $\mathscr{E}_0$ is the ionization energy of the material.

b. Light charged particles

Now let us look at light charged particles, beginning with electrons. These can lose energy owing to elastic collisions with atomic electrons, in which ionization events can occur, just like the heavy charged particles. However, returning to the discussion of the maximum energy loss per interaction, we have from equation (11.23),

$$mv_i^2 - mv_f^2 = mv_e^2 \tag{11.29}$$

where all the masses are now the same. From equation (11.24)

$$v_i = v_f + v_e \tag{11.30}$$

and from equation (11.29)

$$(v_i - v_f)(v_i + v_f) = v_e^2 \tag{11.31}$$

Substituting into equation (11.30) we obtain

$$v_i + v_f = v_e \tag{11.32}$$

Comparing this with equation (11.30) we see that the only solution is $v_f = 0$ and $v_e = v_i$. This means that the maximum energy transfer is the total energy of the initial particle. So in general the electron is stopped and all of its energy is transferred to another electron in one, or at most a few, interactions. For argument, let us take $N = 2$. Then $\delta R/R = 0.7$; in other words the distribution of ranges is 70 percent of the average range. The point is that the range of an electron is not well defined and electrons with the same energy can have vastly different ranges.

Electrons with energies less than about 100 MeV lose energy primarily by the process of ionization. At higher energies a significant energy loss results from *Bremsstrahlung*. Bremsstrahlung is the radiation produced by an accelerated charge. Consider a particle of mass m and charge qe, passing near a nucleus of charge Z, as shown in Figure 11.3. The momentum transferred in this kind of collision is just

$$\Delta p = \int F\,dt \tag{11.33}$$

The force is given by $qeE(r)$, where $E(r)$ is the electric field due to the nucleus. We know $r(t)$ since we know v; we assume that the change in v is relatively small, which it is in reality. From some simple geometry we can write $E(r)$ in terms of E_x and E_y. The integral of E_x vanishes, so we need only consider

$$\Delta v = \frac{\Delta p}{m} = \int_{-\infty}^{\infty} \frac{qeE_y}{m}\,dt \tag{11.34}$$

A straightforward calculation gives

$$\Delta v = \frac{2qe^2 Z}{mvb} \tag{11.35}$$

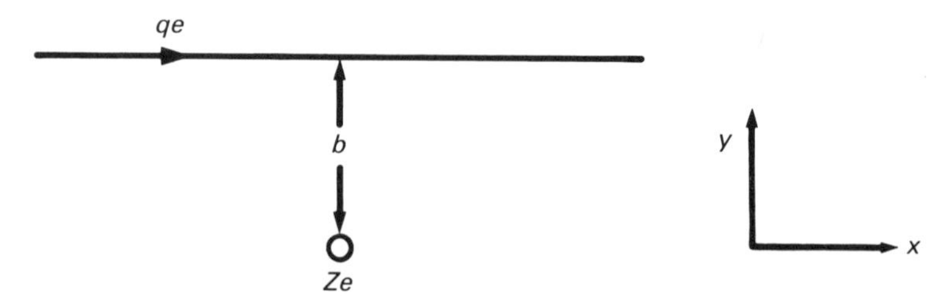

Figure 11.3. The definition of the impact parameter b.

where the impact parameter b is the perpendicular distance between the atom and the path of the electron and is shown in Figure 11.3.

The intensity of the radiation resulting from this velocity change is given by (see J. D. Jackson, *Classical Electrodynamics*)

$$I = \frac{2e^2(\Delta v)^2}{3\pi c^3} \tag{11.36}$$

This is obviously a function of the impact parameter b as shown by equation (11.35) and represents the total radiated energy per unit frequency. The frequency spectrum of I is rather simple. I is a constant in frequency up to some cutoff and is zero above that cutoff. The cutoff frequency is a function of b. In order to find the total energy radiated by a charge over all frequencies per unit length of its path, we must integrate (11.36) over b and ω; that is

$$\frac{d\mathscr{E}}{dx} = N \iint I 2\pi b \, db \, d\omega \tag{11.37}$$

where ω is the frequency, $2\pi b$ is merely a geometric factor and the N is the number density of nuclei in the scattering material. The details of this integration are relatively unenlightening, but the result is of interest:

$$\frac{d\mathscr{E}}{dx} = \frac{16NZ^2e^6q^4}{3\hbar c^3 m} \tag{11.38}$$

All of these calculations are nonrelativistic, so we need to be careful in interpreting them. We see that m appears in the denominator, which means for large m the radiation losses are small. For electrons, m is small and $d\mathscr{E}/dx$ can be large. We note that the radiative loss depends on Z and N of the medium. The nonrelativistic limit for electrons does not extend to very high energy, so a proper approach to this problem would consider the relativistic effects. Such a calculation would yield

$$\frac{d\mathscr{E}}{dx} = \frac{16NZ^2e^4q^2}{3\hbar c^3 m} \mathscr{E} \ln\left(\frac{192\lambda}{Z^{1/3}}\right) \tag{11.39}$$

where λ is a constant of order unity. The point is that in the relativistic case the radiative loss is proportional to the particle's energy. This means that Bremsstrahlung is not an important factor for low-energy electrons, but at higher energies it comes increasingly more significant. For $\mathscr{E} \sim 100$ MeV, losses due to Bremsstrahlung and due to ionization are roughly the same. For higher energies, Bremsstrahlung dominates. We show the relative importance of Bremsstrahlung in Table 11.2. Thus for high-energy electrons the interaction with matter results in a large number of photons—we will see shortly what happens to the photons.

A note on positrons. Positrons are "antielectrons." They mutually

Table 11.2. Energy loss mechanisms: typical values

Particle	Energy (MeV)	$(d\mathscr{E}/dx)_{\text{rad}}/(d\mathscr{E}/dx)_{\text{ion}}$ [a]
e^-	100	1
p^+	1000	5×10^{-6}

[a] rad = radiative; ion = ionization.

annihilate with electrons in matter. However, they cannot do so until they have thermalized—slowed down to a small velocity. While they are moving with a large velocity they lose energy either by ionization or by Bremsstrahlung as do electrons. When they have thermalized they annihilate with electrons, forming two 511 keV γ-rays traveling in opposite directions.

c. *Photons*

Photons have no charge, so they do not interact directly with charges in matter. There are three processes of importance:

1. Photoelectric effect
2. Compton effect
3. Pair production

In the *photoelectric effect* the photon interacts with an atomic electron and gives up all of its energy. The electron is ejected with kinetic energy

$$T_e = \hbar\omega - \phi \tag{11.40}$$

where ϕ is the binding energy of the electron. The photon disappears in the process. Because it is easier to conserve both momentum and energy if the electron involved in the process is bound tightly, the electrons emitted are generally K(inner)-shell electrons. This, of course, leaves a vacancy in an inner electron state that needs to be filled. Two alternative processes for this exist.

1. An outer (L) electron falls into the vacant K-state giving off an X-ray of frequency

$$\omega_x = \frac{\phi_K - \phi_L}{\hbar} \tag{11.41}$$

2. An Auger electron can be given off.

Auger processes require some additional explanation. If an L-shell electron falls into the K-shell vacancy, an energy $\phi_K - \phi_L$ becomes available. Rather than being given off as a photon, this can be used to liberate an L-shell electron from the atom. The kinetic energy of this L-shell electron is thus

$$T_e = (\phi_K - \phi_L) - \phi_L = \phi_K - 2\phi_L \tag{11.42}$$

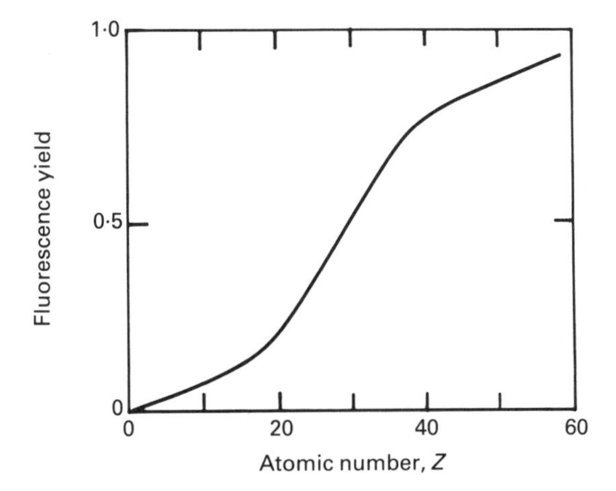

Figure 11.4. The ratio of the number of transitions that proceed via the photoelectric effect to those which proceed via Auger processes (fluorescence yield) as a function of the atomic number.

The ratio of transitions that proceed via the photoelectric effect to those that proceed via an Auger process is a function of the tightness of the binding of the electrons in the atom. More commonly, it is said that this is a function of the atomic number Z. This ratio, usually called the *K fluorescence yield,* is shown as a function of Z in Figure 11.4. We see that Auger processes dominate for low-Z materials but are relatively unimportant for $Z > 40$.

The second process of importance is *Compton scattering*. This is similar to the photoelectric effect, except that in the process of liberating the electron from the atom the photon does not disappear but merely loses some of its energy. We can write equation (11.40) as

$$T_e = \hbar(\omega_f - \omega_i) + \phi \tag{11.43}$$

where ω_i and ω_f are the before and after photon frequencies. We can also write down the equation for the conservation of momentum for Compton scattering:

$$\begin{aligned} \hbar\omega_i - \hbar\omega_f &= cp_e \cos\beta \\ \hbar\omega_f \sin\alpha - cp_e \sin\beta &= 0 \end{aligned} \tag{11.44}$$

where c is the speed of light.

Figure 11.5 defines the angles; p_e is the momentum of the electron after the scattering.

Since we are dealing with electrons here, we should do the problem relativistically to obtain the proper answer. Therefore p is the relativistic momentum of the electron and

$$p_e^2c^2 = T_e(T_e + 2mc^2) \tag{11.45}$$

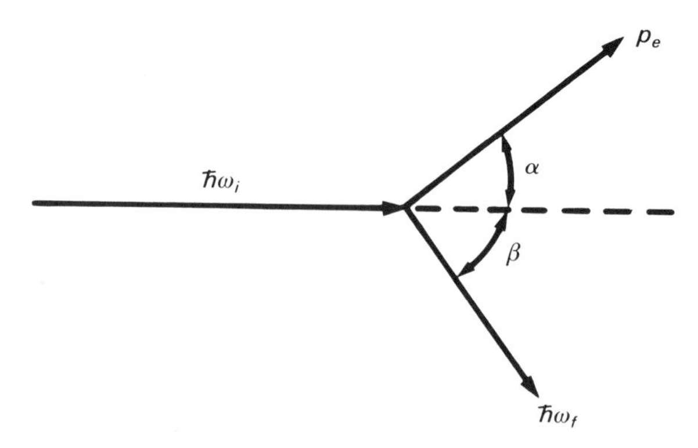

Figure 11.5. Compton scattering.

where T_e is the kinetic energy. A solution to equations (11.43) through (11.45) gives the change in photon wavelength as

$$\lambda_f - \lambda_i = \frac{c}{\omega_f} + \frac{c}{\omega_i} = \frac{(1 - \cos\alpha)\hbar}{mc} \tag{11.46}$$

where we have assumed that $\phi \ll \mathscr{E}_e$. So from equation (11.46) we see that the shift in the photon wavelength is

1. Independent of the photon wavelength. [This means that high-energy (short wavelength) photons lose more energy than low-energy (long-wavelength) photons.]
2. Independent of the type of atoms in the scatterer
3. A function of the scattering angle α

The final important process is *pair production*. This is the simultaneous disappearance of the photon and the production of an electron–positron pair, the opposite of electron–positron annihilation. The geometry is shown in Figure 11.6. We can write down the (relativistic) conservation equations as

$$\hbar\omega = T_e + T_p + 2mc^2 \tag{11.47}$$

$$\left.\begin{aligned} \frac{\hbar\omega}{c} &= p_e \cos\alpha + p_p \cos\beta \\ 0 &= p_e \sin\alpha - p_p \sin\beta \end{aligned}\right\} \tag{11.48}$$

and the relationships for p and T as

$$\left.\begin{aligned} p_2^e c^2 &= T_e(T_e + 2mc^2) \\ p_p^2 c^2 &= T_p(T_p + 2mc^2) \end{aligned}\right\} \tag{11.49}$$

We see from equation (11.47) that T_e and $T_p > 0$, so the minimum energy

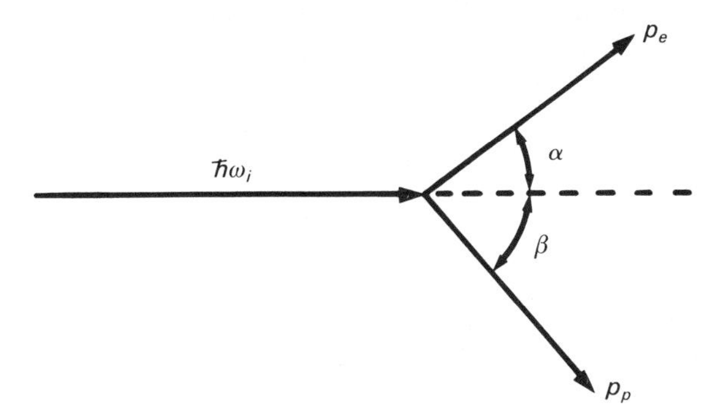

Figure 11.6. Pair production.

necessary for a photon to cause pair production is given by

$$\hbar\omega_{\text{min}} = 2mc^2 = 1022 \text{ keV} \tag{11.50}$$

It can be shown, after some messy algebra, that equations (11.47) through (11.49) *cannot* be solved. So pair production *cannot* occur under the conditions specified. We need another particle to take up some of the momentum. If the particle is a nucleus (which is very heavy compared with the electrons) then a lot of momentum can be taken up with very little energy. So equation (11.47) is still pretty much correct but we have to add an additional term to the right-hand side of equation (11.48). This means that pair production can occur in matter (e.g., inside a gas) but not *in vacuo*.

It is obvious from equations (11.40), (11.42), and (11.47) that the relative probability of energy loss by the three different mechanisms depends on the energy of the initial photon. It will also depend on the type of atoms that are present. Some semiempirical probabilities for these processes as a function of $\mathscr{E}_\gamma = \hbar\omega$ and Z are as follows. For the photoelectric effect the probability, δ_{photo}, is

$$\delta_{\text{photo}} \propto NZ^5\mathscr{E}_\gamma^{-7/2} \tag{11.51}$$

where N is the number of atoms per unit volume. Intuitively, this makes sense as it will be the lower-energy photons that contribute most to this process. For the Compton effect we have

$$\delta_{\text{compton}} \propto \frac{NZ}{\mathscr{E}\gamma} \tag{11.52}$$

and for a pair production

$$\delta_{\text{pair}} \propto NZ^2(\mathscr{E}\gamma - 2mc^2) \tag{11.53}$$

for $\mathscr{E} > 1022$ keV. The total probability is the sum

$$\delta_{\text{total}} = \delta_{\text{photo}} + \delta_{\text{compton}} + \delta_{\text{pair}} \tag{11.54}$$

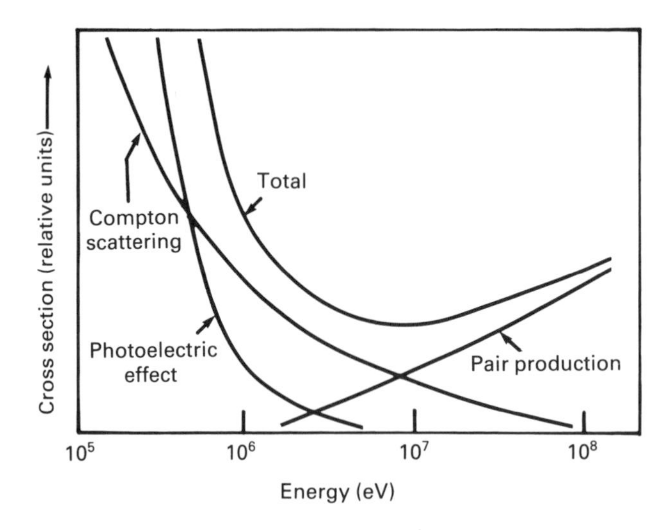

Figure 11.7. The total cross section for γ-rays in a heavy material ($Z \gtrsim 50$) as a function of energy and the contributions due to the various possible processes.

We show the contributions to δ_{total} as a function of $\mathscr{E}_\gamma$ in Figure 11.7. It is interesting to note that in some cases, but not all, δ_{total} shows a minimum at a few million electron volts.

d. Neutrons

Neutrons do not interact directly with charges, because they are not charged. They can, however, interact with nuclei via other mechanisms or with magnetic moments. This latter interaction makes them useful for the study of the magnetic properties of materials, but in the present discussion we concern ourselves only with the former type of interaction.

Unlike the situation with charged particles or even photons, the cross section for reaction of neutrons with nuclei is not a monotonic function of the number of nucleons or the charge of the nucleus. Nuclides with nearly the same Z can have vastly different neutron-reaction cross sections. In fact, even different nuclides of the same element can have enormously different cross sections. However, there are not many isotopes with very large neutron cross sections. (The rare earths are exceptions.) To detect neutrons we need a detector medium that contains a material with a reasonable neutron cross section. Some of the reactions used to detect neutrons are:

$$\begin{aligned} {}^3\text{He} + \text{n} &\rightarrow {}^3\text{H} + \text{p} + \gamma(765) \\ {}^{10}\text{B} + \text{n} &\rightarrow {}^7\text{Li} + \alpha + \gamma(2780) \\ {}^6\text{Li} + \text{n} &\rightarrow {}^3\text{H} + \alpha + \gamma(4780) \end{aligned} \tag{11.55}$$

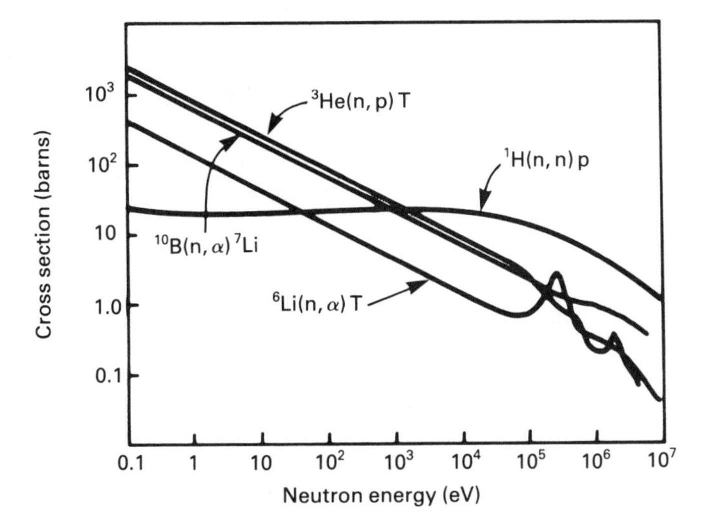

Figure 11.8. Neutron cross section for some reactions as a function of energy. (From P. Nicholson *Nuclear Electronics*. Copyright © 1974 by John Wiley & Sons. Used with permission.)

the $\gamma(\ )$ represents a γ-ray, with the energy given in keV in the parentheses for thermal neutrons. For higher-energy neutrons, the γ-ray energy is larger. For fast neutrons, another reaction,

$$^1\text{H} + \text{n} \rightarrow \text{p} + \text{n} + \gamma \tag{11.56}$$

is sometimes used. Figure 11.8 shows the neutron absorption cross sections as a function of energy for these reactions. You can see that the H(n,n)p reaction is useful above a few thousand electron volts but not for thermal neutrons. For these low-energy neutrons, the ^{3}He(n,p)T or ^{10}B(n,α)^{7}Li reaction is preferred.

e. Ranges

Before we discuss the actual detectors it would be useful to have a feeling for the range of various particles in different types of materials. If we are interested in detecting a particular type of particle with a particular energy using a detector containing a particular medium, it is important to know the range of the particle. If the range is a few millimeters, there is no sense in building a detector 10 cm in diameter. Similarly, if the range is 100 meters we need a different method of detecting the particles.

A simple empirical relationship for the ranges of radiation in different materials in the Bragg–Kleeman rule. This says that if the range of a particle of a given energy in a material of density ρ_1 and atomic number A_1

Table 11.3. Ranges of charged particles in various materials.

		Range (cm) in				
Particle	Energy (MeV)	Air	H_2O	Al	Fe	Pb
Proton	0.1	0.13	—	—	—	—
	0.2	0.3	—	—	—	—
	0.5	0.8	0.001	—	—	—
	1.0	2.3	0.003	0.001	—	—
	2.0	7.14	0.006	0.004	0.002	0.003
	5	34.0	0.03	0.019	0.009	0.012
	10	114.6	0.098	0.063	0.030	0.033
	20	395	0.33	0.21	0.10	0.099
	50	2064	1.7	1.1	0.53	0.46
	100	7138	5.7	3.7	1.78	1.51
α-Particle	0.1	0.097	—	—	—	—
	0.2	0.16	—	—	—	—
	0.5	0.30	—	—	—	—
	1.0	0.50	—	—	—	—
	2	1.0	0.001	—	—	—
	5	3.51	0.0032	0.0024	0.0012	0.0015
	10	10.6	0.0097	0.0064	0.0031	0.0037
	20	34.4	0.032	0.019	0.0092	0.01
	40	116	0.107	0.063	0.030	0.03

Note: The dash indicates less than 10^{-3} cm.

is R_1 then in a material of density ρ_2 and atomic number A_2 the range is

$$R_2 = R_1 \frac{\rho_1 \sqrt{A_2}}{\rho_2 \sqrt{A_1}} \tag{11.57}$$

This expression is to be used *only* for charged particles. Table 11.3 gives some typical average ranges of various particles in matter.

For γ-rays we look at this process somewhat differently. For a flux with initial intensity I_0 we write the flux $I(x)$ as a function of the depth of penetration into some material as

$$I(x) = I_0 e^{-kx} \tag{11.58}$$

where k/ρ (ρ = density) is the mass absorption coefficient. Some values are given in Table 11.4. To give you some idea of how far γ-rays will penetrate different materials, we can find the $1/e$ intensity as

$$x = \frac{1}{k} \tag{11.59}$$

Table 11.5 gives values of x obtained from equation (11.59). From an

Table 11.4. Mass absorption coefficients in various materials

Energy (MeV)	Mass absorption coefficient in Air	H_2O	Al	Fe	Pb
0.01	4.55	4.72	24.3	169	—
0.02	0.71	0.74	3.26	25	95
0.05	0.203	0.221	0.353	1.9	8.5
0.1	0.155	0.171	0.169	0.37	5.5
0.2	0.12	0.14	0.122	0.15	0.95
0.5	0.09	0.097	0.08	0.084	0.152
1.0	0.06	0.07	0.06	0.06	0.07
10.0	0.02	0.022	0.023	0.03	0.05
20.0	0.017	0.018	0.022	0.032	0.062

investigation of this table we can conclude that we should not expect air to attenuate γ-rays; of course if a source is radiating isotropically then the flux density will drop off as $1/r^2$. A couple of centimeters or so of lead, however, will stop nearly any γ-ray and lower-energy ones will be stopped by thin pieces of iron or even a small quantity of aluminum.

11.3 Sources of radiation

A variety of radioactive sources are in common use. These are divided into categories depending on the kind of radiation they produce.

Table 11.5. $1/e$ lengths for γ-rays in some materials as a function of energy[a]

Energy (MeV)	$1/e$ length in Air ($\rho = 0.0012$[b])	H_2O ($\rho = 1$)	Al ($\rho = 2.7$)	Fe ($\rho = 7.9$)	Pb ($\rho = 11.3$)
0.01	180	0.21	0.015	7×10^{-4}	—
0.05	4100	4.5	1.05	0.067	0.010
0.1	5400	5.8	2.2	0.34	0.016
0.5	9600	10.3	4.4	1.51	0.58
1.0	13.0 m	14.3	6.0	2.1	1.26
5	30.9 m	33.3	13.0	4.1	2.06
10	41.7 m	45.5	15.9	4.2	1.77
20	49.0 m	55.6	17.1	3.9	1.43

[a] From Table 11.4 and Equation (11.59), in centimeters.

[b] Units of ρ are g cm^{-3}.

Table 11.6. Pure β^- sources (data from C. M. Lederer *et al.*, Table of Isotopes, 6th ed. John Wiley, New York (1967) and references therein).

Nuclide	$t_{1/2}$	End-point energy (MeV)
^3H	12.3 y	0.019
^{14}C	5730 y	0.156
^{35}S	87.9 d	0.167
^{36}Cl	3×10^5 y	0.71
^{45}Ca	165 d	0.252
^{63}Ni	92 y	0.067
^{90}Sr	27 y	0.546
^{99}Tc	2.1×10^5 y	0.292
^{147}Pm	2.62 y	0.224
^{204}Tl	3.8 y	0.766

a. Electron sources

β^- *Decay sources.* Unstable nuclides that decay via equation (11.12) produce electrons. Often the β-decay process leaves the residual nucleus in an excited state. This then decays to the ground state by γ-emission. This is generally an undesirable effect, as the γ-rays interfere with measurements we would like to make on the electrons. Some β^- processes leave the residual nucleus in the ground state. These are called pure β^- emitters, and a few of the commonly used ones are given in Table 11.6.

The difference in the energy between E1 and E2 (see equation (11.12)) results from differences in the binding energies of the nucleons, and this energy is given up as kinetic energy to the electron and the antineutrino, with the electron having its maximum energy when the antineutrino has zero energy. This maximum electron energy is called the end-point energy and is shown in the table. An example of the energy distribution of β^- decay electrons is shown in Figure 11.9.

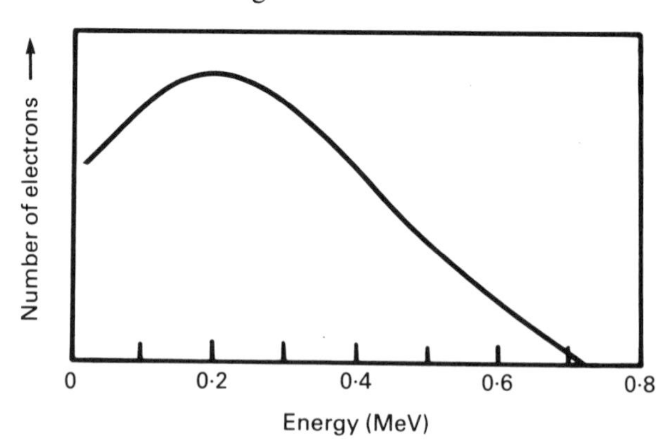

Figure 11.9. Energy distribution of electrons from the β^- decay of ^{36}Cl.

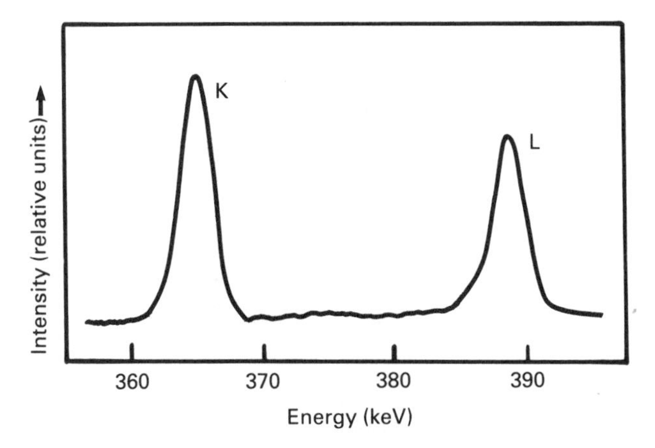

Figure 11.10. Energy spectrum of electrons produced by internal conversion in ^{113m}In.

Internal-conversion sources. The electron spectrum shown in Figure 11.9 represents "white" electrons. For some (or many) applications we require *monochromatic* electrons. One way of obtaining these is by the use of an internal-conversion source. The energy of the electron is given by equation (11.16). The energy levels of the nucleus are quantized for a particular transition $\mathscr{E}_\gamma$, and hence $\mathscr{E}_\gamma + \phi$ is well defined. The binding energy ϕ of the electron depends on the shell from which it comes, as well as the kind of atom. Different energies can come from different electron shells in the same source. An example is shown in Figure 11.10. The peaks in the spectrum are very well defined, although they are usually close together. This is because the spacing of the inner electron energy levels is usually much less than the spacing of the nuclear energy levels. This kind of source is the most convenient for producing monochromatic electrons in the range of a 100,000 or so up to a few million electron volts.

Auger electron sources. The previous arguments concerning the monochromaticity of internal-conversion electrons applies here as well. However, since Auger processes are prevalent only in low-Z atoms, the binding energies are low (a few thousand electron volts) and the electron energies are in this range. As a result, they are mostly absorbed inside the source before they are emitted.

Accelerator sources. Electrons can be produced by a hot filament. These electrons can then be accelerated across an electrical potential. A conventional cathode ray tube is an example of a device in which the electrons are accelerated in this manner to 10–20 keV. Large accelerators can produce charged particles with energies up to 5×10^{11} eV. Protons or ions can be accelerated in these types of machines if an appropriate ion source is used.

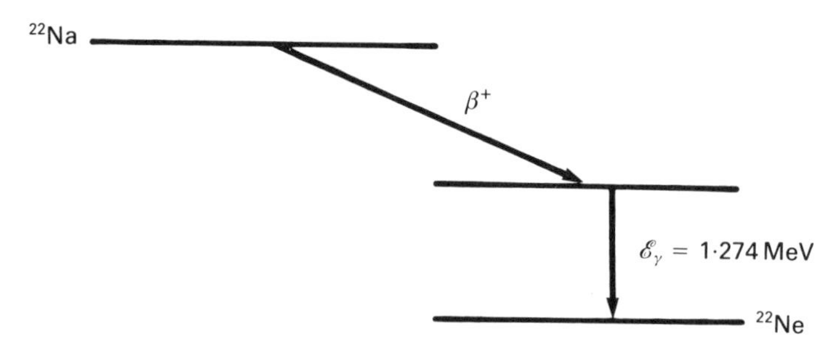

Figure 11.11. The decay scheme of ^{22}Na.

β^+ decay sources. β^+-Decay produces positrons rather than electrons. If we are actually interested in using the positrons, then the source and any covering material must be very thin. Positron sources with a thick covering will produce 511 keV annihilation radiation, but you can do a number of interesting experiments with this radiation as well. In positron sources there are generally some γ-rays, which result because the β^+-decay leaves the daughter nucleus in an excited state. This is illustrated in the simplified decay diagram for ^{22}Na shown in Figure 11.11. ^{22}Na is probably the most commonly used source of this kind because it is long-lived and is fairly inexpensive to produce. However, ^{64}Cu is frequently used at research facilities with neutron reactors. It is produced by the (n,γ) reaction

$$^{63}\mathrm{Cu} + \mathrm{n} \rightarrow {}^{64}\mathrm{Cu} + \gamma \tag{11.60}$$

Natural (nonenriched) copper can be used, since ^{63}Cu is 69 percent naturally abundant. Characteristics of useful positron sources are given in Table 11.7.

b. α-Particle sources

Alpha particles may be produced via the reaction (11.10). As their range in matter is small, the sources prepared from these kinds of radionuclides must be thin and any covering over the source must be kept to a minimum. The energy in reaction (11.10) results from the difference in the binding energies of the nuclide on the left-hand side of the equation and the nuclide plus the α-particle on the right-hand side. This energy is given up as a kinetic energy

Table 11.7. Some useful positron sources

Nuclide	$t_{1/2}$	Other γ-radiation (MeV)
^{22}Na	2.6 y	1.274
^{64}Cu	12.9 h	1.348
^{68}Ge	287 d	1.077
^{58}Co	71.3 d	0.811

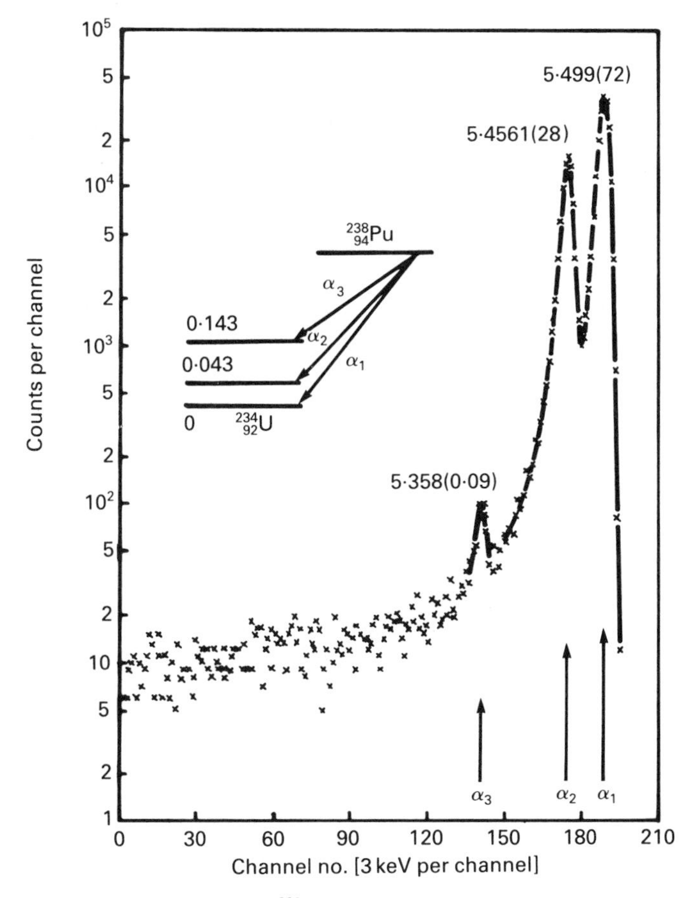

Figure 11.12. α-Particle spectrum of ^{238}Pu. Each peak is identified by its energy in MeV and its relative intensity in parentheses. (From G. Knoll, *Radiation Detection and Measurement.* Copyright © 1979 by John Wiley & Sons. Used with permission.)

to the α-particle as well as some recoil energy to the daughter nucleus. The daughter nucleus may be left in the ground state or one of several excited states with a corresponding decrease in the α-particle kinetic energy. An example of this is the decay of ^{238}Pu as shown in Figure 11.12.

A variety of α-particle sources have been commonly used and these are listed in Table 11.8.

α-Particles of high energies may also be produced by stripping the electrons from helium atoms (i.e., doubly ionizing them) and then accelerating the ions in an accelerator using electric and/or magnetic fields.

c. γ-*Ray sources*

γ-Rays are produced whenever a reaction or decay occurs that leaves a nucleus in an excited state. Most common γ-ray sources are nuclides that

Table 11.8. Some useful α-particle sources (data from A. Rytz, Atomic Data and Number Data Tables 12 (1973) 479).

Nuclide	$t_{1/2}$	Energy (MeV)	Intensity
^{148}Gd	93 y	3.1827	1
^{210}Po	138 d	5.3045	1
^{230}Th	7.7×10^4 y	4.6210	0.76
		4.6875	0.23
^{231}Pa	3.2×10^4 y	4.9517	0.23
		5.0141	0.25
		5.0297	0.20
		5.0590	0.11
^{232}Th	1.4×10^{10} y	3.953	0.23
		4.012	0.77
^{234}U	2.5×10^5 y	4.7220	0.72
		4.7739	0.28
^{235}U	7.1×10^8 y	4.219	0.06
		4.365	0.12
		4.374	0.06
		4.401	0.56
		4.598	0.05
^{236}U	2.4×10^7 y	4.445	0.26
		4.494	0.74
^{238}U	4.5×10^9 y	4.149	0.23
		4.196	0.77
^{238}Pu	86.4 y	5.452	0.28
		5.495	0.72
^{239}Pu	2.4×10^4 y	5.1046	0.12
		5.1429	0.15
		5.1554	0.73
^{240}Pu	6.5×10^3 y	5.1238	0.24
		5.1683	0.76
^{241}Am	433 y	5.4430	0.13
		5.4857	0.85
^{242}Pu	3.79×10^5 y	4.858	0.24
		4.898	0.76
^{242}Cm	16.25 d	6.066	0.26
		6.110	0.74
^{243}Cm	30 y	5.7415	0.12
		5.7847	0.73
		5.992	0.06
		6.067	0.02
^{244}Cm	18 y	5.759	0.23
		5.801	0.77
^{253}Es	20.5 d	6.592	0.08
		6.633	0.90
^{254m}Es	276 d	6.4288	0.93

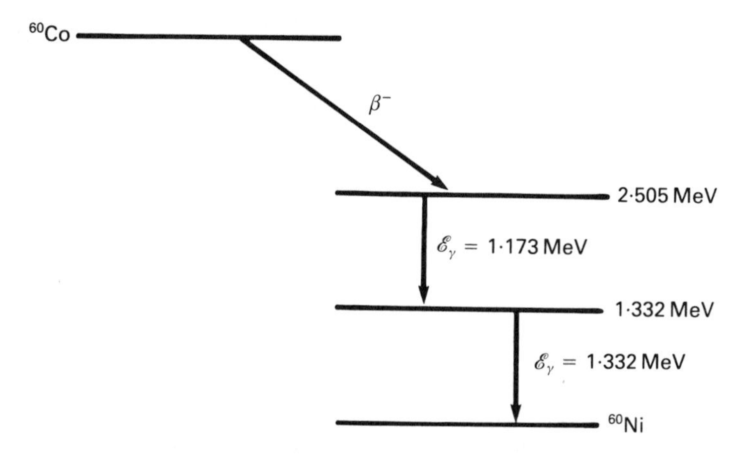

Figure 11.13. The decay scheme of ^{60}Co.

decay via β^+ or β^- decay. The γ-rays are emitted as a result of the decay of the daughter nucleus to the ground state. A common example, ^{60}Co, is shown in Figure 11.13. The source is labeled ^{60}Co because this is the radioactive nuclide; however, as you can see in the figure, the γ-rays result from the de-excitation of the daughter ^{60}Ni nucleus. Table 11.9 gives some of the more common γ-ray sources that decay via β^+ or β^-. The sources listed in Table 11.9 all have fairly short half-lives and are produced artificially using an appropriate reaction on a stable isotope. For example, ^{60}Co is produced from natural ^{59}Co (100 percent naturally abundant) by neutron irradiation via the reaction

$$^{59}\mathrm{Co}(\mathrm{n},\gamma)^{60}\mathrm{Co} \tag{11.61}$$

Table 11.9. Some common γ-ray sources

Source	Decay[a]	Daughter	$t_{1/2}$	γ-Ray Energies (MeV)
^{22}Na	β^+	^{22}Ne	2.6 y	0.511, 1.274
^{24}Na	β^-	^{24}Mg	15 h	1.368, 2.754
^{54}Mn	EC[b]	^{54}Cr	313 d	0.834
^{57}Co	EC	^{57}Fe	270 d	0.014, 0.122, 0.136
^{60}Co	β^-	^{60}Ni	5.26 y	1.173, 1.332
^{109}Cd	EC	^{109}Ag	450 d	0.088
^{131}I	β^-	^{131}Xe	8.1 d	0.080, 0.284, 0.364, 0.637
^{137}Cs	β^-	^{137}Ba	30 y	0.661
^{203}Hg	β^-	^{203}Tl	46 d	0.279
^{207}Bi	EC	^{207}Pb	30 y	0.569, 1.063, 1.769
^{208}Tl	β^-	^{208}Pb	3.1 min	0.510, 0.583, 0.860
^{212}Pb	β^-	^{212}Bi	10.6 h	0.238, 0.300

[a] Principal mode.
[b] EC = electron capture.

γ-Rays can also be produced by nuclear reactions. A common one that yields γ-rays of high energy (4–6 MeV) is the (α,n) reaction. Sources of this type, however, are more commonly used for the production of neutrons.

d. X-Ray sources

X-Rays can be produced in two distinctly different ways:

1. As a result of Bremsstrahlung
2. From decay processes involving orbital electrons (characteristic X-rays).

Bremsstrahlung. X-Rays produced as the stopping radiation for electrons in matter have a continuum of energies up to the energy $\mathscr{E}_e$ of the incident electron. This is shown in Figure 11.14. The electrons passing through a material also produce characteristic X-rays: these are monochromatic X-rays superimposed on the Bremsstrahlung continuum.

Characteristic X-rays. Characteristic X-rays result from atomic transitions. Two processes can be used: electron capture and external radiation.

In the process of electron capture, one of the inner electrons is taken up by the nucleus, converting a proton to a neutron. The atom still has the correct number of orbital electrons for the type of element but now one of the inner energy states is vacant. X-rays are produced by the de-excitation of electrons that fill the vacancy. In order to produce X-rays without interfering γ-rays, the EC decay must be to the ground state of the daughter nucleus. Table 11.10 gives examples of some sources that are relatively free from other radiation in an energy range that would interfere with the X-rays.

Internal-conversion processes can produce X-rays in a similar manner, although here γ-rays are generally present and interfere with the X-rays.

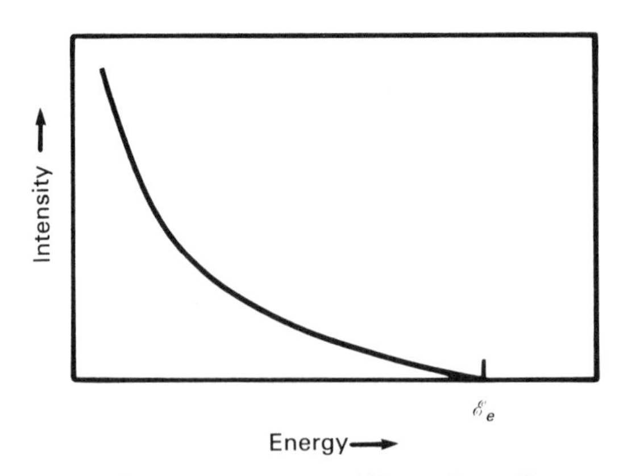

Figure 11.14. The energy spectrum of X-rays due to Bremsstrahlung.

Table 11.10. Electron-capture decays that produce X-rays

Parent nuclide	Daughter	$t_{1/2}$	K-shell energy (keV)
^{37}Ar	^{37}Cl	35 d	2.957
^{41}Ca	^{41}K	8×10^4 y	3.690
^{44}Ti	^{44}Sc	48 y	4.508
^{49}V	^{49}Ti	330 d	4.949
^{55}Fe	^{55}Mn	2.6 y	5.895

External irradiation can also be used to excite the orbital electrons to higher energy levels. These electrons then de-excite, yielding characteristic X-ray frequencies. A broadband X-ray spectrum produced by Bremsstrahlung can be used to excite these transitions and in this case the process is referred to as *X-ray fluorescence.* Electrons and α-particles can also be used as the source of radiation to excite these transitions. The technical details of the production of monochromatic X-rays by this method will be discussed in Chapter 13.

e. Neutron sources

(α,n) sources. Neutrons can be obtained via the (α,n) reaction by irradiating an appropriate material with α-particles. The most commonly used reaction to obtain neutrons in this way is

$$^{9}\mathrm{Be} + \alpha \rightarrow {}^{12}\mathrm{C} + \mathrm{n} \tag{11.62}$$

The α-particles may be provided by any of a number of sources (see Table 11.8). The more commonly used ones for the reaction (11.62) are given in Table 11.11.

Sources are generally prepared by mixing powdered source material with powdered beryllium. The powdered material is then compressed into a "pellet" and sealed in a can. The "standard" commercial small neutron sources are sealed in a double walled stainless steel can as shown in Figure

Table 11.11. Properties of ^{9}Be (α, n) sources (data from G. F. Knoll, Radiation Detection and Measurement, John Wiley, New York (1979) and references therein).

α-Source	$t_{1/2}$	$\mathscr{E}_\alpha$(MeV)	Neutrons per 10^6 α particles
^{239}Pu	2.4×10^4 y	5.14	65
^{210}Po	138 d	5.30	73
^{238}Pu	87 y	5.48	79
^{241}Am	433 y	5.48	82
^{244}Cm	18 y	5.79	100
^{242}Cm	162 d	6.10	118

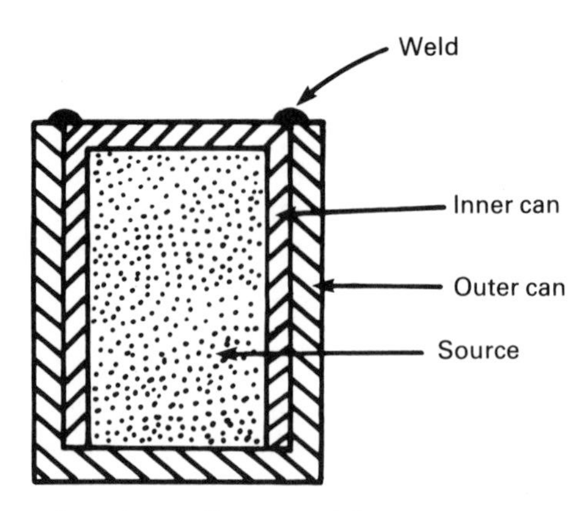

Figure 11.15. Construction of a small commercial neutron source, shown at approximately actual size.

11.15. These types of neutron sources produce "white" neutrons that range in energy from thermal (low energy—a fraction of an electron volt) up to around 10 MeV). Several other (α,n) reactions are sometimes used although not so commonly. Some of these are

$$\begin{aligned} &^{10}B + \alpha \rightarrow {}^{13}N + n \\ &^{11}B + \alpha \rightarrow {}^{14}N + n \\ &^{19}F + \alpha \rightarrow {}^{22}Na + n \\ &^{13}C + \alpha \rightarrow {}^{16}O + n \\ &^{7}Li + \alpha \rightarrow {}^{10}B + n \end{aligned} \tag{11.63}$$

Photoneutron sources. Two reactions are of interest here:

$$\begin{aligned} &^{9}Be + \gamma \rightarrow {}^{8}Be + n \\ &^{2}H + \gamma \rightarrow {}^{1}H + n \end{aligned} \tag{11.64}$$

The γ-rays cause the liberation of the neutron from the nucleus. The energy of the γ-ray must be at least equal to the binding energy of the neutron in the nucleus. This is 1.67 MeV for the first reaction and 2.23 MeV for the second reaction. Additional energy of the γ-ray is given to the neutron as kinetic energy. Hence monoenergetic γ-rays (as we would generally have) will yield monoenergetic neutrons. These sources are prepared by covering the γ-ray emitter with either a beryllium or deuterium oxide (D_2O) target. Some common sources are given in Table 11.12. Although this is a convenient way of producing monoenergetic neutrons, there are two difficulties associated with it.

1. The half-life of the γ-ray source is relatively short (see the table).
2. The neutron yield is about 1 per 10^6 γ-rays, so the γ-ray flux from the source is much larger than the neutron flux.

Table 11.12. Some useful photoneutron sources (data from K. H. Beckurts and K. Wirtz, Neutron Physics, Springer-Verlag (1964)).

γ-Source	$t_{1/2}$	γ-Ray energy (MeV)	Target	Neutron energy (keV)
^{24}Na	15 h	2.757	Be	969
			D_2O	265
^{140}La	40 h	2.51	Be	747
			D_2O	140
^{124}Sb	60 d	1.691	Be	23

Fission sources. The radioactive decay of some heavy nuclides proceeds via spontaneous fission in which neutrons are given off. These sources have the advantage that they produce large quantities of neutrons per unit mass of source. As they must be produced in high-energy reactors, they are not as commonly used as other types of neutron sources. ^{252}Cf has gained the widest acceptance as a fission neutron source. This decays via α-emission ($t_{1/2} = 22$ y) and by spontaneous fission ($t_{1/2} = 66$ y). About one fission event occurs for each 32 α-decays and each fission yields about four neutrons. The neutron energy spectrum has a peak at about 1 MeV but has a high-energy tail extending to about 10 MeV.

12
Nuclear instrumentation

12.1 Radiation detectors

a. Gas-filled detectors

There are three basic types of gas-filled detectors: (1) ionization chambers, (2) proportional counters, and (3) Geiger–Müller tubes. Their modes of operation have a number of points in common, so we will start with a generalized gas-filled detector.

The basic design consists of a chamber to hold the gas, a window that is transparent to whatever kind of radiation we want to detect, and an arrangement of electrodes for producing an electric field. A simple parallel-plate arrangement is illustrated in Figure 12.1.

The net result of any sort of ionizing radiation entering the chamber will be to produce electron–ion pairs. An electric field is set up in the region between the plates in Figure 12.1 by applying a voltage between the electrodes. This attracts the electrons to one plate and the (+) ions to the other. The charge collected by the (+) and (−) electrodes (anode and cathode, respectively) is measured by electronics that we discuss later. For each "particle" of ionizing radiation that enters the chamber, the number of electron–ion pairs that results depends on the energy of the particle. Thus the charge collected on the electrodes is a measure of the energy of the particle. Let us look at this process more quantitatively.

Consider an electron–ion pair being formed at some location between the plates. This is illustrated in Figure 12.2. We denote the distance of the ionization from the cathode as a fraction f of the total plate separation d. The energy gained by a charge q traveling through a potential ΔV is

$$\mathscr{E} = q\,\Delta V \tag{12.1}$$

We make the assumption here that the ion is singly ionized, so that both charges are of magnitude e. Also we know that in a parallel-plate capacitor the electric field is a constant and V varies linearly with position between the plates. So we write the energy gained by the (+) and (−) charges in falling through the potential to the cathode and anode, respectively as

$$\mathscr{E}_+ = efV \qquad \mathscr{E}_- = e(1-f)V \tag{12.2}$$

We can calculate the change in the voltage across the plates due to the

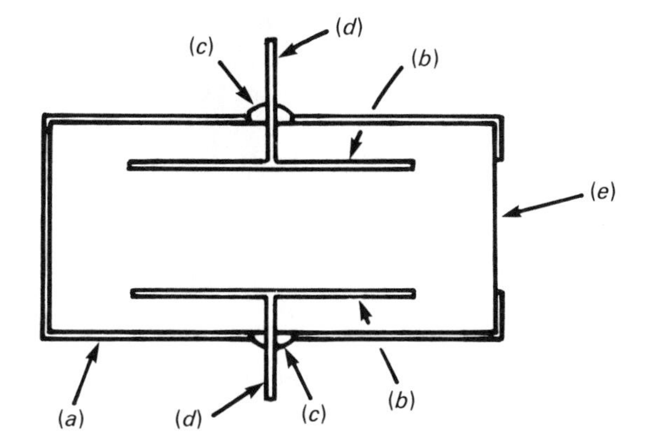

Figure 12.1. A parallel-plate gas-filled radiation detector: (*a*) chamber; (*b*) plates; (*c*) electrical feedthroughs; (*d*) electrical connections; (*e*) window.

arrival of one electron at the anode. The energy of a capacitor with a voltage V across it is

$$\mathscr{E} = \frac{CV^2}{2} \tag{12.3}$$

where C is its capacitance. We can write

$$\frac{\partial \mathscr{E}}{\partial V} = CV \tag{12.4}$$

then

$$\Delta \mathscr{E} = CV \, \Delta V \tag{12.5}$$

If $\Delta \mathscr{E}$ results from one electron, then equations (12.2) and (12.5) give

$$e(1-f)V = CV \, \Delta V_- \tag{12.6}$$

or

$$\Delta V_- = \frac{e(1-f)}{C} \tag{12.7}$$

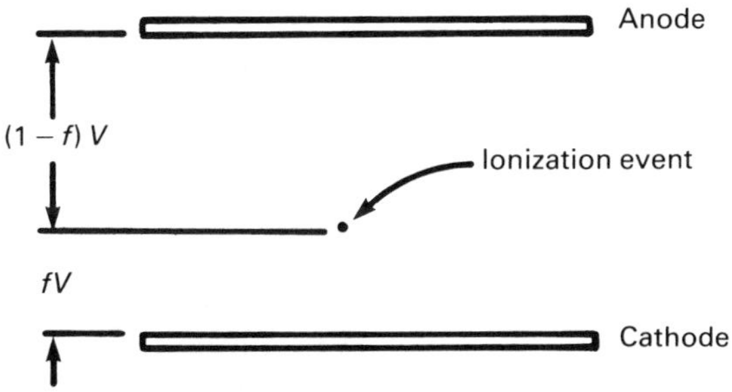

Anode
(1 − f) V
Ionization event
fV
Cathode

Figure 12.2. Ionization in the gap of a parallel-plate radiation detector.

It is interesting to note that ΔV is independent of V. The size of the voltage is, however, due to both the electron and to the ion. For the ion we can easily calculate from equations (12.2) and (12.5)

$$\Delta V_+ = \frac{fe}{C} \tag{12.8}$$

so the total size of the voltage charge is given by

$$\Delta V = \Delta V_- + \Delta V_+ = \frac{e}{C} \tag{12.9}$$

This means that the size of the voltage pulse detected at the output of the tube due to the production of an electron–ion pair is independent of the location between the plates where the event occurs. This is an encouraging result, because we have no way of knowing where between the plates the event took place. So for a particle of ionizing radiation of energy $\mathscr{E}_\gamma$ we should see a voltage pulse ne/C, where n is the number of electron–ion pairs. If the gas has an ionization energy $\mathscr{E}_i$, then $n = \mathscr{E}_\gamma / \mathscr{E}_i$ and

$$\Delta V = \frac{e}{C}\frac{\mathscr{E}_\gamma}{\mathscr{E}_i} \tag{12.10}$$

This would be very nice if it worked, but there is a problem. Let us calculate how long it takes between the formation of the electron–ion pair and the detection on the voltage change on the plates. We make the following assumptions.

1. The gas in the tube is krypton (at.wt = 84).
2. The separation d of plates is 4 cm (a reasonable size).
3. The ionization event is in the middle of the tube ($f = 0.5$).
4. The voltage drop between plates is 1000 V.

First we consider the electrons. The force acting on the electron is

$$F = ma = -eE = \frac{eV}{d} \tag{12.11}$$

so the time to travel a distance $d/2$ (2 cm) is given by

$$t_- = \sqrt{da} = \left(\frac{md^2}{eV}\right)^{1/2} = 3 \text{ ns} \tag{12.12}$$

For the calculation for the ion we need only substitute a different mass:

$$t_+ = [(84)(1.6 \times 10^{-27})(0.04)^2/(1.6 \times 10^{-19})(1000)]^{1/2} = 1\ \mu\text{s} \tag{12.13}$$

or about 300 times longer than the time for the electron. Suppose a radioactive particle entered the gas; what would the voltage change across the plates look like? This is illustrated in Figure 12.3: note that the time scale is highly nonlinear. It is quite probable that another ionizing particle

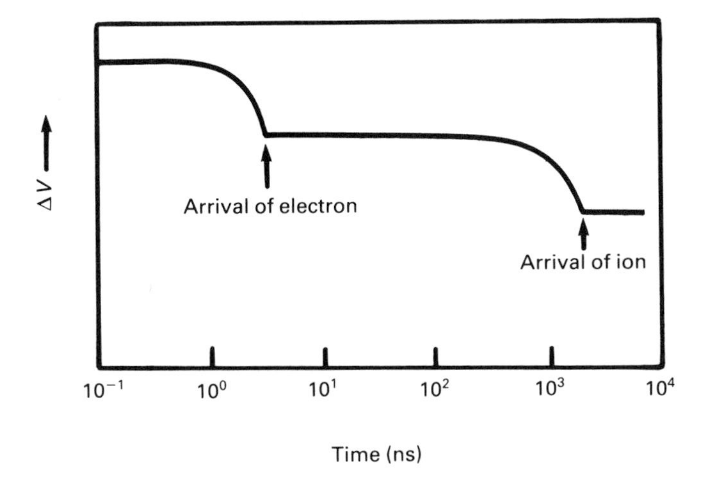

Figure 12.3. Voltage detected as a function of time for the parallel plate chamber of Figure 12.1 after the occurrence of an ionization event.

will enter the chamber within a millisecond or so after the first one. In this case the electrons from this second event will be detected before the ions from the first one.

Unless the rate of particles entering the chamber is very low, we will get a confused mess of voltage pulses at the anode and cathode and we will have no way of knowing which anode pulse corresponds with which cathode pulse. There are a couple of ways of dealing with this problem. We can look only at the electrons arriving at the anode and ignore the (+) ions arriving at the cathode. In any case the time response is better for the electrons than for the ions. Unless the rate of particles entering the counter is very high, all the electrons from one event will be counted before electrons from another event begin to arrive and we will get a well-defined pulse for each particle. This is, in fact, part of the solution but not all of it. The problem is equation (12.7). If we look only at the electrons, the size of the voltage pulse will be a function of where in the chamber the electron–ion pair was formed. The rest of the solution to this problem lies in the geometry of the electrodes. Two methods are available.

Consider first the case of a coaxial capacitor: see Figure 12.4. The cathode is formed by an outer metal cylinder with a window to allow the radiation to enter the chamber. Commonly the chamber is made of stainless steel, as this is quite strong and durable, while the window is made of beryllium, as this has a low atomic weight. The anode consists of a thin wire in the center of the cylinder. In a cylindrical arrangement like this, the electric field between the anode and cathode is not constant and hence the potential does not vary linearly with spatial position. The potential as a function of the radial distance from the centre of the anode is given in terms

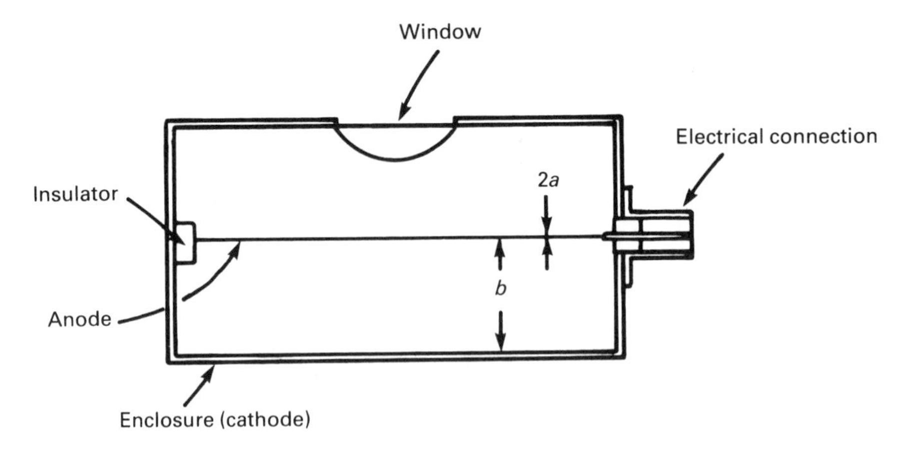

Figure 12.4. The coaxial ionization chamber. The radius of the anode wire is a and the radius of the cathode (chamber) is b.

of the total potential drop V_0 (note that the cathode is grounded) as

$$V(r) = V_0 \frac{\log(b/r)}{\log(b/a)} \tag{12.14}$$

This is shown in Figure 12.5.

It is also important here that the volume of the gas in the tube per unit radius is proportional to the distance from the center of the tube. That is,

$$\frac{dV}{dr} = 2\pi r L \tag{12.15}$$

where V is the volume and L is the length of the cylinder. Now we can look at only the electrons. On the average they are the result of ionization events that occur uniformly in volume inside the tube but not uniformly as a

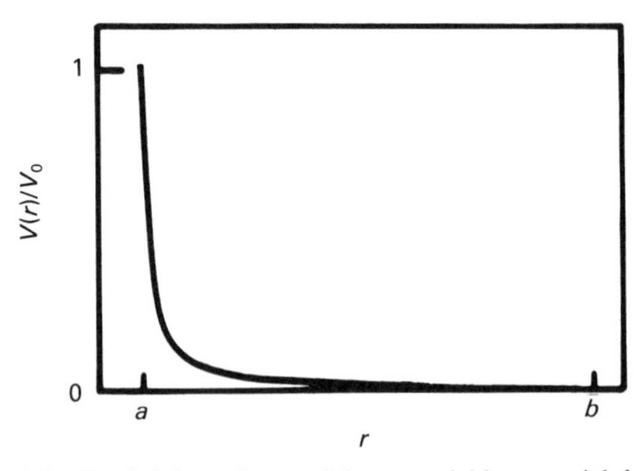

Figure 12.5. Radial dependence of the potential in a coaxial detector.

function of r—see equation (12.15). From Figure 12.5 we can see that most of the electrons are going to fall through the potential V_0. It is only those that are created very close to the anode that fall through a potential drop that is measurably less than V_0. You can see from equation (12.14) that we can make this region even smaller if we decrease the radius of the anode wire.

However, things are even better than Figure 12.5 would suggest. From equation (12.15) we see that the volume of gas contained within an element dr at small r is much less than the volume of gas in dr at large r. Thus the probability of an ionization taking place very close to the anode is extremely small. The result is that essentially all electrons fall through a potential V_0; similarly, essentially all ions fall through a potential ~0. Although it is the total energy, that due to the electrons plus that due to the ions, that we are interested in, the ions contribute essentially nothing and we can therefore look only at the electrons. This coaxial arrangement, or something similar, is used for most gas-filled detectors.

Another solution to the problem of building an energy-sensitive ionization chamber retains the parallel-plate geometry. This is the gridded ionization chamber. A diagram of this detector is shown in Figure 12.6. The volume of gas between the cathode and the grid is large compared with the volume between the grid and the anode, so that most of the electron–ion pairs are produced in the region between the cathode and the grid. The electrons that are collected (and detected) at the anode are those that are produced in this larger region and make it through the grid. The energy of these electrons when they reach the anode is governed primarily by V_2; that is, they all fall through the same potential. Thus the ΔV detected at the anode can be related to the particle energy.

Ionization chambers are often of the gridded type, although they may be

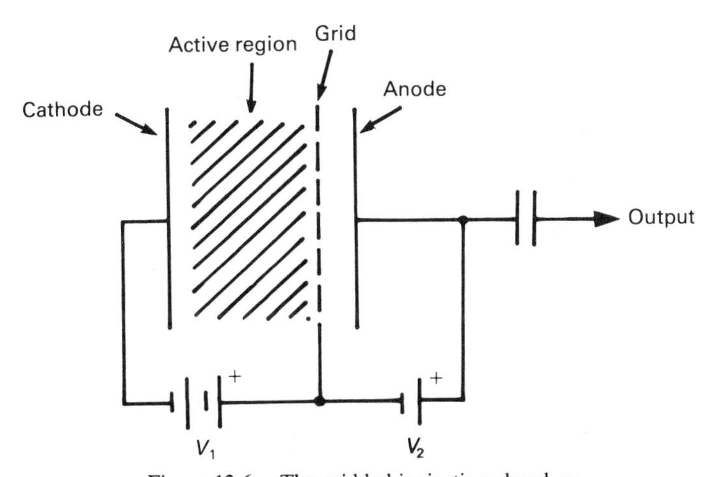

Figure 12.6. The gridded ionization chamber.

of the coaxial geometry as well. Proportional counters are generally of the coaxial geometry. Geiger–Müller tubes are often coaxial although in this case the geometry is not important, as we shall see.

The device described thus far is actually an *ionization chamber.* The size of the voltage pulse at the output from such a device is small. It can, however, be used successfully to detect high-energy particles—lower-energy particles do not create a sufficient ΔV to make the ionization chamber very useful. The proportional counter is a device that works along similar lines but overcomes this problem of low gain. We could say that the Geiger–Müller tube has too high a gain. The operation of these devices is contrasted below.

Proportional counters and Geiger–Müller tubes. Consider the plot in Figure 12.7. For low bias voltages the output is independent of bias voltage but increases with increasing particle energy: this is what we expect from equation (12.10).

Region *b* is the proportional counter region. What happens here is as follows. If the bias voltage is large enough then the electrons falling to the anode gain enough energy to cause secondary ionization. This is rather like the avalanche effect that occurs in Zener diodes. Thus the total number of charges collected is much larger than we would expect from equation (12.10). It can in fact be as much as 10^4 times as large. We modify equation (12.10) as

$$\Delta V = \frac{M\mathscr{E}_\gamma e}{\mathscr{E}_i C} \tag{12.16}$$

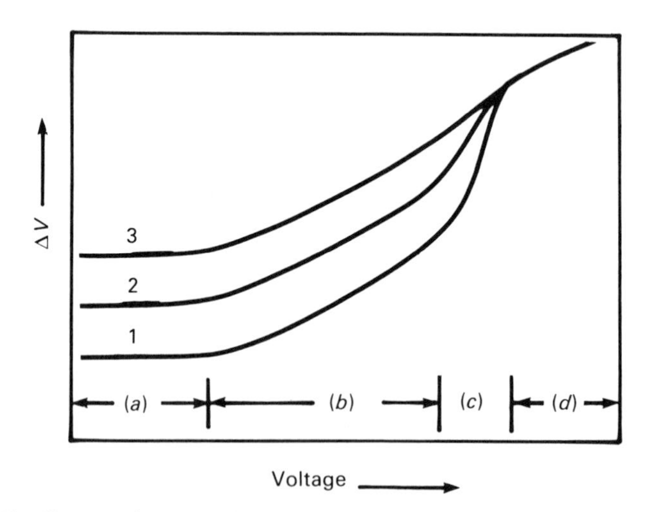

Figure 12.7. Output voltage as a function of bias voltage for a gas-filled detector. The curve represent the results for incident particles of three different energies, $\mathscr{E}_1 < \mathscr{E}_2 < \mathscr{E}_3$. The regions are (*a*) ionization region; (*b*) proportional region; (*c*) nonproportional region; (*d*) Geiger–Müller region.

where M is referred to as the *gas multiplication factor*. The criterion in the proportional counter region is that M is independent of the particle energy. This means that in equation (12.16) $\Delta V \propto \varepsilon_\gamma$, the other factors all being energy independent. In the nonproportional region, M is no longer independent of $\mathscr{E}_\gamma$. This is not an especially useful range of voltages for a detector, because the proportionality of the lower voltage region is lost and the gain is lower than in the Geiger–Müller region.

In this Geiger–Müller region the avalanche-like effect becomes uncontrolled and any particle above some small energy will cause the same large charge to be collected by the anode. Hence ΔV is no longer a function of $\mathscr{E}_\gamma$ and this means that the gain of the tube is very high. However, all possibility of determining the energy of the particle is lost. The Geiger–Müller tube is a very sensitive counter of ionizing radiation but it cannot be used for measuring the energy spectrum. There is a particular problem with the Geiger–Müller tube: the positive ions move much more slowly than the electrons, so that long after the electrons have been counted the large pulse of (+) ions arrives at the cathode. Since so much ionization has taken place, this ion current is large enough to cause an emission of electrons from the cathode. These electrons are accelerated toward the anode and cause more ionization. This whole process can occur over and over again and since the (+) ions move more slowly than the electrons there is a rapid build up of positive charges in the tube. This, of course, interferes with its further operation. We must, therefore, have some means of returning the tube to the condition before the initial ionization occurred so that it will be ready to detect another particle. This process is referred to as *quenching* and two methods are in common use: the use of a quenching gas, and electronic quenching.

In the electronic method the bias voltage on the tube is reduced for about 300 μs to allow the (+) ions to be collected in such a way that further ionization is avoided. After this the tube is returned to its original bias voltage and it is ready to detect another particle.

In the other method a small percentage of another gas is added to the tube. This is usually something organic, but halogens have been used as well. The energy of the ions is used by this gas to dissociate the molecules rather than to cause emission of electrons from the cathode. Ethyl alcohol is sometimes used but this becomes totally dissociated after about 10^9 counts and must be replaced. Halogens seem to regenerate (i.e., recombine) and do not get used up so rapidly.

The choice of gas used to fill these detectors depends on what we are interested in detecting. For most applications one of the noble gases argon, krypton, or xenon is generally used as the *major* component. For the purpose of detecting neutrons, BF_3 is often used because boron has a large neutron corss section.

Uses of gas-filled detectors. Ionization chambers can be used for high

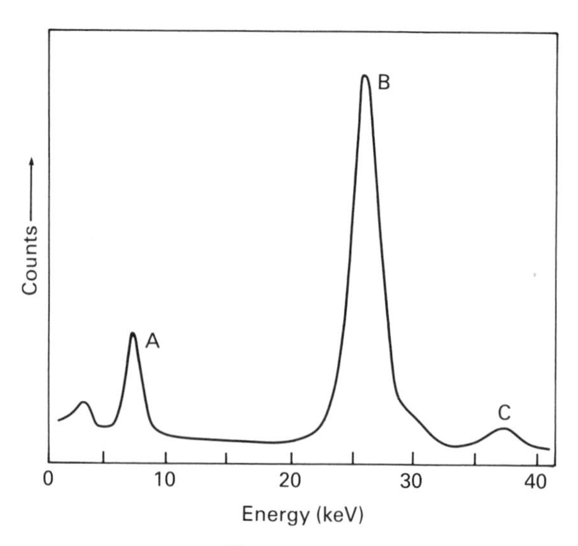

Figure 12.8. Energy spectrum for a ^{121m}Sn source obtained with a $Xe\text{–}CO_2$ proportional counter.

energies but the efficiency is generally not good because the range of the particles in the gas is too large. Detectors that use denser detector media are generally preferred, at least for the detection of γ-rays.

The additional gain obtained with the proportional counter makes it particularly suited for looking at low-energy particles. Energy resolution of about 10 percent or so can be obtained from a few thousand electron volts up to 50–100 keV with reasonable efficiencies. An example of an energy spectrum obtained with a proportional counter is shown in Figure 12.8. The details of how to obtain such a spectrum will be discussed in the next section. Peak C in Figure 12.8 results from γ-rays and peak B from electrons. Peak A demonstrates a problem with some detectors: this peak originates from processes in the gas *inside* the proportional counter caused by the radiation from the source but does not come from the radiation source directly. It is therefore important to choose a gas for the detector so that any such peaks will not interfere with the measurement of interest.

b. Scintillation detectors

A scintillator is a material that produces optical photons when struck by ionizing radiation. A photomultipler tube is then used to detect the light. The operation of the PMT was discussed in detail in Chapter 10, so here we will concentrate on the nature of the scintillation material. The requirements for a suitable scintillation material are as follows.

1. The material must fluoresce when struck by the radiation of interest.

2. The wavelength of the fluorescent radiation must be compatible with the spectral response of the PMT (see Figure 10.4).
3. The material must be transparent to the radiation it produces.

The PMT and the scintillator must be covered in such a way that only light produced by the scintillation makes its way into the PMT; that is, so that all ambient light is excluded. In addition, the covering over the scintillator must be transparent to the type of radiation to be detected. We divide scintillator materials into five categories: (1) inorganic crystals, (2) organic crystals, (3) plastics, (4) liquids, (5) gases. We look at these in detail.

Inorganic crystals. Inorganic crystals are the most common scintillators and the most popular is NaI. The room-temperature luminescent efficiency is improved by adding a small amount of thallium. The thallium is called an *activator,* and we designate the material as NaI(Tl). Most NaI crystals you will find around a nuclear physics laboratory will be activated with about 0.1 per cent Tl. It is interesting to note, however, that the efficiency at 77 K (liquid-nitrogen temperature) is higher for pure NaI than for Tl-activated NaI. NaI is very good for high-energy particles because the large atomic number of iodine means it has a large absorption cross section. One disadvantage of NaI is that it is highly deliquescent and must be encapsulated in an airtight container. An example of a mounted NaI crystal is shown in Figure 12.9. The crystal is often packed in a layer of powder inside the aluminum can; MgO is the commonly used material. This has three possible advantages:

1. It absorbs any moisture which may be present in the can.
2. It helps to reflect light towards the scintillator.
3. It provides some mechanical isolation for the crystal.

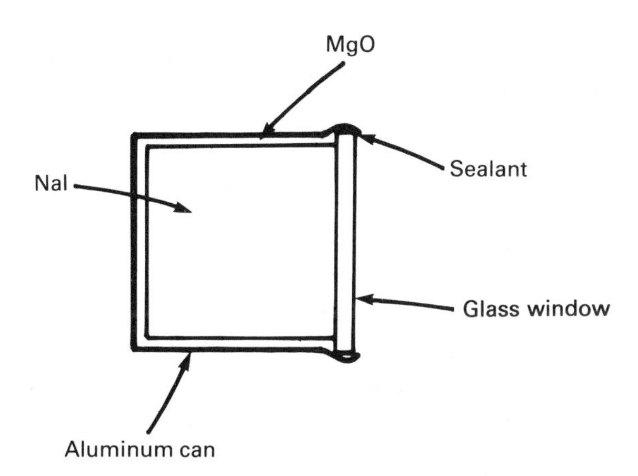

Figure 12.9. Cross section of a mounted NaI scintillator.

The glass window is coupled optically to the photocathode of the PMT using a suitable type of grease. When the crystal cannot be attached directly to the PMT because of the geometry of the experiment, the light is conveyed from the scintillator to the PMT by means of a plastic (Perspex) light pipe. It is important that the NaI scintillator be a single crystal so that it is sufficiently transparent. Crystals can be grown up to about 20 cm in diameter.

Although NaI is by far the most common scintillator used, there are a few other inorganic materials as well as some materials from other categories that have advantages in some specialized applications. The other inorganic materials which are sometimes used are as follows.

CsI activated with either thallium, or less frequently sodium, can be used as a scintillator. It has some advantages over NaI:

1. Slightly higher average atomic number for better efficiency
2. Higher mechanical strength
3. The fact that it is not deliquescent

LiI(Eu) is of particular interest because the lithium is effective in detecting neutrons according to the reaction in (11.55). Since ^{6}Li is only 7.4 percent naturally abundant, lithium enriched in ^{6}Li improves the neutron cross section of the detector considerably.

ZnS(Ag) produces large quantities of light compared with other scintillation material. However, it can be manufactured only as a polycrystalline powder and is therefore opaque to the light it produces if very large quantities are used. Only thin layers 0.1 mm or less thick can be used. As such, it is most effective for particles with a small range, namely α-particles or heavier ions.

$CaF_2(Eu)$ the scintillating properties of this material are only moderate, but it has the advantage that it is extremely strong and impervious to adverse environmental conditions, so it can be used in places where other materials cannot.

$Bi_4Ge_3O_{12}$ (bismuth germanate) this is of interest because of its high atomic number component (bismuth). As a result it is highly efficient at stopping γ-rays. Unfortunately, its light output is not as great as most other scintillator materials.

CsF is used when fast response time is required. Although its efficiency is low, it is better than the organics.

Organic crystals. The organic crystals that have been used as scintillators are primarily anthracene and stilbene. These are hydrocarbons with a benzene-like ring structure and can be prepared as single crystals up to a few centimeters on a side. They have much better time resolution than the inorganic crystals but, because they are made up only of elements with low atomic number, their efficiency is very low. They have been replaced for essentially all applications by plastic detectors.

Plastics. The plastic detector consists of a plastic host material to which some complex organic impurities have been added. The organic molecules absorb the radiation to be detected and de-excite via photoemission, yielding light that is detected by the PMT. Popular host plastics are polyvinaltoluene and polystyrene. The impurity, typically in the order of a few percent, can be any of a variety of organic materials. Common ones are *p*-terphenyl; (2,5)-diphenyloxazole; and 2-phenyl-5-(4-biphenyl)-1,3,4-oxadiazole. Often, however, you will see these materials referred to by a manfacturer's name, such as NE102. These materials have several advantages.

1. They are inexpensive.
2. They can be made quite large.
3. They can be formed into any required shape.
4. They are not easily broken.
5. They are very fast (i.e., they have good time resolution).

They have one serious disadvantage: they are very inefficient, for the same reason as are the organic crystals.

Liquids. Liquid scintillators are essentially the same as plastic scintillators except that the host material is a liquid rather than a plastic. The impurities are generally the same as for plastic detectors. Common host liquids are toluene; benzene; xylene; and 1,4-diazole. Their advantages and disadvantages are essentially the same as for the plastic detectors, plust the fact that they must be contained in a liquid-tight container with an optically transparent window to couple to the PMT. One unusual use of liquid scintillators is what is called internal counting. Low-energy β-particles can be detected without the problem of attenuation due to the windows on the detector by dissolving the β-source in the detector medium.

Liquid and plastic detectors can be constructed using a variety of impurities. The inclusion of boron or lithium makes neutron detection possible. The addition of such impurities is *generally* not possible in inorganic crystals, although ZnS has been prepared with boron or lithium impurities for the purpose of neutron detection, and LiI has also been used.

Gases. Among gases, xenon is sometimes used as a detector for heavy charged particles. As with the liquids, it must be contained appropriately. It has the disadvantage that the photons produced are in the ultraviolet and cannot be readily detected by a PMT. A wavelength shifter can be used to shift the wavelength to the visible region. One of the best shifters is diphenylstilbene.

Uses of scintillation detectors. Scintillators are useful in the range of 50 keV up to a few million electron volts; the higher energies require larger crystals. A percentage resolution in energy comparable to that of the

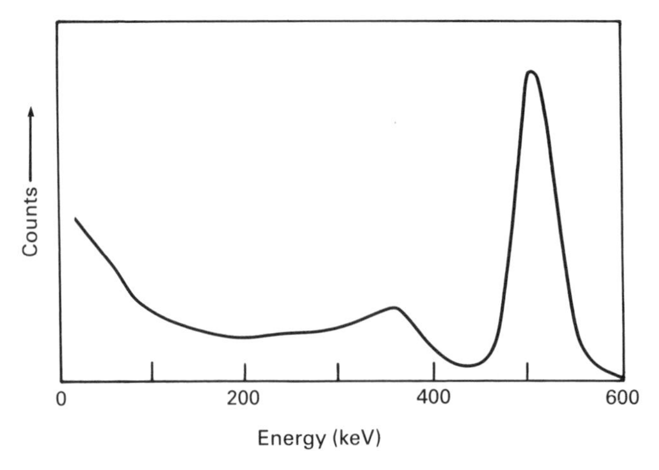

Figure 12.10. Energy spectrum of a ^{22}Na source measured with a NaI(Tl) scintillator coupled to a photomultiplier tube.

proportional counter is possible; an example of a spectrum is shown in Figure 12.10. It is important to point out that the time resolution of scintillation detectors is much better than that of proportional counters so they are the choice when this is a requirement.

c. *Solid-state detectors*

A simple solid-state detector might consist merely of a parallel-plate capacitor with a piece of pure (intrinsic) semiconductor as the dielectric (see Figure 12.11): it would work as follows. Ionizing radiation causes the production of electron–hole pairs. This is analogous to the formation of electron–ion pairs in a gas-filled detector. The electrons and holes are then attracted to the anode and cathode, respectively, and serve as an indication of the presence of ionizing radiation. The number of electron–hole pairs detected indicates the energy of the radiation. Geometry problems would not be as serious, as they were for the proportional counter, since the

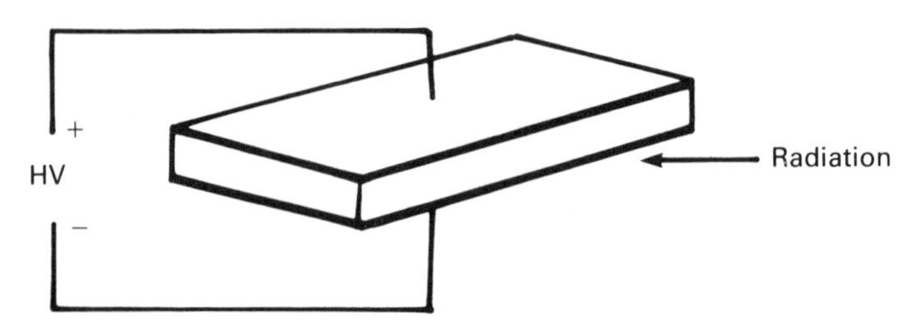

Figure 12.11. The geometry of a simple solid-state radiation detector.

mobilities of the electrons and holes are very similar. However, there are some problems with this kind of a detector. Assume, first of all, that we wish to use the detector at room temperature. We want a semiconducting material with a small energy gap in order to detect as many electron–hole pairs as possible: for this we would prefer germanium to silicon. However, the smaller energy gap in germanium means that at room temperature we will have a lot of electron–hole pairs that are caused by thermal excitations. That is to say the tail of the F–D distribution will overlap the valance and conduction bands considerably. Thus a current will flow that has nothing to do with the detection of radiation. In the semiconducting detector this is called leakage current and is analogous to the dark current in a PMT or photodiode. Silicon would be better so far as leakage current is concerned. The solution to this problem is to use germanium but to cool it down (usually to liquid-nitrogen temperature) to reduce the leakage current. Even so, there is a problem existing. To understand this, let us turn to Figure 6.7. Any semiconductor, no matter how pure we try to make it, will contain some impurities. So any semiconductor will have a conductivity or carrier concentration that is dominated at low temperature by the impurity carriers. To eliminate the problem of leakage current in germanium we would have to cool it to a temperature sufficiently low that the current would be dominated by the impurity carriers. There are solutions to this problem and it is possible to make a detector out of intrinsic germanium. However, there is another way of doing things—to use a semiconducting junction. We will discuss this approach first and then get back to intrinsic detectors. We discuss three types of junction detectors: the diffused junction detector; the surface barrier detector; and the lithium-drifted detector.

Diffused junction detectors. Diffused junction detectors are diodes constructed in a very unusual way. We begin with a piece of silicon that has been doped with boron. Since the boron is an acceptor, the semiconductor is a p-type material. One surface of this p-type material is exposed to a vapor containing phosphorus. Since the phosphorus has a valence of 5 (silicon's is 4), it is a donor impurity. The phosphorous diffuses into the surface and, in the region where the diffusion occurs, the concentration is greater than the concentration of boron and therefore converts the p-material to an n-material. The depth of the diffused region is typically 0.1 to about 2.0 μm. Enough phosphorus is diffused so that the concentration of majority electrons in the n-type region is much greater than the concentration of the majority holes in the p-type region. This means that the depletion region extends much further into the p-type region than into the n-type region; see Figure 12.12. Electrical contact is made to either side of the diode through a thin layer of metal (usually gold). deposited on the surface.

These detectors are operated with a reverse bias for the same reasons that photodiodes are reverse biased. They have some practical use as detectors

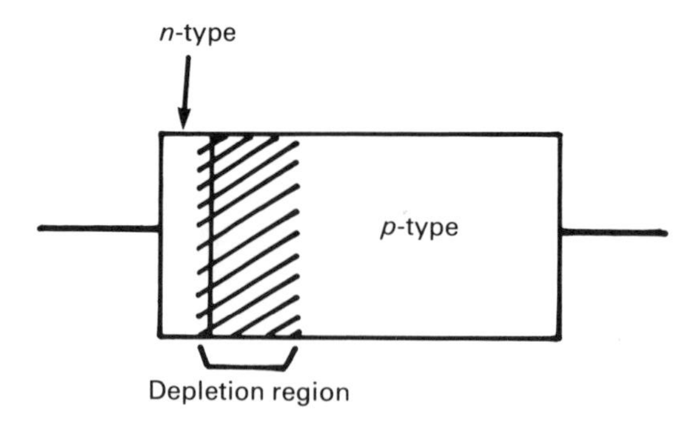

Figure 12.12. The diffused junction semiconducting radiation detector.

of α-particles or other heavy charged particles. As their depletion region is small, they are of extremely low efficiency in detecting γ-rays. As the depth of the diffused layer is somewhat difficult to control in the fabrication of these devices, there is a potential difficulty. The layer of n-type material that is not part of the depletion region is not sensitive to ionizing particles, but the particles must pass through this layer before being detected. They will lose some energy in the process and it is somewhat difficult to relate the detected energy to the incident energy. For this reason, the surface barrier detector is often used.

Surface barrier detectors. A semiconductor with a depletion layer essentially at the surface can be formed as follows. A piece of n-type silicon is allowed to oxidize on the surface. The oxidized p-type material behaves like an n-type material (recall the the methods of making thermistors). Electrical contact is made by a thin layer of gold on the oxidized surface. The gold contributes in some way to the formation of the depletion region, but the process is not well understood. As with the junction detector, the surface barrier detector is adequate for heavy charged particles, but the small detector volume results in an extremely low efficiency for γ-rays. Two alternatives are discussed below.

Lithium-drifted detectors. Lithium-drifted detectors are similar to diffused junction detectors but are nearly always made from germanium because its higher atomic number than that of silicon makes it more suitable for detecting γ-rays. "Pure" germanium is a p-type material because it is impossible to remove all of the boron and aluminum acceptor impurities. We begin with a piece of "pure" p-type germanium. Lithium is diffused into it from one end (as the phosphorus was in the diffused junction detector). Lithium has extremely high mobility in most solids (10^7 times that of boron or phosphorus) and in germanium (or silicon) it behaves as a donor, creating an n-type region. After diffusion of the lithium into the p-type germanium, the crystal looks like that shown in Figure 12.13*b*. The initial

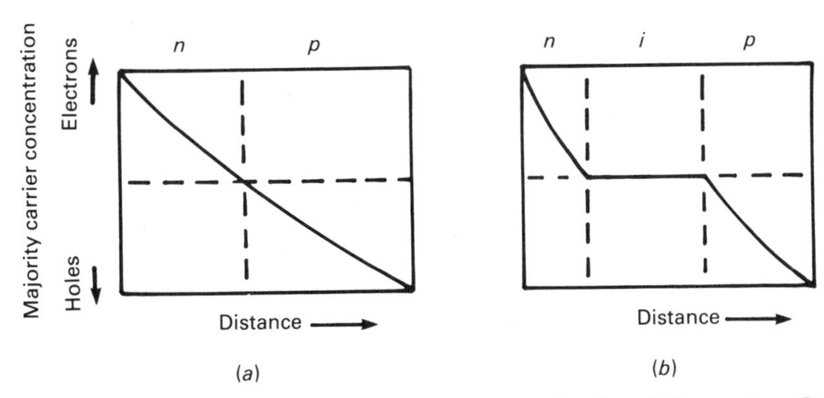

Figure 12.13. Majority charge carrier concentration as a function of distance in a Ge(Li) detector (*a*) before and (*b*) after drifting.

diffusion is generally done at an elevated temperature (say, 500°C). The crystal is then returned to room temperature, although even here the mobility of the lithium is still quite high. A large reverse bias is now applied to the semiconductor. A large electric field is set up across the depletion region and the lithium diffuses in such a way that a large charge-free region is formed around the junction. This acts rather like an intrinsic region and for this reason this is sometimes called a *pin* detector (recall the *pin* photodiodes). It is more commonly called a Ge(Li) detector (pronounced Jelly!). The I region is the radiation-sensitive part and can be made large enough to yield a good γ-efficiency. These detectors must be used at low temperature (usually liquid-nitrogen temperature) in order to reduce the leakage current due to thermally excited ionization. These detectors must be kept at low temperature not only when they are being used, but at all times, to keep the lithium from diffusing and destroying the *pin* configuration. This requirement is an inconvenient feature of these detectors. The intrinsic detector has eliminated this problem.

Intrinsic germanium detectors. Intrinsic germanium detectors are diode-like devices that have a very large depletion-like region formed by high-purity "intrinsic" germanium. Germanium can be purified to an impurity level of about 10^{-6} p.p.m.; this amounts to about 10^{10} impurities per cubic centimeter. The pure germanium is slightly p-type owing to presence of residual acceptor impurities. The general arrangement of the detector is shown in Figure 12.14: the intrinsic material is designated p-type while the contacts are made by p^+ and n^+; the "+" indicating a heavily doped material. The n^+ and p^+ regions are diffused into the p-type germanium or are merely deposited onto the surface. The device is reverse biased as shown in Figure 12.14 and the intrinsic "p" region acts as the detector. The device has a major advantage over the Ge(Li) detector—we do not have to keep it at low temperature to avoid diffusion of the

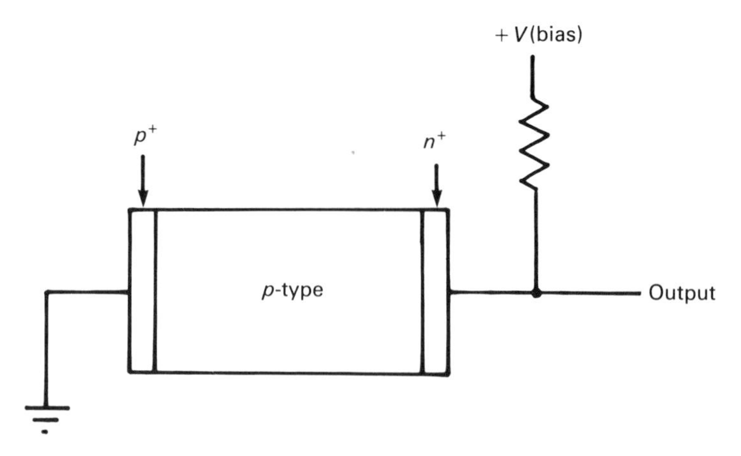

Figure 12.14. The intrinsic germanium radiation detector.

impurities; we can store it at room temperature. However, we still have to cool it (usually with liquid nitrogen) in order to use it; this is necessary to reduce the leakage current to an acceptable level. These detectors tend to be more expensive because of the cost of producing high-purity germanium.

These types of detectors generally have exceptional energy resolution in comparison with the other types of detectors we have discussed (an example is illustrated in Figure 12.15), so when energy resolution is a crucial factor and absolute efficiency is not, a solid-state detector is the clear choice.

While the Ge(Li) and the intrinsic germanium detectors have become the most widely used, a few other semiconductors are sometimes used to make detectors similar to the intrinsic type. Some of these other materials have the advantages that they can be used at room temperature and that they are more efficient. The first feature results from a larger energy gap (i.e., less leakage current) and the second from a higher atomic number. Some such semiconducting materials are described in Table 12.1. As you can see, the disadvantage of such materials is a much poorer energy resolution.

12.2 Nuclear instrumentation electronics

Here we discuss the basics of some of the more common instrumentation used to process signals from nuclear detectors. We begin with a brief discussion of the different kinds of signals.

a. Types of signals

Linear signals. Linear signals contain information, usually about particle energy, in the form of their amplitude. There are a variety of types of linear

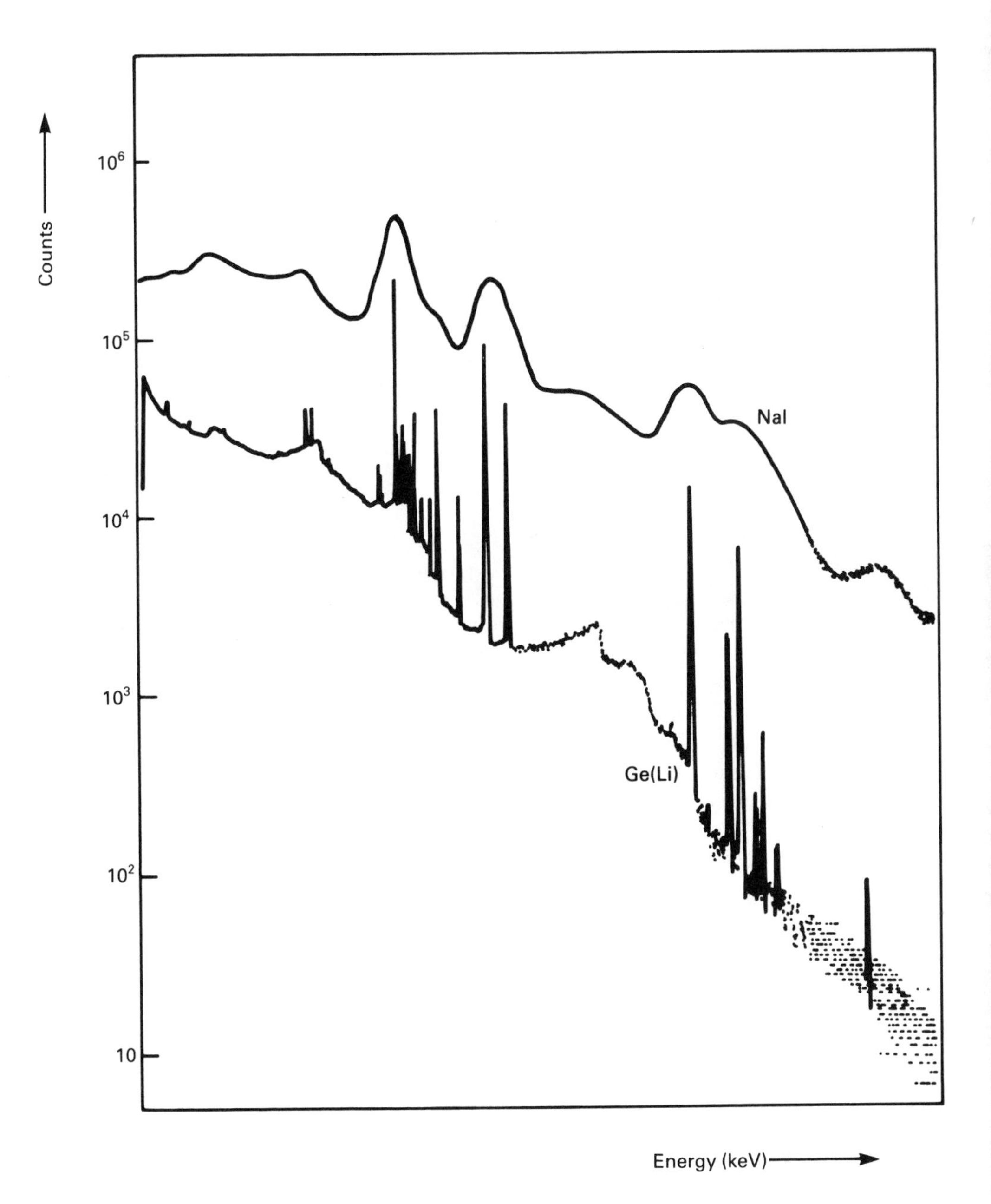

Figure 12.15. Energy spectra for the decay of ^{108m}Ag and ^{110m}Ag. A comparison of the spectra obtained using a NaI scintillator and a Ge(Li) solid state detector. (From G. Knoll, *Radiation Detection and Measurement*. Copyright © 1979 by John Wiley & Sons. Used with permission.)

Table 12.1. Some semiconducting detector materials used in detector

Material	Average Z	$\mathscr{E}_g$ (eV)	Resolution (% at 100 keV)
Ge	32	0.74	0.4
CdTe	50	1.47	3.1
HgI_2	62	2.13	2.9
GaAs	32	1.43	2.1

pulses that are distinguished by their pulse shape. For the present discussion we will not worry about the details of the pulse shape. Linear pulses are usually in the range of 0 and 10 volts with 10 volts being the amplitude of the pulse representing the maximum value of the quantity (e.g., energy) of interest.

Logic signals. Logic signals are pulses that merely represent an ON or OFF state, that is, a "logical 1" or "logical 0." As long as the amplitude of the pulse is within certain limits, the pulse is classified as either "1" or "0": its amplitude represents no more information that that. There are two kinds of logic pulses used in standard nuclear instruments: standard positive pulses and fast negative pulses. The voltage levels for these are given in Table 12.2. In addition to their voltage levels, these pulses are distinguished by their rise time and duration.

Nuclear electronics components, even when made by different manufacturers, are generally compatible with one another. They generally share their source of DC power and most pieces of equipment will plug into a standard power supply (preamplifiers are usually an exception). There are two standard systems of nuclear instruments:

1. The NIM (nuclear instrument module) system
2. The CAMAC (computer automated measurement and control) system

The power supply unit into which the instruments are plugged and from which they obtain their source of DC power is called a bin for the NIM system or a crate for the CAMAC system. Although the CAMAC system is newer, you are more likely to encounter the NIM system in a teaching laboratory.

Each NIM module connects to the NIM bin by means of a 42-pin

Table 12.2. Standard logic levels (volts)

Pulse type	"Logical 0"	"Logical 1"
Standard positive	−2 to +1.5 V	+3 to +12 V
Fast negative	−0.2 to +1.1 V	−0.6 to −1.8 V

Table 12.3. NIM connections

Pin no.	Voltage
10	+6 VDC
11	−6 VDC
16	+12 VDC
17	−12 VDC
28	+24 VDC
29	−24 VDC
33	117 VAC (hot)
34	AC ground
41	117 VAC (neutral)
42	DC ground

connector. Only 10 of these pins are necessary to supply power to the modules; the others are spares and can be used to carry a variety of signals. The identities of the various pins are given in Table 12.3.

The CAMAC system is more complex and we will not go into further detail here. The interested reader is referred to *IEEE Trans. Nucl. Sci.* **NS-20,** 2 (1973).

b. Nuclear instrumentation

We can classify all nuclear instruments according to the kind of signals they require for input and the kind of output they produce. The more common components are given in Table 12.4 along with a description of their input and output signals. We consider the operation of each in turn.

Preamplifiers. Preamplifiers generally do not plug into a NIM bin but are situated close to the output of the detector. They do, however, normally obtain their DC voltage from the bin, generally via a cable plugged into a connector on the back of the amplifier. The purpose of the preamplifier is

Table 12.4. Input/output signal specifications for common nuclear instruments

Module	Input	Output
Preamplifier	Linear	Linear
Amplifier	Linear	Linear
Delay	Linear	Linear
Linear gate	Linear + logic	Linear
Discriminator	Linear	Logic
Single-channel analyser	Linear	Logic
Coincidence (fast) gate	2 or more logic	Logic
Anticoincidence gate	2 or more logic	Logic
Time–amplitude converter	2 logic	Linear

primarily to increase the size of the voltage pulse from the detector. As the output from most detectors is a charge pulse rather than a voltage pulse *per se,* most nuclear preamplifiers are charge-sensitive devices and convert this charge to a voltage by means of a capacitor on the input. Some pulse-shaping can be provided by the preamplifier as well. It is important to have a very high input impedance for the preamplifier and most modern devices of this type have a FET input stage.

Two types of preamplifiers—biased and unbiased—are commonly encountered in nuclear physics instrumentation. In biased preamplifiers, the high voltage supply for the detector is connected *through* the preamplifier. These are used for gas-filled detectors and solid-state detectors. For solid state detectors the preamplifier is often connected permanently to the detector, to eliminate noise. In fact, it is not uncommon for the input FET circuit to be attached to the same cold-finger as the detector crystal in order to reduce thermal noise in the input stage. Unbiased preamplifiers are used in cases where the detector bias is connected directly to the detector, as in the case of photomultiplier tubes.

Amplifiers. Amplifiers accept pulses from a preamplifier. Most amplifiers can be set to accept either negative or positive linear input pulses (or both) but will produce only positive output pulses. Some pulse-shaping is provided as well. The gain of the amplifier is adjustable (that of the preamp generally is not) and is generally set so that the region of energy (or whatever) of interest produces output pulses in the range of 0 to +10 volts. The particular gain required varies greatly, depending on the particular situation, but voltage gain values in the range 10 to 1000 are typical.

Delays. Delay boxes provide a delay for the pulse in a particular part of the circuit. With circuits that involve precision time measurements it is essential to consider the delay introduced by the finite velocity of signal propagation in the wires. For standard coaxial cables the rule of thumb is "1 ns per foot." Delay boxes can provide an adjustable delay. Delays of up to a microsecond can be provided by passive delay boxes. These are essentially just coils of wire. A variety of active circuits are used for longer delays.

Linear gates. Linear gates have two inputs: one is a linear signal, the signal of interest, and the other is a logic signal. The linear gate allows the linear pulse to pass from the input to the output *unaltered* when a "logical 1" pulse is applied to the gate input at the same time. If there is a "logical 0" at the gate input then any linear pulses applied to the input will not be passed to the output.

Discriminators. Discriminators accept a linear input. If the voltage of an input pulse is above some specified level then the discriminator will produce a single "logical 1" pulse at the output. The level at which the discriminator will trigger can be adjusted and on some units may be as low as ~10 mV, although ~100 mV is more typical. This is a particularly useful device for producing standard logic pulses that are synchronized with linear pulses, and for eliminating low-voltage noise pulses.

Single-channel analyser (*SCA*). The SCA consists essentially of two discriminators: one, as just described, is the *lower-level discriminator* (LLD) and the second, which blocks all signals above a maximum settable voltage, is the *upper-level discriminator* (ULD). Hence all input linear signals between some lower value V_L and some upper level V_U trigger the SCA and a standard logical 1 pulse is output. Values of V_L and V_U can typically be adjusted between ~100 mV and 10 V. Two methods of setting these levels are used. The first is the straightforward method of setting V_L and V_U independently via separate potentiometers. The second method is to set V_L with one potentiometer; a second potentiometer sets the width of the voltage window ΔV so that $V_U = V_L + \Delta V$. In this way both the lower threshold and the width of the window can be adjusted independently.

(*Fast*) *coincidence gates and* (*fast*) *anticoincidence gates*. Fast coincidence gates are essentially multiple-input logical AND gates. Most commonly, coincidence and anticoincidence gates are incorporated into a single unit. There are generally several inputs (say, four) and each can be specified as "coincidence" or "anticoincidence." A "logical 1" pulse is produced at the output when all of the gate "conditions" are met. An example of a four-input fast coincidence/anticoincidence gate is illustrated in Figure 12.16.

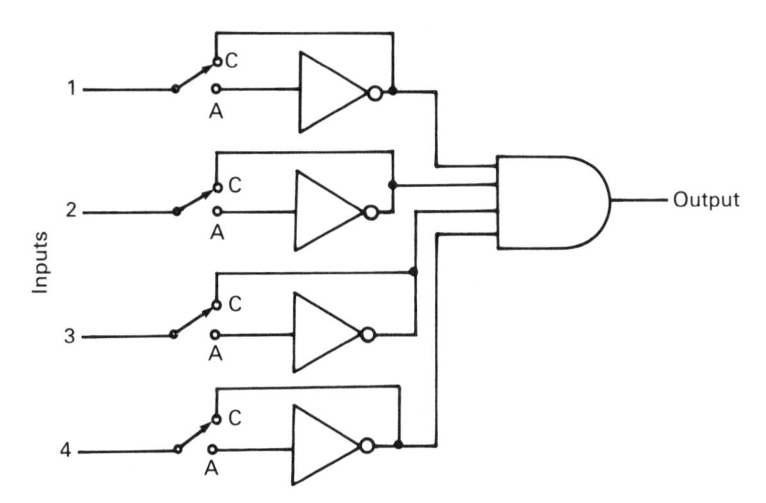

Figure 12.16. Schematic diagram for a four-input fast coincidence/anticoincidence gate. Each input can be set to coincidence (C) or anticoincidence (A) mode.

Time-to-amplitude converters. Time-to-amplitude converters are also known as TACs, T/As or time-to-pulse-height converters (TPHC). These are the most elaborate of the nuclear instruments described here but are particularly useful. A logical start pulse is input into the "start" input. This begins a clock in the TAC. Then when a logic pulse is input into the "stop" input, the clock is stopped, and a single linear pulse is output. The voltage of this linear pulse is proportional to the length of time elapsed between the start and stop pulses. The maximum time interval of interest, τ_{max}, can be chosen on the TAC; τ_{max} might be adjustable between, say, 100 ns and 100 ms. The output will be in the range of 0 to 10 V, with a time difference between input "start" and "stop" pulses of τ_{max} producing an output linear pulse of +10 V; shorter times will produce linear outputs that scale proportionately. A useful circuit for testing the linearity of a TAC and for determining unknown time delays in an experiment is illustrated in Figure 12.17. The voltage of the linear TAC output pulse will be proportional to the introduced delay (τ_d) from the delay box plus any unknown time delay in the circuit (τ_u). This relationship is given as

$$V_{out} = \frac{10(\tau_d + \tau_u)}{\tau_{max}} \text{ volts} \tag{12.17}$$

The MCA in the PHA mode collects counts only in the channel corresponding to this voltage. Typically we would have (ν = total number of MCA channels)

$$X(\text{Ch}) = \frac{\nu V_{out}}{10} = \frac{\nu(\tau_d + \tau_u)}{\tau_{max}} \tag{12.18}$$

An example of the results of this kind of experiment for measuring X(Ch) as a function of τ_d is given in Figure 12.18.

We should note that if a *stop* pulse is not observed after $t > \tau_{max}$ then the TAC resets itself without outputting a pulse and waits for another *start* pulse.

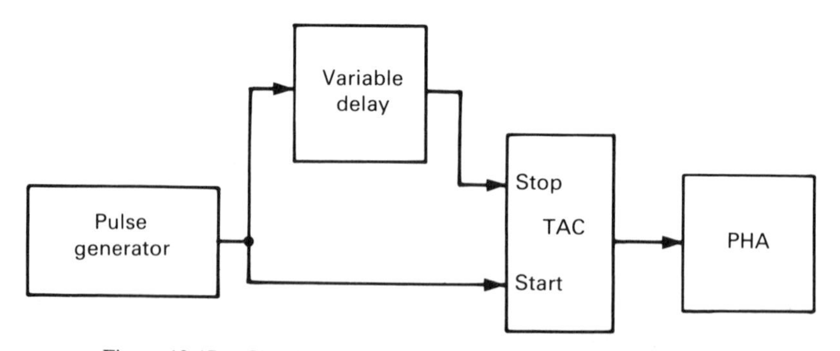

Figure 12.17. Circuit for calibrating a time-to-amplitude converter.

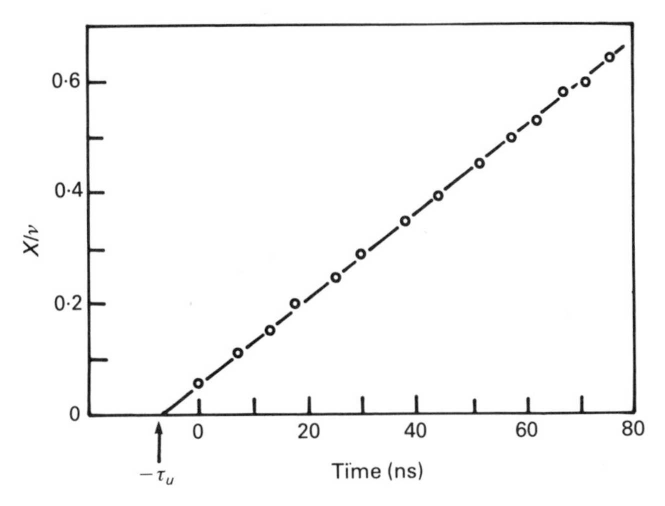

Figure 12.18. Data obtained from the linearity check of the TAC circuit in Figure 12.17. The graph shows the method of determining the unknown time delays in the circuit.

12.3 Nuclear measurement systems

Numerous experiments can be done with the instruments described in the previous section. Here we describe only two of the more useful ones. They will give you ideas about how other experiments might be set up. The ORTEC manual (see the Bibliography) has some good experiments and explains their operation in detail.

a. Nuclear counting with a proportional counter

Figure 12.19 shows a typical system for counting radiation from a source. The pulse from the proportional counter is amplified and passed through an SCA before counting. A standard frequency counter may be used here. The

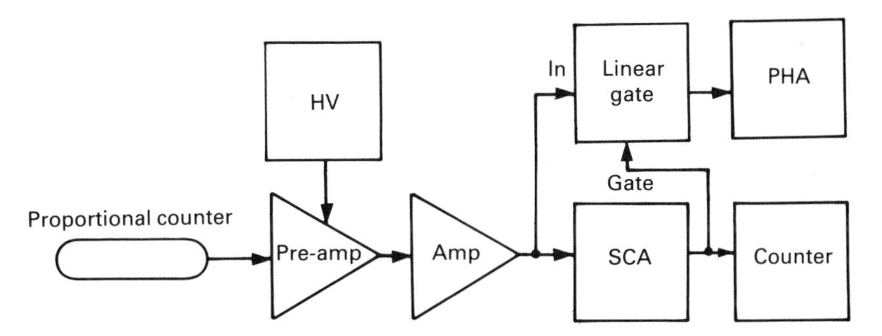

Figure 12.19. A nuclear counting system using a proportional counter. A SCA and a linear gate are used in conjunction to allow for the counting of only those energies of interest.

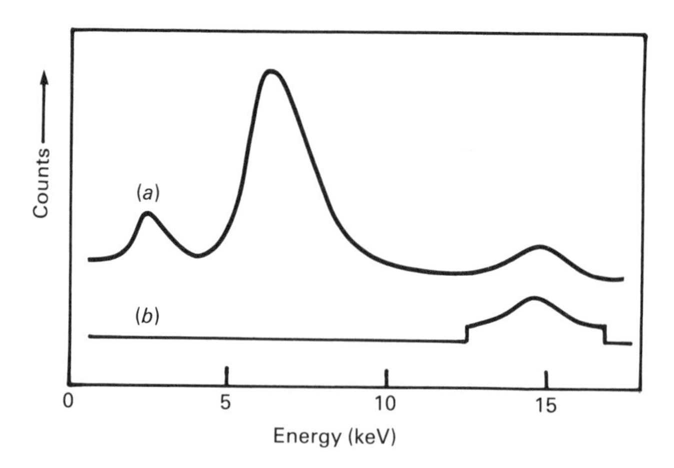

Figure 12.20. Low energy portion of the energy spectrum of ^{57}Co obtained using the circuit in Figure 12.19 and a Kr–CO_2 proportional counter. See Figure 12.22 for the energy-level diagram of this source.

output of the counter will tell us the total number of particles detected within an energy range specified by the SCA. The energy spectrum can be obtained on an MCA in the PHA mode by looking directly at the amplifier output. Various energies in the PHA spectrum may be selected by using a linear gate in conjunction with the SCA. The SCA produces gate pulses when the energy lies within a specified range that allows the linear gate to pass the linear pulses from the amplifier to the MCA. Figure 12.20 illustrates some spectra of the ^{57}Co (EC)^{57}Fe transition. In spectrum (*a*) the SCA was set to allow all energies to pass, and in spectrum (*b*) the SCA was set to allow only energies around the 14 keV peak to be recorded. Many MCAs have this kind of discrimination (LLD and ULD) built into their input.

b. Nuclear lifetime measurements

Chapter 11 explained how the lifetimes of nuclear states with moderately long lives (a few minutes to a few months) could be measured in a simple manner. Often, however, nuclear states have lifetimes of much less than one second. The method described here can measure these. The experimental apparatus is illustrated in Figure 12.21. The basic principle is as follows. As an example we consider the transition in ^{57}Co (EC)^{57}Fe and specifically the lifetime of the 14 keV state. From Figure 12.22 we see that a γ-ray of $136 - 14 = 122$ keV is produced each time the 14 keV state becomes populated. Some time after that (t_{mean} on the average) the 14 keV state will decay to the ground state, giving off another γ-ray. We set detector system 1 to look for the 122 keV γ-ray. We do this by properly adjusting the amplifier and SCA so that only the energies near 122 keV are

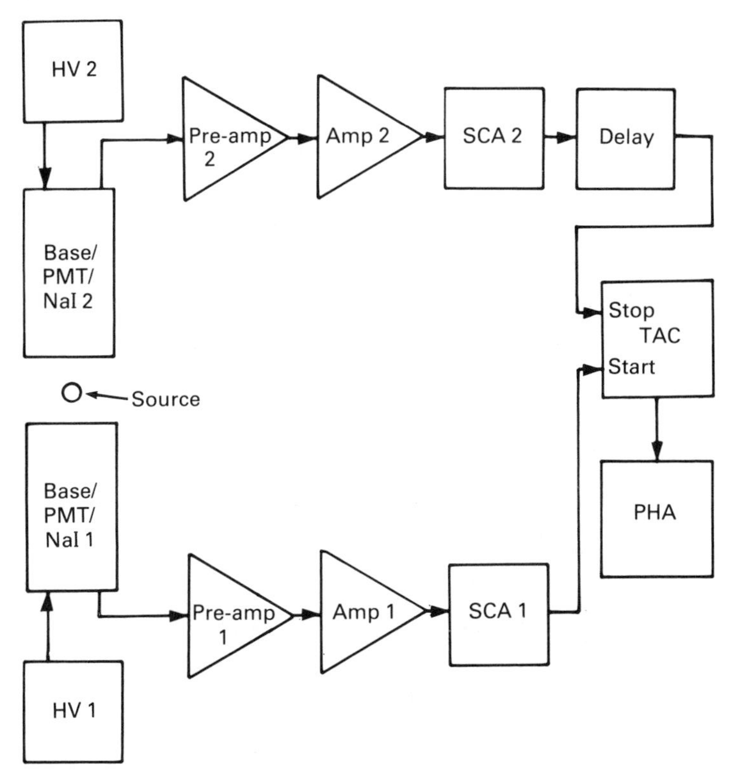

Figure 12.21. Circuit diagram for a coincidence experiment used to measure nuclear half-lives.

passed. A system like that in Figure 12.19 can be used to do this. The presence of a pulse tells us that the 14 keV state has been populated and we use it to start our TAC. Detector circuit 2 is set to observe only the 14 keV γ-rays and is used to stop the TAC. The time recorded by the TAC is the time between the detection of the γ-ray of these two energies and is the

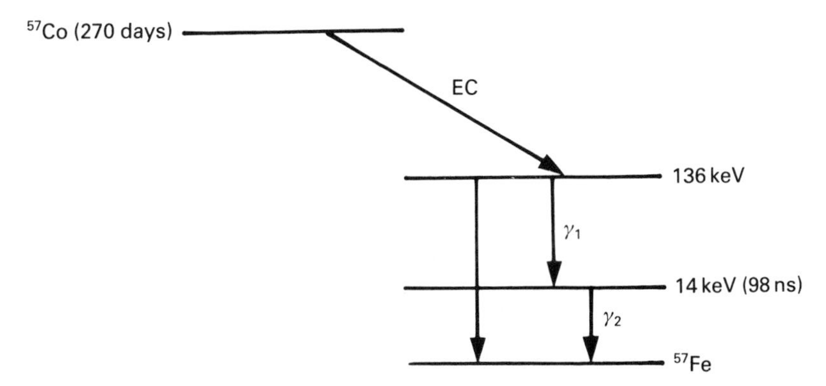

Figure 12.22. Energy-level diagram of a ^{57}Co source.

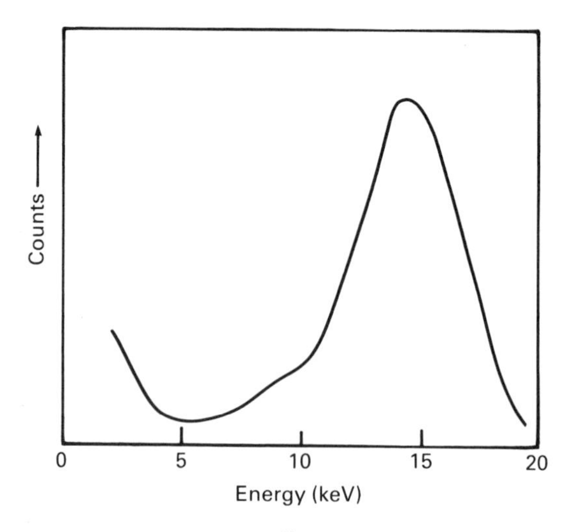

Figure 12.23. The low-energy spectrum of a ^{57}Co source obtained using a 1-inch diameter by $\frac{1}{4}$-inch thick NaI scintillator.

time that the nucleus remained in the 14 keV state. Figure 12.21 shows that NaI(Tl) scintillators coupled to PMTs are used as detectors. Because of the low energy, a thin ($\frac{1}{4}$-inch thick) NaI crystal is used to detect the 14 keV γ-ray and a larger (2×2)-inch crystal is used to detect the 122 keV γ-rays. Spectra obtained from these detectors are shown in Figures 12.23 and 12.24. As you can see, the proportional-counter spectrum (Figure 12.20) for the low-energy region is better resolved than the scintillator spectrum in Figure 12.23. It is generally not possible in this kind of an experiment to use a proportional counter because the response time is not sufficiently small.

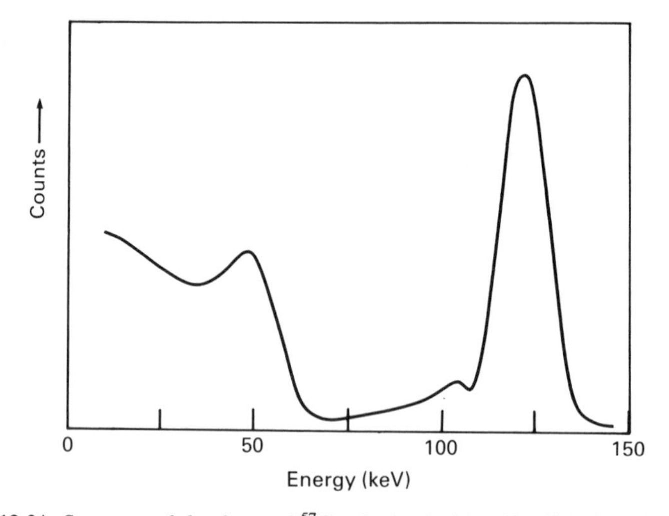

Figure 12.24. Spectrum of the decay of ^{57}Co obtained with a (2×2)-inch NaI scintillator.

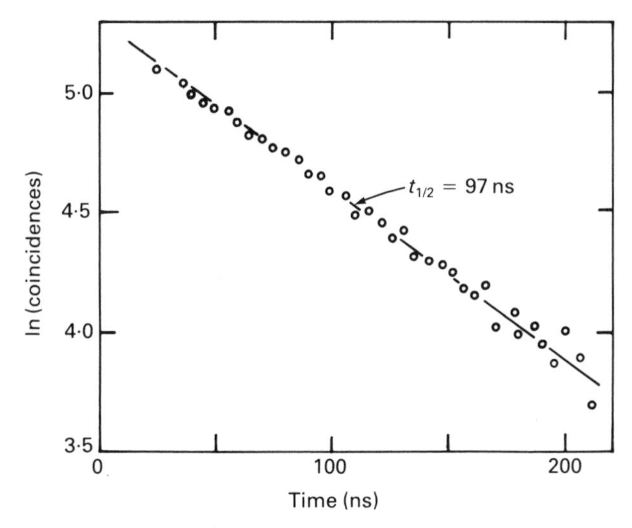

Figure 12.25. Count rate as a function of time obtained from the PHA in Figure 12.21 for the decay of the 14.4 keV state of ^{57}Fe.

Figure 12.25 shows the kind of data that would be recorded on the PHA. For the purpose of determining the mean lifetime, the location of $t = 0$ is not important. It is necessary, however, to find the proportionality between PHA channel number and time. This can be done using the method described in Figure 12.18. This proportionality is given by the slope of the line. The x intercept gives $t = 0$ but is not necessary for this experiment. For the data shown in Figure 12.25 it was found that one channel on the PHA corresponded to 0.73 ns (this, of course, will depend primarily on the setting of the TAC). The half-life is determined by the location on the curve at which the count rate drops by a factor of 2. Locating two such points on the data in Figure 12.25 yields, as the figure shows, $\tau_{1/2} = 97$ ns, in good agreement with the literature value of 98 ns.

13
X-ray diffraction methods

13.1. The production of x-rays

High-energy electrons produce high-energy photons by the process of Bremsstrahlung when they interact with matter. There will be a distribution of energies (or wavelengths), but there will be a high-energy cutoff determined by the kinetic energy of the electrons. See Figure 13.1 for an illustration of an X-ray continuum produced by the bombardment of a metal target with high-energy electrons. It is convenient to think of the wavelength of the X-rays rather than their energy. This gives the cutoff in terms of the energy $\mathscr{E}$ (in electron volts) of the bombarding electrons as

$$\lambda\ (\text{Å}) = 1.24 \times 10^{-4} \mathscr{E}^{-1}$$

If the X-rays produced are sufficiently energetic (of short enough wavelength) some of the energies will correspond to electronic transitions in the target metal that produce X-rays. Thus, superimposed on the continuum spectrum of Figure 13.1 we will see sharp lines corresponding to the energies of these atomic transitions. Figure 13.2 shows the situation for 35 keV electrons incident on a molybdenum target.

The transitions of interest in the transition metals (which we would usually use for targets) are shown in Figure 13.3. K_α and K_β transitions are those from the L and M shells, respectively to the K shell. These typically have wavelengths of about 1 Å; L lines have $\lambda \sim 5$ Å and are not often used. We confine further discussion to the K lines.

The K level consists of a single energy state. The L level consists of three closely spaced substates. Thus there are three different but very close K_α energies. In order of increasing wavelength (decreasing energy) they are referred to as $K_{\alpha 1}$, $K_{\alpha 2}$, and $K_{\alpha 3}$. Similarly, the M level has five substates, which result in five closely spaced X-ray lines referred to as $K_{\beta 1}$, $K_{\beta 2}$, $K_{\beta 3}$, $K_{\beta 4}$, and $K_{\beta 5}$. Because of the relative intensities, the only ones that are commonly used are $K_{\alpha 1}$, $K_{\alpha 2}$ and $K_{\beta 1}$. Table 13.1 gives wavelengths of these lines for common target materials. Often it is the K_α lines that are of interest for the measurement and it would be convenient to be able to eliminate the K_β radiation. A filter can be used with a sharp cutoff in its absorption coefficient at some wavelength between those of the K_α and K_β lines. An example of filtering copper radiation with a nickel filter is

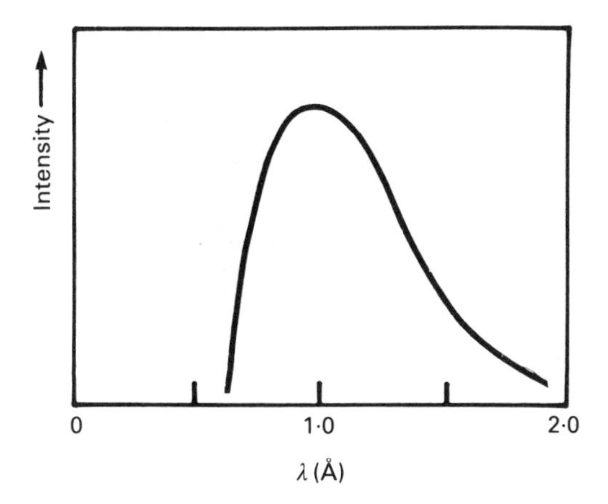

Figure 13.1. X-ray spectrum of 20 keV electrons incident upon a molybdenum target.

illustrated in Figure 13.4. Suitable filters for removing K radiation for common target materials are given in Table 13.2. The thickness is that required to reduce the K_β intensity to less than 0.2 percent of the K_α intensity.

Let us now consider the design of X-ray tubes. X-rays are produced by first producing electrons, usually from a filament, and accelerating them across a large potential (~30–50 kV) after which they are incident upon the target. A typical design is illustrated in Figure 13.5. As about 99 percent of the electron energy is converted not into X-rays but into heat, it is essential to water-cool the anode target. The X-rays are liberated from the anode in all directions but are only allowed to escape from the tube in specific

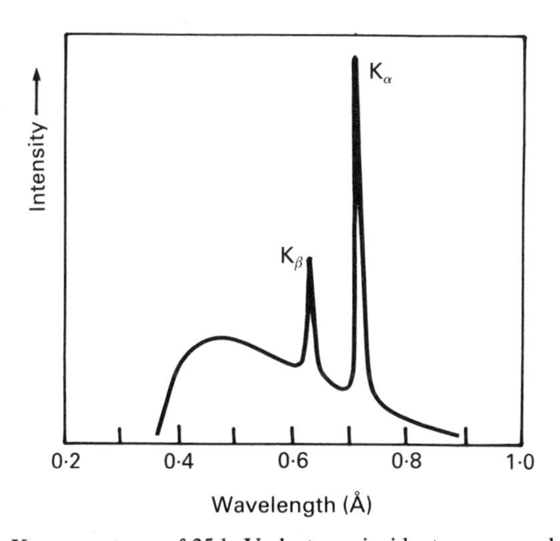

Figure 13.2. X-ray spectrum of 35 keV electrons incident upon a molybdenum target.

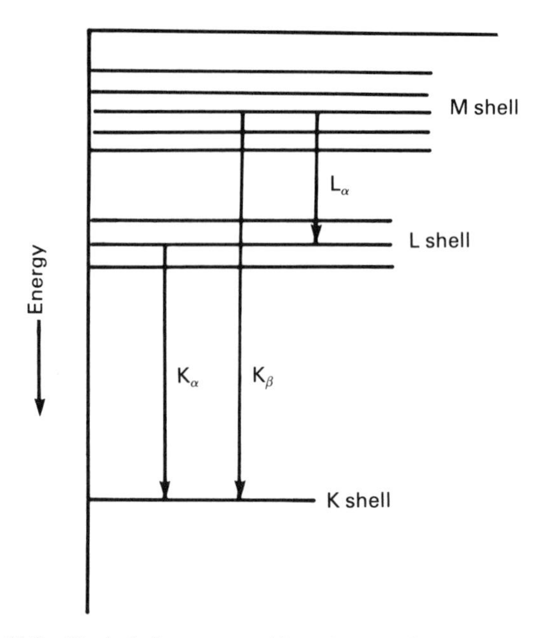

Figure 13.3. Typical electron transitions that can fill a K-shell vacancy.

directions determined by the location of beryllium windows. It is possible to use a rotating anode to increase the power output of an X-ray tube (see examples in Figure 13.6). This allows the production of X-rays from one area of the anode while the remainder of the anode is allowed to cool.

13.2 Diffraction of x-rays by crystalline materials

A crystalline material consists of atoms arranged in a periodic array. The set of space coordinates that define the locations of the atoms is referred to as the lattice. There are a variety of kinds of lattices, distinguished by their

Table 13.1. Wavelength in angstroms of characteristic X-rays

Target	$K_{\alpha 1}$ (Å)	$K_{\alpha 2}$ (Å)	K_α weighted average (Å)	$K_{\beta 1}$ (Å)
Ag	0.55941	0.56380	0.56084	0.49707
Mo	0.70930	0.71359	0.71073	0.63229
Cu	1.54056	1.54439	1.54184	1.39222
Co	1.78897	1.79285	1.79026	1.62079
Fe	1.93604	1.93998	1.93735	1.75661
Cr	2.28970	2.29361	2.29100	2.08487

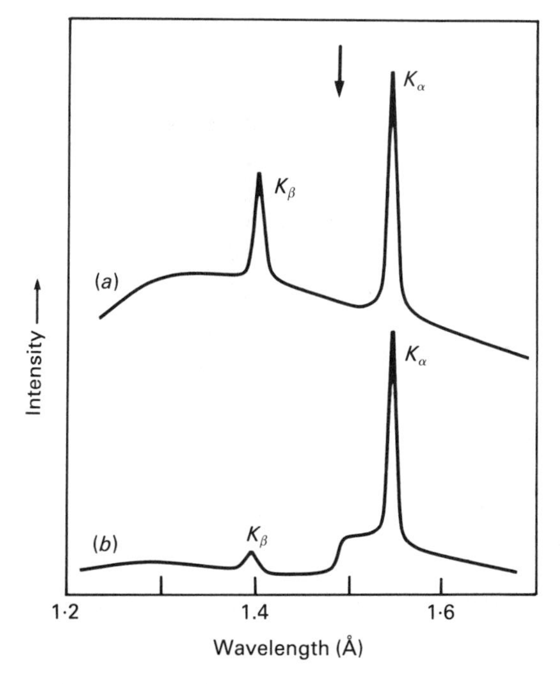

Figure 13.4. Comparison of copper X-ray spectra (*a*) without a filter and (*b*) with a nickel filter. The arrow indicates the position of the absorption edge of nickel.

symmetry. In three dimensions there are 14 different basic lattice symmetries called the 14 Bravais lattices; the simplest is the cubic structure. The 14 different structures are shown in Figure 13.7. The structures are distinguished by the relative distances between the atoms along three different (not necessarily orthogonal) axes and by the angles between these directions. Table 13.3 shows the conditions on these quantities for the different Bravais lattices. The a, b, and c are called the lattice parameters and the relationship between these directions and the angles α, β, and γ given in Table 13.3 are shown in Figure 13.8. The Bravais lattice type is also determined by the locations of atoms: for example, in the cubic case the

Table 13.2. Use of filters to remove K_β radiation

Target	Filter	Incident $I(K_\alpha)/I(K_\beta)$	Thickness (mm)	Percent K_α transmitted
Mo	Zr	5.4	0.109	30
Cu	Ni	7.5	0.020	42
Co	Fe	9.4	0.015	47
Fe	Mo	9.0	0.015	47
Cr	V	8.5	0.015	48

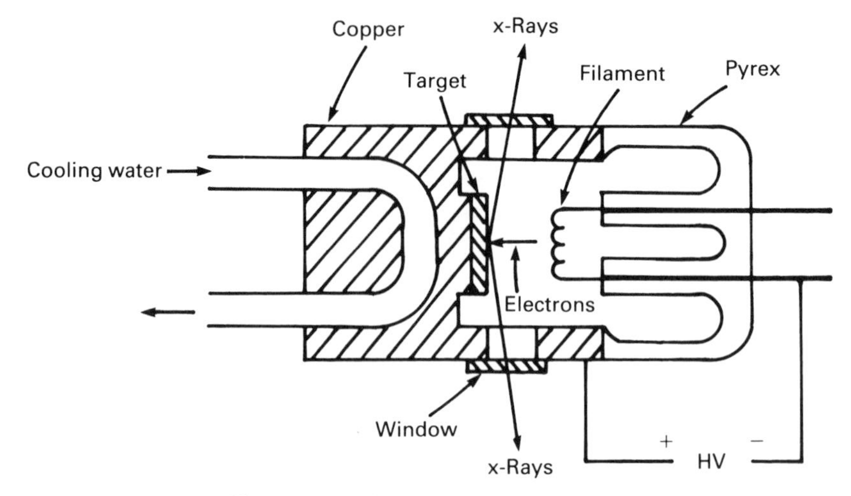

Figure 13.5. Construction of an X-ray tube.

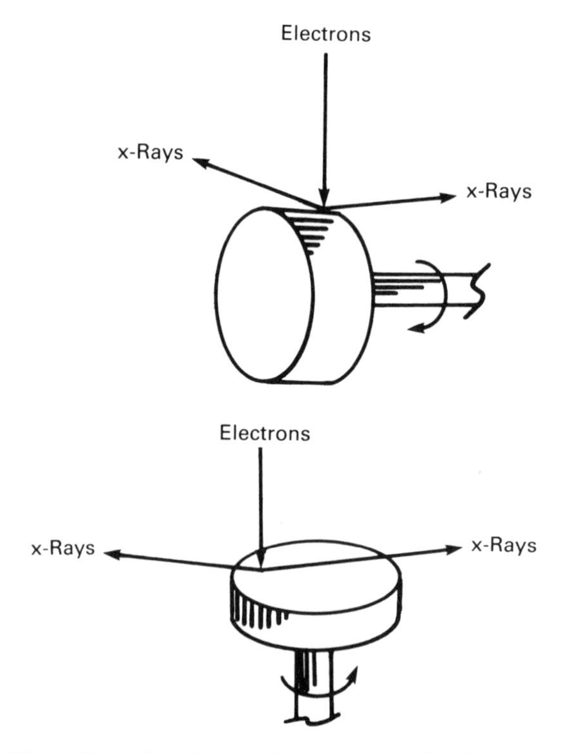

Figure 13.6. Rotating-anode arrangements for X-ray tubes.

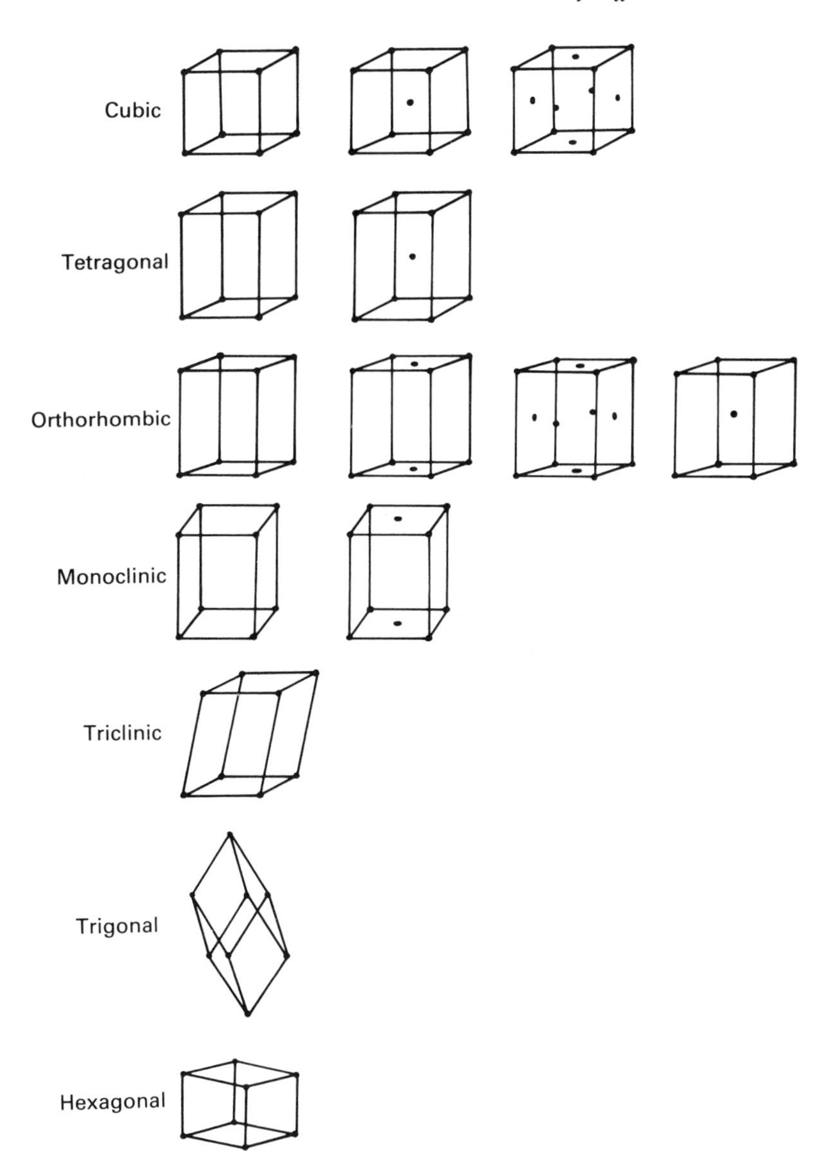

Figure 13.7. The fourteen Bravais lattices in three dimensions.

lattice can be simple, body-centered, or face-centered. For simplicity, the subsequent discussion of diffraction considers only the simple cubic lattice.

It is important to be able to locate directions within the crystal structure. To do this we establish a coordinate system within the structure with the origin located at one of the lattice points (since the structure repeats, the choice of lattice point is irrelevant). A right-hand coordinate system is set up in which x, y, and z are unit vectors along the three crystallographic

Table 13.3. Conditions on the lattice parameters and the angles for the various crystal systems in three dimensions

System	Conditions
Cubic	$a = b = c$ $\alpha = \beta = \gamma = 90°$
Tetragonal	$a = b \neq c$ $\alpha = \beta = \gamma = 90°$
Orthorhombic	$a \neq b \neq c$ $\alpha = \beta = \gamma = 90°$
Monoclinic	$a \neq b \neq c$ $\alpha = \beta = 90° \neq \gamma$
Triclinic	$a \neq b \neq c$ $\alpha \neq \beta \neq \gamma$
Trigonal	$a = b = c$ $\alpha = \beta = \gamma < 120°, \neq 90°$
Hexagonal	$a = b \neq c$ $\alpha = \beta = 90°, \gamma = 120°$

directions. The direction of a plane is designated by the three components of a vector drawn from the origin and normal to the plane. Figure 13.9 shows some examples of how to identify crystallographic planes. The three vector components are enclosed in parentheses and are referred to as Miller indices. A negative index is indicated by a *bar* over the appropriate quantity. Indices are given as integers, so we would refer to (211) rather than $(1\frac{1}{2}\frac{1}{2})$. The designation (200) would refer to planes parallel to (100) but half as far apart. The simple rules for determining the Miller indices of a plane are as follows.

1. Locate the intercepts of the plane on the x, y, and z axes (call them x_0, y_0 and z_0).

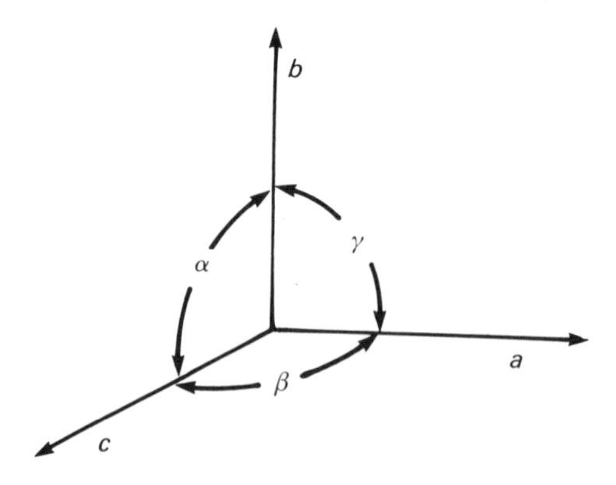

Figure 13.8. The relationship of lattice directions and angles for Table 13.3.

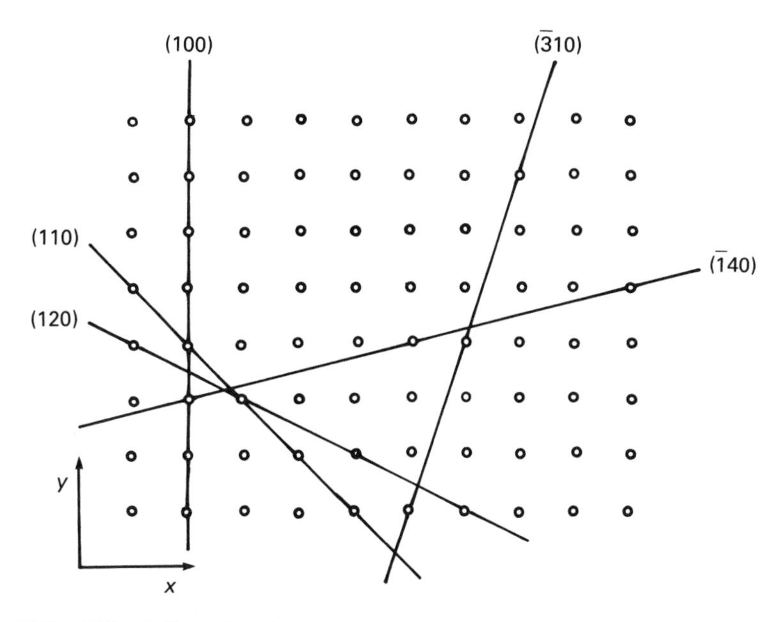

Figure 13.9. Miller indices of various lattice planes parallel to the z axis in three dimensions.

2. Take the reciprocals $1/x_0$, $1/y_0$ and $1/z_0$.
3. Reduce these to the smallest three integers by multiplying each by an appropriate quantity ξ.
4. The Miller indices of the plane are

$$(hkl) = \left(\frac{\xi}{x_0}\frac{\xi}{y_0}\frac{\xi}{z_0}\right) \tag{13.1}$$

Each crystal structure will contain a large number of planes (designated by different Miller indices) in which a regular array of lattice points can be found.

Now we are able to look at the X-ray diffraction from lattice planes. Consider two planes of atoms separated by a distance d as shown in Figure 13.10. We consider X-rays of wavelength λ incident upon the lattice at an angle θ to the planes. X-rays scatter from the two planes of atoms as shown in the figure. The beam that reflects from the second plane is phase-shifted relative to the beam that reflects from the first plane by an amount

$$\phi = 4\pi\frac{s}{\lambda} \tag{13.2}$$

where the distance s is given by

$$s = d \sin\theta \tag{13.3}$$

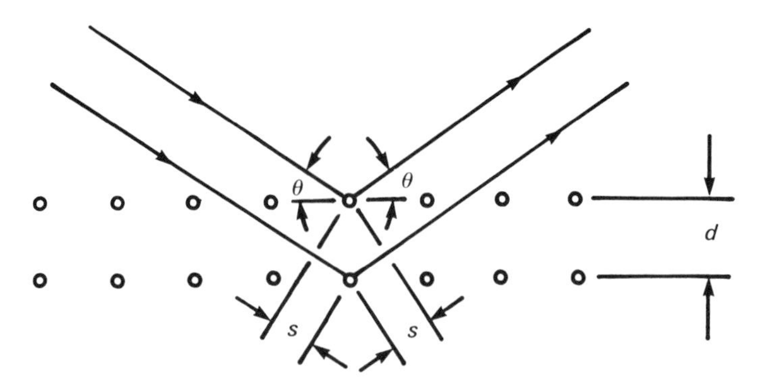

Figure 13.10. The scattering of X-rays from crystal planes.

Constructive interference occurs when

$$\phi = 2\pi n \tag{13.4}$$

for integer values of n. Equations (13.2) through (13.4) yield Bragg's Law for scattering of X-rays

$$n\lambda = 2d \sin \theta \tag{13.5}$$

These reflections can occur only for $n \geqslant 1$ and $\sin \theta \leqslant 1$ so

$$\lambda \leqslant 2d \tag{13.6}$$

For typical lattice spacing of a few angstroms, the necessary wavelengths are in the X-ray region. As you can see from Figure 13.9, any particular lattice structure has a variety of planes of atoms separated by different distances when the crystal is viewed from different angles. Thus, as we look at reflections at different angles we will see large X-ray intensities at angles that satisfy equation (13.5), while at angles that do not satisfy (13.5) we will see low X-ray intensities. Measuring the X-ray intensity as a function of angle will tell us the spacing of lattice planes in different directions. From this information the symmetry of the lattice and the lattice parameters can be determined. We describe experimental methods for doing this in the next section.

13.3 X-ray measurement techniques

a. Debye–Scherrer method

A powdered sample is used in this method. Monochromatic X-rays are incident upon the sample and both forward-scattered and back-scattered beams are observed, generally by means of a photographic film. The sample is packed inside a thin glass capillary tube or is attached with glue to the outside of a glass fiber. This sample holder is located along the axis of a

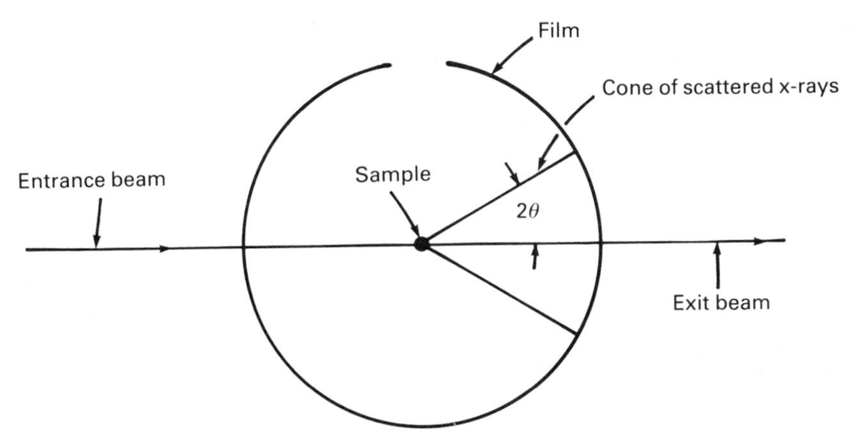

Figure 13.11. The geometry of the scattered X-ray beam in Debye–Scherrer camera.

cylindrical camera body. The film is in the form of a strip wrapped around the inside of the camera. The X-rays enter through a collimator and constructive scattering occurs in the shape of cones illustrated in Figure 13.11.

The crystallites in the sample are, in general, distributed in all directions.

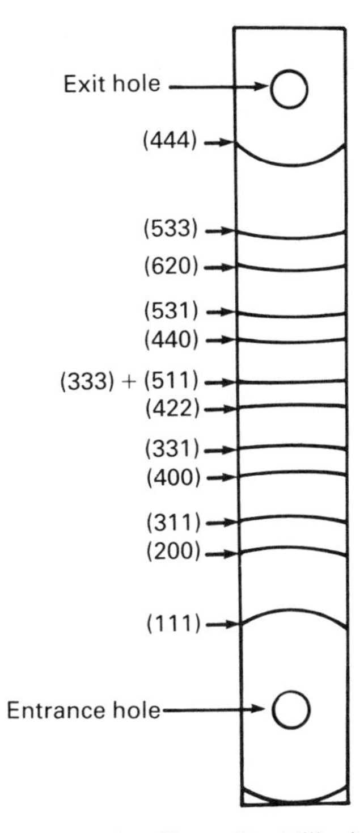

Figure 13.12. Debye–Scherrer pattern for silicon. The Miller indices of the various reflections are indicated.

The incident beam is in a fixed direction. Those crystallites with planes that are oriented at some direction θ relative to the beam direction, so that equation (13.5) is satisfied, will produce reflections of high intensity. Because of the symmetry, the reflections occur in the form of cones as shown in Figure 13.11.

Figure 13.12 shows the pattern on a typical Debye–Scherrer photograph. Note the two holes in the film to accommodate the entrance and exit collimators.

The type of crystal structure is determined by the existence of particular reflections and by their intensities. The lattice parameters are determined by the spacing of the lines. The location of the lines is determined from the Bragg Law, equation (13.5)); this is written in the form

$$\lambda\sqrt{h^2 + k^2 + l^2} = 2a \sin \theta \tag{13.7}$$

where h, k, and l are the Miller indices (hkl).

b. Laue photography

The Laue method uses single crystals and can have either transmission or backscatter geometry. Figure 13.13 shows the geometry for these two configurations. The transmission pattern is recorded by film B and the backscattered pattern by film A. The incident beam is directed through a hole in the backscatter film by a collimator (usually a stainless-steel tube). Newer cameras often use Polaroid film. In this case there is no actual hole in the film but rather the center portion is blocked off by a metal collar around the collimator.

As the incident beam direction is fixed, the reflections occur only along specific directions as narrow beams. Thus the diffraction pattern recorded on the film is a series of well-defined points. Figure 13.14 shows the set of

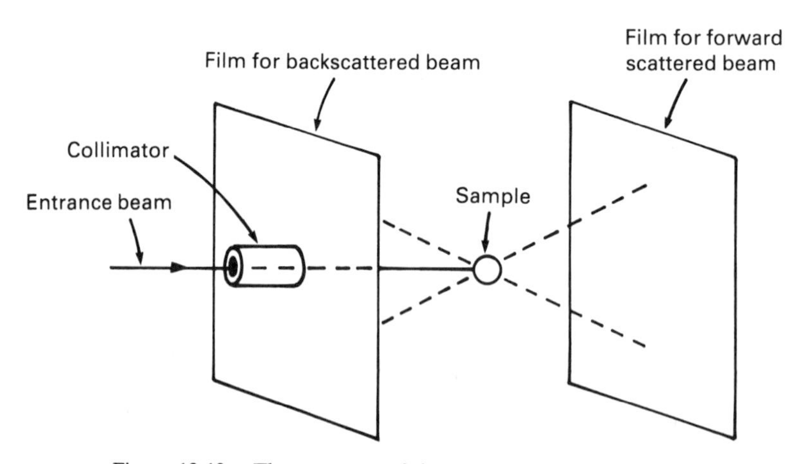

Figure 13.13. The geometry of the Laue single-crystal method.

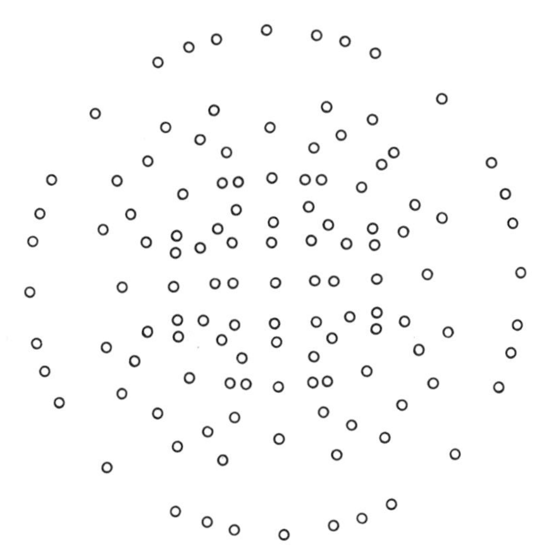

Figure 13.14. Pattern of lattice reflections arising from the simple cubic structure in the Laue method. The incident beam is normal to the (001) plane.

reflections for a simple cubic structure along the (001) direction. The size of the image is determined primarily by the distance between the crystal and the film. This is an exremely powerful method for the determination of crystal structure. The analysis of the photographs is usually done by a graphical means using a comparison of the photograph with points on a curved coordinate system similar to that in Figure 13.14. This requires knowing the crystal-to-film distance and this is generally fixed to 5 cm to allow for use of standard graphic charts.

Laue photography is the standard method for orienting single crystals. In order to do this, it is necessary to position the crystal precisely in the X-ray beam and to be able to rotate it between photographs to observe the diffraction pattern produced. To do this the sample crystal is mounted on a *goniometer.* Such a device allows for the precise rotation of the sample about three orthogonal axes.

c. *The scanning diffractometer*

The scanning diffractometer method uses a polycrystalline sample and is similar to the Debye–Scherrer method except that an X-ray detector (proportional counter or scintillation detector) is used to scan the X-ray intensity as a function of angle. This is the most accurate method of determining lattice parameters and the relative intensities of diffraction peaks. The basic geometry is shown in Figure 13.15. The detector is rotated about angle θ_2. It is important to note that θ_1 must be equal to θ_2, so in fact the sample is rotated relative to the X-ray tube by some angle $\Delta\theta$ while the detector is rotated relative to the X-ray tube by an angle $2\,\Delta\theta$.

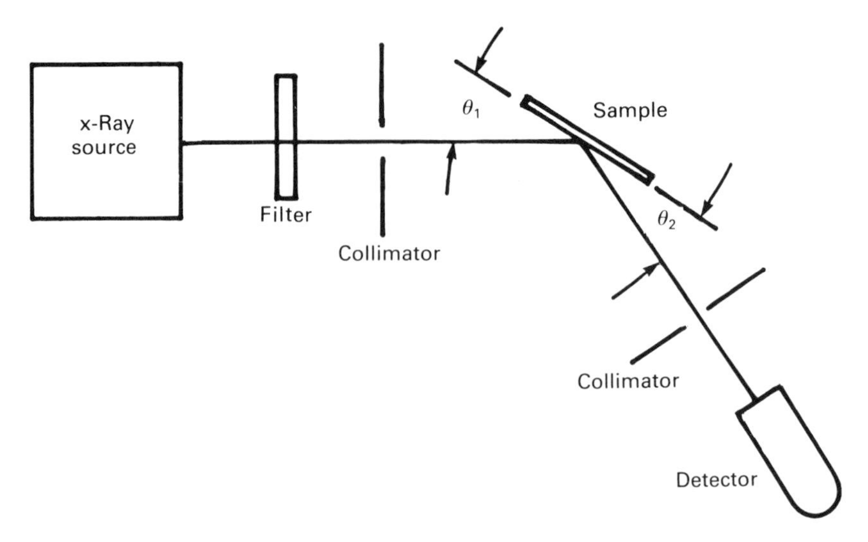

Figure 13.15. Schematic diagram of the scanning diffractometer.

The sample is generally a polycrystalline powder but can also be a polycrystalline foil. This is mounted on a glass or plastic plate (or sheet) and is held in a vertical plane in the sample holder (A in the figure). The particular machine illustrated has a vertical rotation axis. Some machines rotate the sample and detector about a horizontal axis, and would have the sample mounted accordingly.

A block diagram of the electronics for a scanning diffractometer is shown in Figure 13.16. The scanning of the angle is synchronized to a chart recorder timebase. The detected X-ray intensity is recorded on the chart recorder *Y* axis.

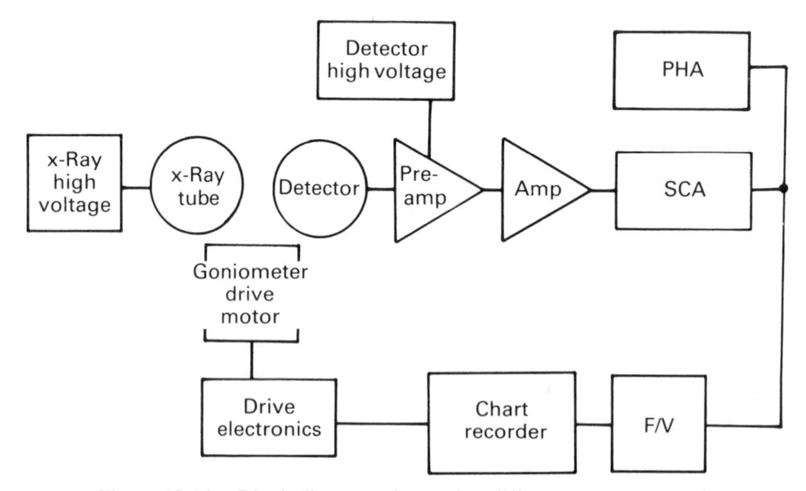

Figure 13.16. Block diagram of scanning diffractometer electronics.

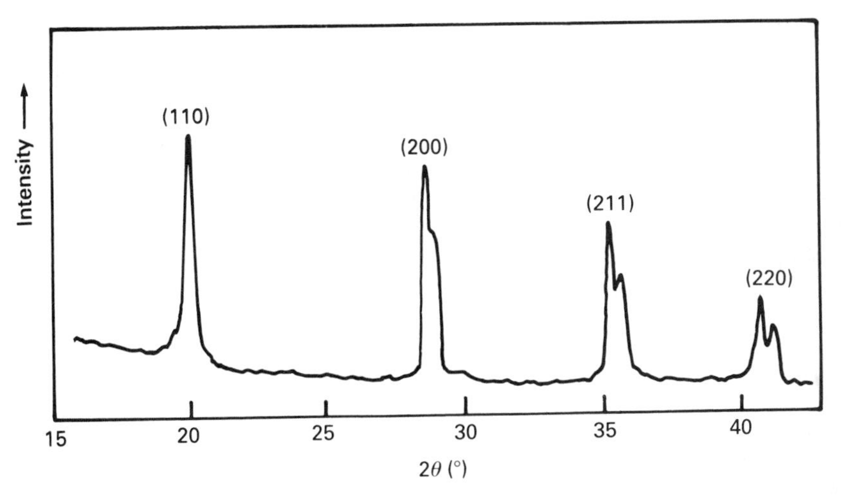

Figure 13.17. X-ray diffraction scan for polycrystalline α-Fe using Mo K_α radiation. The Miller indices for the reflections are given in the figure.

Diffraction lines can be measured for angles θ_2 (see Figure 13.15) down to about 6° or so. For angles less than this, the straight-through (unreflected) beam intensity is too large.

Figure 13.17 illustrates the kind of result that can be obtained from this type of machine. For high angles, the "twinning" of the peaks is due to the presence of the two slightly different wavelengths K_{α_1} and K_{α_2}.

14
Magnetic materials and measurements

14.1 Classification of magnetic materials

We can gain an understanding of the magnetic properties of materials by an investigation of electrons in solids. We know that moving electric charges produce magnetic fields. As the electron is charged and, in the classical sense, it moves, it produces a magnetic field. There are, in fact two motions to consider for each electron: its orbital motion and its spin. Quantum-mechanically speaking, we write these angular momenta of an atom as **L** and **S**, respectively, the total angular momentum being given by the vector sum

$$\mathbf{J} = \mathbf{L} + \mathbf{S} \tag{14.1}$$

The magnetic field produced by the electrons in atom is thus related to **J**. In real atoms we need to think about how all the magnetic fields due to the different electrons will add up. There are some simple rules. These rules stem, in part, from the Pauli exclusion principle and are called Hund's rules.

1. The total value of **S** will be the maximum allowed by the Pauli principle.
2. The total value of **L** will be the maximum consistent with rule 1.
3. The value of $|\mathbf{J}|$ is equal to $|\mathbf{L} - \mathbf{S}|$ when the shell is less than half-full and it is equal to $|\mathbf{L} + \mathbf{S}|$ when the shell is more than half-full. If the shell is exactly half-full, then $|\mathbf{L}| = 0$ and $\mathbf{J} = \mathbf{S}$.

One important manifestation of these rules is that filled shells do not contribute to the magnetic field. Of course, a complete picture of what is happening to the electrons would have to include the band structure discussed in Chapter 1. A simple analogy, however, would seem to indicate that elements with filled shells are not good conductors and that these elements would also not have strong magnetic properties. This is certainly true, if we confine our arguments to the elements.

From a magnetic standpoint, the transition metals are among the most interesting. As an example let us look at the 3d transition series. We will consider iron in detail. Iron has an outer electron configuration of $4s^2\,3d^6$. The 4s orbital is full and the 3d orbital has a total capacity of 10 electrons.

Table 14.1. The six 3d electrons in iron

Electron no.	L	S
1	−2	↑
2	−1	↑
3	0	↑
4	+1	↑
5	+2	↑
6	+2	↓
Total	+2	$4/2 = 2$

Note: The spin of an electron is $\frac{1}{2}$.

Let us see where the six 3d electrons go. By the Pauli principle there are five possible energy states for the 10 electrons. These five levels have corresponding values of L of −2, −1, 0, 1, 2. So the first five electrons go one each into the five different L states all with their spins parallel. Table 14.1 shows the situation. This is a result of the 1st rule. The last electron must go into one of the already occupied states with its spin antiparallel to the other electron in that state. According to rule 2 it will go into the $L = 2$ state. Since the shell is more than half-filled rule 3 tells us how L and S add up. We find from the table that $J = 4$.

Quantum mechanics tells us that the resultant magnetic moment of the atom is given by

$$|\boldsymbol{\mu}| = \mu_B g(J(J+1))^{1/2} \tag{14.2}$$

where μ_B is a constant known as the Bohr magneton and g is a factor near 2 that tells us something about the strength of the interaction coupling L and S.

We should point out here that the valence state of the atom is very important in determining its magnetic moment. Let us look at the simple case of free atoms, and assume $g = 2$. Now we can plot μ as a function of atomic number across the 3d series. This is shown in Figure 14.1. Details of the calculations are given in Table 14.2.

We see that the elements manganese through nickel, according to this simple model, have the largest magnetic moments. Experimental evidence also shows that this is the case.

Let us consider the magnetic properties of a large piece of material. If each atom of the material possesses a moment $\boldsymbol{\mu}$, then the piece of material as a whole possesses a magnetic moment, referred to as the *magnetization* **M**, which is the vector sum of the $\boldsymbol{\mu}$,

$$\mathbf{M} = \frac{1}{\mathrm{V}} \sum_{\text{all atoms}} \boldsymbol{\mu}_i \tag{14.3}$$

where V is the volume of the material. There is, as well, a small magnetic

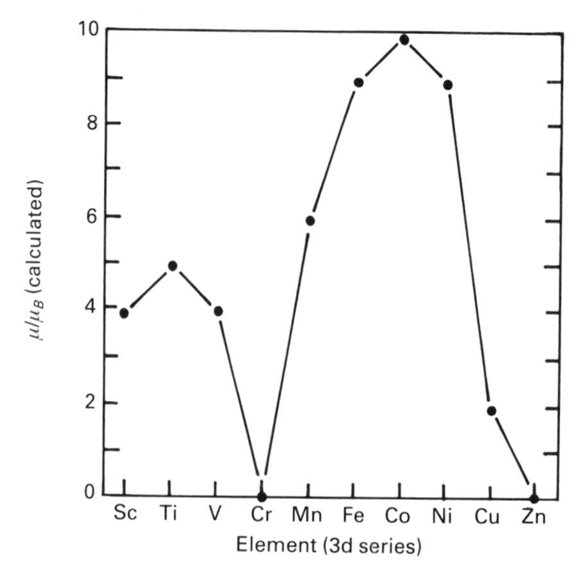

Figure 14.1. Magnetic moments of the 3d transition series calculated using Hund's rules.

moment associated with the nuclei for most atoms. In this simple discussion we will not be concerned with this small contribution.

We can now look in detail at the differences in the various kinds of magnetic materials.

a. Diamagnets

Diamagnets are truly nonmagnetic materials. They are nonmagnetic because each atom in the material, taken individually, has no magnetic

Table 14.2. The application of Hund's rules to 3d transition metals

Element	Z	Outer electrons	L	S	J	μ/μ_B
Sc	21	$3d^1$	2	$\frac{1}{2}$	$\frac{3}{2}$	3.9
Ti	22	$3d^2$	3	1	2	4.9
V	23	$3d^3$	3	$\frac{3}{2}$	$\frac{3}{2}$	3.9
Cr	24	$3d^4$	2	2	0	0
Mn	25	$3d^5$	0	$\frac{5}{2}$	$\frac{5}{2}$	5.9
Fe	26	$3d^6$	2	2	4	8.9
Co	27	$3d^7$	3	$\frac{3}{2}$	$\frac{9}{2}$	9.9
Ni	28	$3d^8$	3	1	4	8.9
Cu[a]	29	$3d^{10}\,4s^1$	0	$\frac{1}{2}$	$\frac{1}{2}$	1.7
Zn	30	$3d^{11}$	0	0	0	0

[a] For copper it is the 4s electron we need to consider.

moment. Hence

$$|\mathbf{M}| = 0 \qquad \text{because} \qquad |\boldsymbol{\mu}_i| = 0 \text{ for all } i \tag{14.4}$$

An important aspect of the behavior of any magnetic material is what happens to it when we place it in an external magnetic field. What happens to the diamagnet? Practically nothing. There is a very small induced moment that results from the interaction of the electron cloud with the applied field. This is called the diamagnetic effect and we stress that it is a very small effect. Zinc is an example of a diamagnetic material—not surprising on the basis of Figure 14.1. However, if you guessed that chromium is diamagnetic you were wrong! The reasons are rather complex.

b. Paramagnets

In a paramagnet each atom has a moment, $\boldsymbol{\mu}_i$. However, $|\mathbf{M}| = 0$, because the $\boldsymbol{\mu}_i$ point in random directions. The reason is that the magnetic interaction between the moments is not sufficiently strong to order them and thermal energy keeps the direction of each $\boldsymbol{\mu}_i$ moving around all the time. Hence the vector sum of the $\boldsymbol{\mu}_i$ is always zero.

Now what happens when we place a paramagnet in an external magnetic field $\mathbf{H}$? The interaction between the external field and the individual $\boldsymbol{\mu}_i$ causes at least some of them to line up in the direction of $\mathbf{H}$. Hence $\sum \boldsymbol{\mu}_i$ is no longer identically zero but now depends on the degree of alignment of the moments (as well as the size of $|\boldsymbol{\mu}_i|$). We define the *magnetic susceptibility* as

$$\chi = \frac{M}{H} \tag{14.5}$$

We can measure M by methods described later as a function of H; the situation for a paramagnet is shown in Figure 14.2. The susceptibility is derived from the *slope* of the curve.

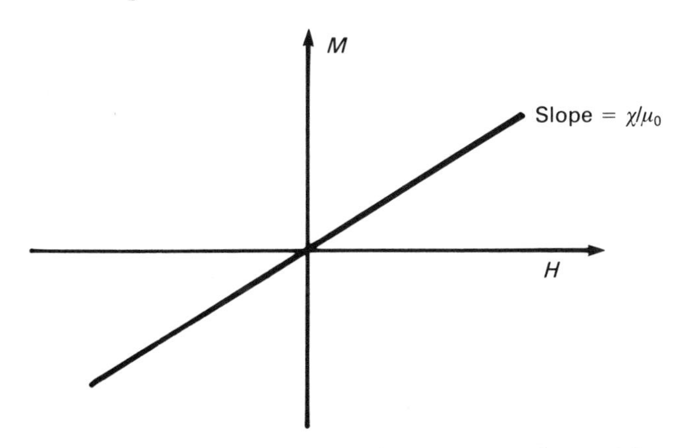

Figure 14.2. Magnetization curve for a paramagnetic material.

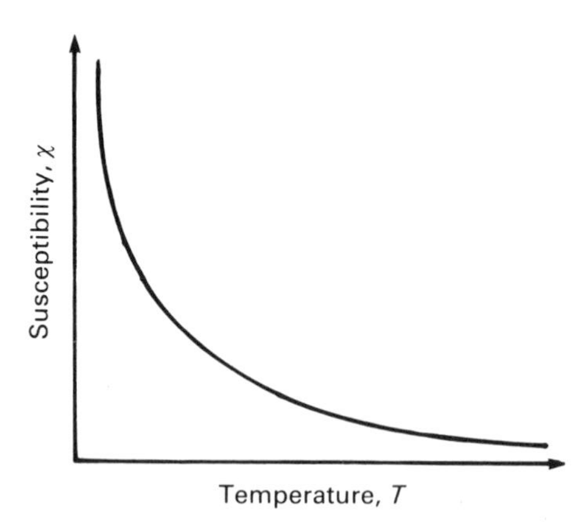

Figure 14.3. Temperature dependence of the paramagnetic susceptibility as predicted by the Curie law.

We can increase the paramagnetic susceptibility by lowering the temperature. Thus, in a given field H the ability of the moments to align is improved as a result of the decrease in the thermal energy. For a paramagnet, the temperature dependence of the susceptibility is given by Curie law (temperature in kelvins):

$$\chi(T) = (\text{const})T^{-1} \tag{14.6}$$

(see Figure 14.3). The constant depends primarily on the $|\boldsymbol{\mu}_i|$. Manganese is a well-known paramagnet.

c. *Ferromagnets*

In a ferrogmagnet each atom has a moment $\boldsymbol{\mu}_i$ and the magnetic interations between the moments are sufficiently large (as compared with the thermal energy) to make at least some of the moments align parallel. Thus $M \neq 0$ even in the absence of an applied magnetic field, because $\sum \boldsymbol{\mu}_i \neq 0$. We can, of course, force the ferromagnet into a paramagnetic state by increasing the thermal energy (i.e., raising the temperature), and we can increase M by decreasing the temperature. Figure 14.4 shows the magnetization M as a function of temperature. At T_c (known as the Curie temperature) $M = 0$ and the material is in the paramagnetic state. A simple theory (the mean-field theory) gives the temperature dependence of M for a ferromagnetic as

$$M = (\text{const})(T_c - T)^{1/2} \qquad T < T_c \tag{14.7}$$

Above the Curie temperature the material has a susceptibility that behaves according to the Curie–Weiss law:

$$\chi(T) = (\text{const})(T - T_c)^{-1} \qquad T > T_c \tag{14.8}$$

This is illustrated in Figure 14.5. Iron is the most widely known ferromag-

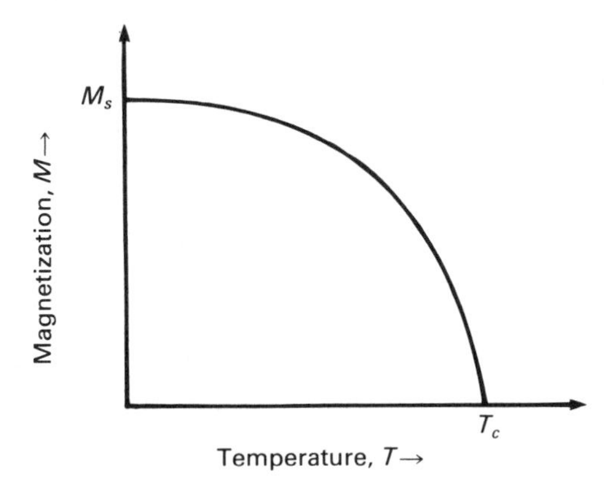

Figure 14.4. Magnetization-versus-temperature for a ferromagnet for $H = 0$.

netic material. Other elements that are ferromagnetic are shown in Table 14.3 along with their T_c. Various alloys and compounds are also ferromagnetic.

d. Magnetic versus magnetized

A few words of explanation are required concerning these terms. If we were to look at the atoms in a piece of iron on a very small scale (say 2 nm; this is ~10 lattice constants so a cube of that size would contain ~10^3 atoms), we would find that at some temperature a particular percentage of the iron magnetic moments (as given by Figure 14.4) would be spontaneously aligned in the same direction. Hence, iron is magnetic. If we were to look at

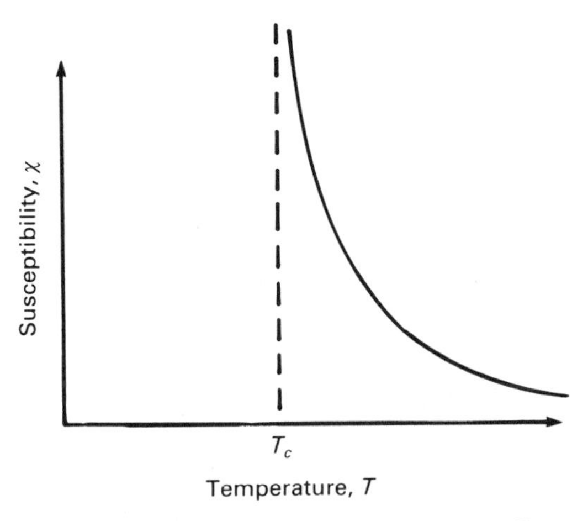

Figure 14.5. Paramagnetic susceptibility of a ferromagnet above T_c as predicted by the Curie–Weiss law.

Table 14.3. Ferromagnetic elements

Element	T_c (K)	M at 0 K (Tesla)
Fe	1043	0.1740
Co	1388	0.1446
Ni	627	0.0510
Gd	293	0.2060
Dy	88	0.2920

another 2 nm section in the same piece of iron we would find the same thing with one exception. The direction of M would (most likely) be different. In fact, a piece of iron would normally comprise a number of areas; each may be 10 nm on a side and each with its own particular magnetization direction. These areas are called *domains*. This is illustrated in Figure 14.6; the arrows show the predominant magnetization direction within each domain. Thus, on a *macroscopic* scale a truly unmagnetized ferromagnetic would have $\sum \boldsymbol{\mu}_i = 0$ where the sum is over all moments in the material.

The material becomes *magnetized* if there is a preferred orientation of the domain magnetizations. A simple method of magnetizing ferromagnet is to place it in a magnetic field. This situation is shown in Figure 14.7. Beginning with an unmagnetized material ($M = 0$ at $H = 0$) we apply a magnetic field in a particular direction. This forces the magnetization directions of some of the domains to line up. For sufficiently strong applied fields, all of the domains line up. This produces the *saturation magnetization* M_s which is an indication of μ_i as well as the fraction of moments that are aligned within each domain. It is, in fact, this M_s that is plotted as a function of temperature in Figure 14.4. If we now reduce the applied field to zero, some of the domains will remain lined up and produce a residual magnetization M_r—see Figure 14.7. It will actually require an applied field in the opposite direction (H negative) to reduce M to zero. Increasing and decreasing the

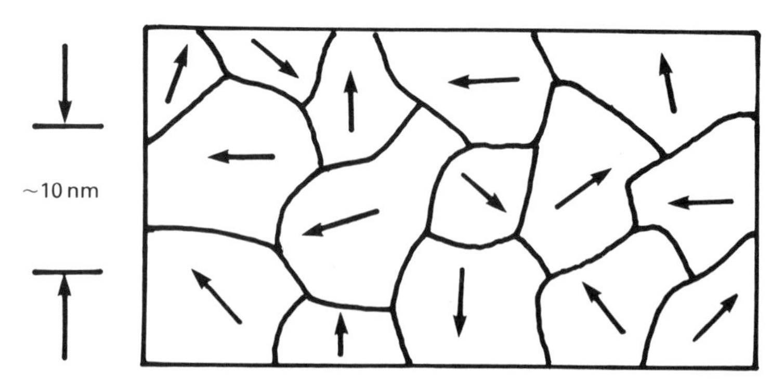

Figure 14.6. The spontaneous magnetization of domains in a ferromagnetic material.

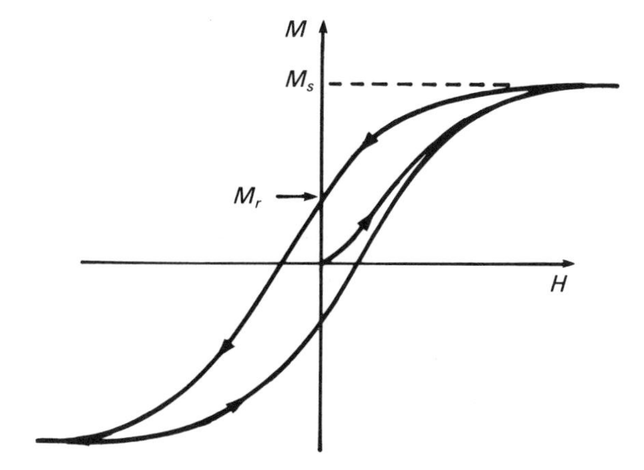

Figure 14.7. The magnetic hysteresis loop of a ferromagnet.

applied field from large positive to large negative values will cause the magnetization to follow the *hysteresis loop* shown in the figure.

There are numerous other classes of magnetic materials each with its own particular magnetization-versus-temperature curve, susceptibility-versus-temperature curve and magnetization-versus-applied field curve. The three we have described above are the most fundamental and important.

14.2 Production of magnetic fields

We consider three methods here; the use of permanent magnets, electromagnets, and superconducting magnets.

a. Permanent magnets

Permanent magnets are ferromagnetic materials that have been permanently magnetized. This means they have a large value of M_r. We can redraw Figure 14.7 as in Figure 14.8. Here, rather than plotting M on the vertical axis we plot flux density B to make things more compatible with the language used in the field of permanent magnets. We can relate B to M as

$$\mathbf{B} = \mu_0(\mathbf{M} + \mathbf{H}) \tag{14.9}$$

In order to be a good permanent magnet it is important that a material have both a large B_r and a large coercive force, H_c, because the product BH is the measure of the magnetic energy. The maximum magnetic energy $(BH)_{\max}$ in a material is given by the point in the second quadrant of the hysteresis loop in Figure 14.9 where this product is a maximum.

The hysteretic properties of a ferromagnetic depend not only on its composition but also on the method by which it has been prepared (e.g., by heat treatment).

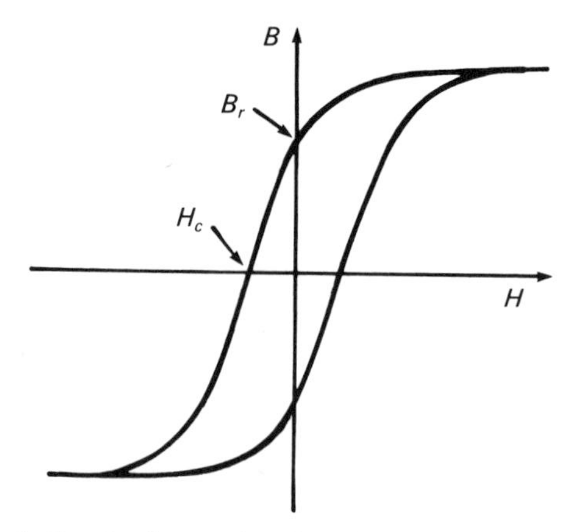

Figure 14.8. Magnetic flux density as a function of applied magnetic field for a ferromagnet.

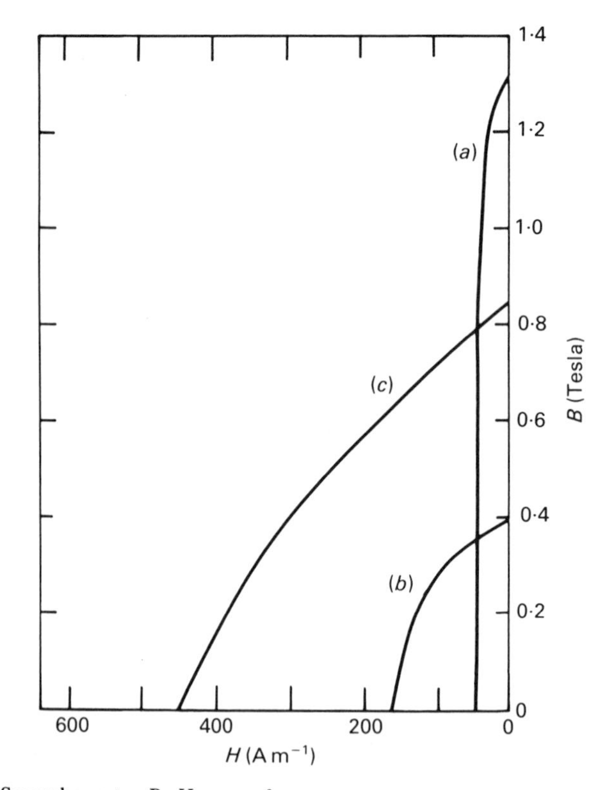

Figure 14.9. Second quarter B–H curves for some common permanent magnet materials, (a) Alnico 5, (b) typical ceramic and (c) Co–Sm.

We divide permanent magnets into three categories: (1) transition-metal alloys, (2) ceramics, and (3) transition-metal–rare-earth alloys.

Transition-metal alloys. Alloys comprised of transition metals form the traditional strong magnetic materials. Alnico alloys are the best known of these. These are Fe–Al–Ni–Co alloys, sometimes also including other elements. Alnico alloys are prepared by melting the elements together, followed by slow cooling in an applied magnetic field.

Other transition metal alloys each have their own particular heat treatment that produces the best permanent magnets. One of the best transition-metal permanent magnet materials is 76.7 wt.% Pt + 23.3 wt.% Co. As you might guess, however, it is very expensive.

Ceramics. The magnetic element in ceramic magnets is iron, but it is in the form of an oxide. Various heavy-metal oxides in the powdered form are mixed with ferric oxide powder. These are then heated under pressure in a magnetic field. The resulting alloys have fairly large H_c but a rather small B_r. The net $(BH)_{max}$ is comparable to those of some of the average transition-metal alloys. They are inexpensive and are easily produced in a variety of shapes.

Transition-metal–rare-earth magnets. Transition-metal–rare-earth alloys are of interest here because of the large magnetic moment carried by some rare-earth elements. The cobalt–samarium alloy Co_5Sm is the most commonly used at present. Some new alloys (e.g., Nd–Fe–B) have better properties (BH_{max} up to 400,000 T A m^{-1}) and may in time, replace Co_5Sm. Co_5Sm is prepared in the following way. Co_5Sm powder is mixed with a powder of the composition 60 wt.% Sm + 40 wt.% Co. This is sintered at a high temperature in an applied magnetic field. The resultant alloy has superior magnetic properties but is, unfortunately, somewhat brittle and expensive to make.

The properties of some commonly encountered permanent magnetic materials are given in Table 14.4. Figure 14.9 shows some second-quarter B–H curves from which B_r, H_c and $(BH)_{max}$ can be determined.

b. Electromagnets

Electromagnets consist of coils of conducting wire (or in some cases ribbon) through which a current is passed. We discuss some of the commonly encountered types.

Solenoids. Solenoids are cylindrically symmetric coils, usually without a core. They are suitable for moderate fields up to a few hundred gauss and provide good field homogeneity over a limited volume in their center. A straightforward derivation (given in nearly any first-year physics text) gives

Table 14.4. Permanent magnetic materials

Name	Composition	B_r (T)	H_c (10^3 A m^{-1})	$(BH)_{max}$ (10^3 T A m^{-1})
Alnico 1	12 Al, 22.5 Ni, 5 Co, 50.6 Fe	0.66	43.0	11.1
Alnico 2	10 Al, 18 Ni, 13 Co, 6 Cu, 53 Fe	0.70	51.7	13.5
Alnico 3	12 Al, 26 Ni, 3 Cu, 59 Fe	0.64	44.6	10.7
Alnico 4	12 Al, 28 Ni, 5 Co, 55 Fe	0.55	58.1	10.7
Alnico 5	8 Al, 15 Ni, 24 Co, 3 Cu, 50 Fe	1.20	57.3	39.8
Alnico 5DG	8 Al, 14 Ni, 24 Co, 3 Cu, 51 Fe	1.33	54.5	51.7
Alnico 8	7.5 Al, 14 Ni, 38 Co, 3 Cu, 8 Ti, 29.9 Fe	0.71	159	43.8
Alnico 9	7 Al, 15 Ni, 35 Co, 4 Cu, 5 Ti, 34 Fe	1.04	127	67.7
Cunife 1	60 Cu, 20 Ni, 20 Fe	0.57	47.0	14.7
Cunife 2	50 Cu, 20 Ni, 2.5 Co, 27.5 Fe	0.73	20.7	6.2
Cunico 1	50 Cu, 21 Ni, 29 Co	0.34	56.5	6.8
Cunico 2	35 Cu, 24 Ni, 41 Co	0.53	35.8	7.9
Remalloy	20 Mo, 12 Co, 68 Fe	0.88	25.5	10.3
Vicalloy 1	52 Co, 38.5 Fe, 9.5 V	0.90	23.9	8.0
Vicalloy 2	52 Co, 35 Fe, 13 V	1.00	35.8	23.9
Pt–Co	76.7 Pt, 23.3 Co	0.65	342	75.6
Ceramic	$BaO_6Fe_2O_3$	0.39	191	27.9
	$SrO_6Fe_2O_3$	0.34	263	24.7
Co–Sm	Co_5Sm + 60 Sm, 40 Co[a]	0.84	557	127

Note: Unless otherwise indicated, compositions are wt.%.
[a] Co–Sm is a mixture of Co_5Sm and an alloy of 60 wt.% Sm and 40 wt.% Co.

the axial field at the center of a solenoid as

$$H = \mu_0 nI \tag{14.10}$$

where n is the number of turns per unit length. The radial variations of the field in a solenoid are relatively small and the variation along the length is shown in Figure 14.10.

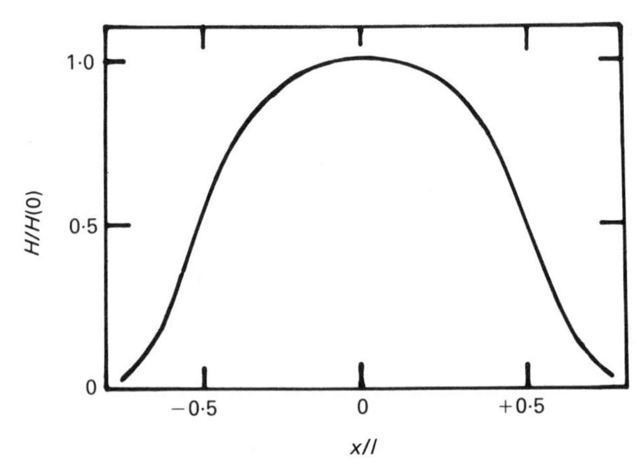

Figure 14.10. Variation of the magnetic field along the axis of a solenoid. The length l, is eight times the radius. x is the distance from the center along the axis and $H(0)$ is the field at the center of the solenoid.

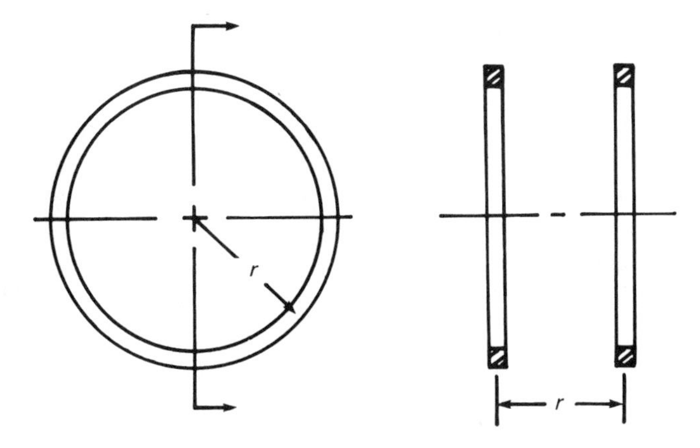

Figure 14.11. The geometry of Helmholtz coils.

Helmholtz coils. Helmoltz coils are of a particular geometry that will produce a very uniform axial magnetic field over a fairly large volume. The geometry consists of two thin solenoids of radius r parallel to each other and separated by a distance r, as shown in Figure 14.11. The axial field at the center of the coils is given by

$$H = \frac{8\mu_0 NI}{5^{3/2} r} \tag{14.11}$$

where n is the number of turns on one of the coils. Be sure when you connect these that the currents in the two coils are flowing in the *same* direction otherwise the fields from the two will cancel. The variation of the axial component of the magnetic field as a function of position is illustrated in Figure 14.12.

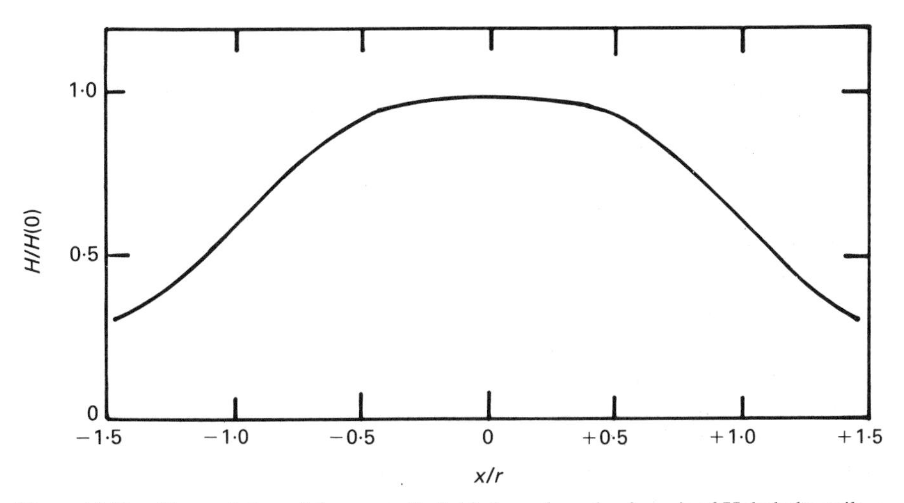

Figure 14.12. The variation of the magnetic field along the axis of a pair of Helmholtz coils. x is the distance from the center along the axis and $H(0)$ is the field at the center of the coils.

Electromagnets with armatures. Some electromagnets use a soft magnetic material (e.g., 97 wt.% Fe + 3 wt% Si) with a large saturation magnetization to concentrate the flux lines in a particular region. The field lines are concentrated in the region between the pole pieces. A small magnet of this type might weigh 20–30 kg and have pole pieces ~7 cm in diameter. It might produce a magnetic field of a few hundred amps per meter, which would be fairly homogeneous over the diameter of the pole pieces and over the distance between them, typically 2 cm or so.

A large laboratory electromagnet might weight several hundred kilograms and produce fields of up to $\sim 10^6$ A m^{-1}. This magnet might have 25 cm diameter pole pieces with a gap between them of 5 cm or more. In general, the smaller the pole gap the larger the field that can be produced by a particular magnet. The coils can be either copper or aluminum and can be varnish-insulated wire or layers of flat ribbon separated by insulating layers. Larger magnets will be water-cooled to prevent overheating, with the water pipes embedded in the coil windings.

Superconducting magnets. Superconductivity is a field in which there have been recent and very remarkable advances. In 1987 a new class of high-critical-temperature superconductor (materials based on rare-earth–Ba–Cu–O formulations) has been developed. Critical temperatures in excess of 150 K have now been reported and it is not clear when or where the present race for higher and higher critical temperatures will end. For sure the commercial application of these materials is close at hand. However, this recent development is, as yet, an unfinished story and it will presumably be some time before magnets made of these new materials will be commonplace in the physics laboratory. For the present, therefore, we will confine the discussion of specific superconducting materials in this chapter to the established lower-critical-temperature materials.

Let us begin with a review of properties of superconductors. We will approach this phenomenologically and explain the behavior of different kinds of superconductors. We will not delve into a quantum-mechanical explanation of why superconductivity exists—theoretical treatments of superconductivity can be found in some of the references.

In a superconductor the electrical resistivity drops to zero at some temperature. This temperature is called the *critical temperature* (see Figure 14.13). A number of elements show this behavior and they are listed in Table 14.5.

Something else important that happens to a superconductor is the behavior of magnetic flux lines. Figure 14.14 shows what happens. Above T_c, magnetic flux lines penetrate the material. If the material is cooled below T_c, the flux lines are expelled. This is known as the Meissner effect. This occurs if the field is not too strong. Thus if we were to plot the magnetization as a function of applied field it would look like that of a diamagnet (i.e., a small negative susceptibility).

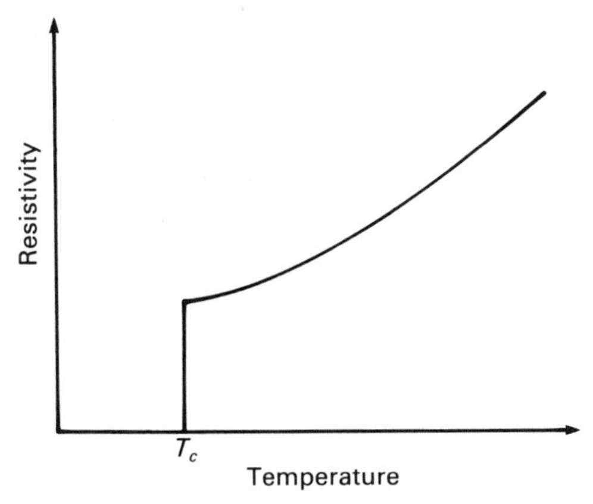

Figure 14.13. The resistivity of a material near the superconducting transition temperature T_c.

Table 14.5. Superconducting elements (all of type I) with their critical temperatures and critical fields

Element	T_c (K)	H_c (T)
Al	1.19	0.0099
Cd	0.56	0.003
Ga	1.09	0.0051
In	3.4	0.0294
Ir	0.14	0.002
La-β	6.0	0.16
Pb	7.18	0.08
Hg-α	4.15	0.041
Hg-β	3.95	0.034
Nb	9.46	0.194
Os	0.66	0.0065
Pr	1.4	—
Re	1.698	0.0198
Ru	0.49	0.0066
Ta	4.48	0.083
Tc	7.75	0.14
Tl	2.39	0.017
Th	1.37	0.016
Sn	3.722	0.031
Ti	0.39	0.01
W	0.012	0.107
U-α	0.68	0.2
V	5.41	0.137
Zn	0.875	0.0053
Zr	0.55	0.0047

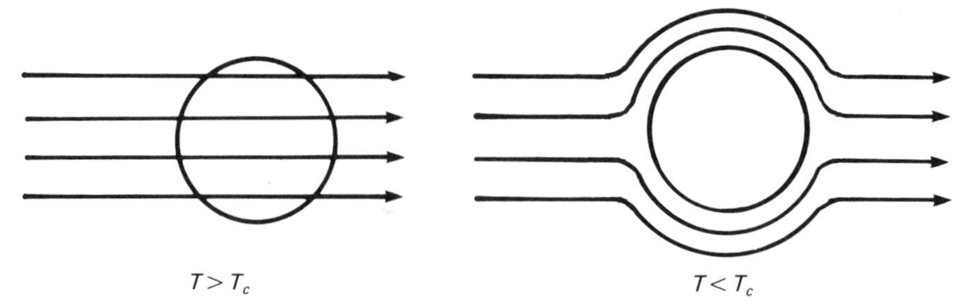

Figure 14.14. The behavior of a superconductor in a magnetic field, illustrating the expulsion of the flux lines below T_c—the Meissner effect.

This is illustrated in Figure 14.15 and, as you can see, this occurs only up to some critical field H_c. At this point the superconductor becomes normal (nonsuperconducting). This kind of superconductor is known as Type I. The idea that superconducting coils could be used to produce magnetic fields without the expenditure of energy (normally needed in electromagnets because of the resistance) was fine as long as the field required was less than H_c. However, the largest field we could hope to produce would be $H_c(T = 0)$ as shown in Figure 14.16. Table 14.4 gives values of this quantity for elemental superconductors. As we can see the critical fields are all less than 0.2 T. This is not very high. (As the convention in the field of superconductivity is to refer to H_c in gauss or tesla we will follow this convention; however, strictly speaking you should realize that in SI units $H = B/\mu_0$.)

Fortunately, in the 1950s another kind of superconductor was discovered: the Type II superconductor. These were alloys or intermetallic compounds that showed M–H curves like that in Figure 14.17. Up to some field H_{c1} these behave like Type I superconductors. At H_{c1} however, flux lines begin to creep into the material a little at a time. These are called fluxoids. The flux does not completely penetrate the sample until a field H_{c2} is reached.

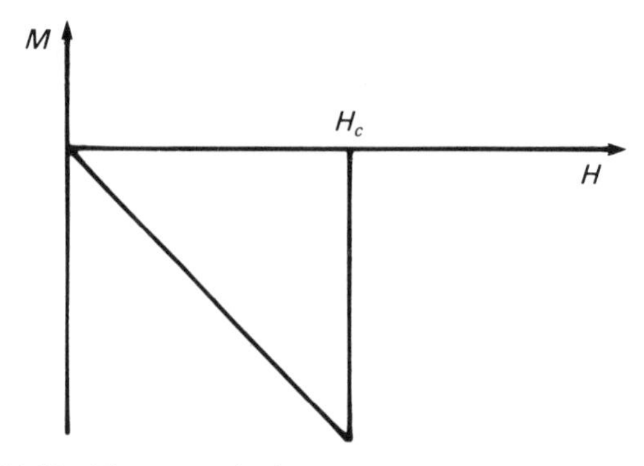

Figure 14.15. The magnetization curve for a type I superconductor.

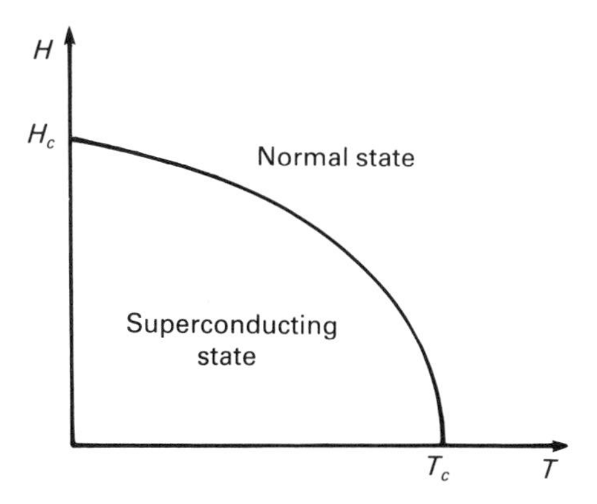

Figure 14.16. The phase diagram for a type I superconductor.

Electrically speaking the sample is superconducting (has zero resistance) up to H_{c2} when it becomes normal. H_{c2} can be quite a lot larger than H_{c1} (see Figure 14.18) and the use of such materials allows the manufacture of magnets that produce magnetic fields much larger than those attainable with normal magnets—and they do not dissipate any heat.

Figure 14.19 shows the H_{c2} vs T curves for some intermetallic compounds and alloys. Although it is clear that Nb_3Sn and NbTi are not the best materials from this standpoint, they have the best combinations of superconductivity and mechanical properties and are most commonly used. In cases where fields in excess of about 10 T are required, Nb_3Sn (an intermetallic compound) is used. If fields above 10 T are not required Nb–Ti (an alloy) is normally used, despite its lower T_c, because of its superior mechanical properties and ease of manufacture.

Nb_3Sn is extremely brittle and it is not simple to produce forms suitable for magnet fabrication. Most commonly this material is available as ribbons, laminated together with copper and niobium. Figure 14.20 shows a typical configuration. Wires are also generally available in the form of Nb_3Sn in a

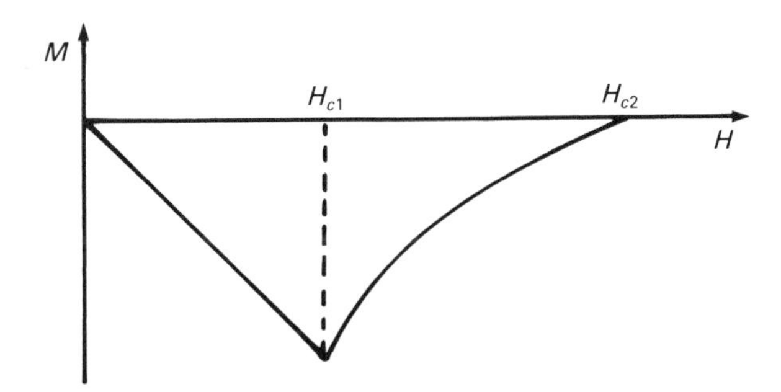

Figure 14.17. The magnetization curve for a type II superconductor.

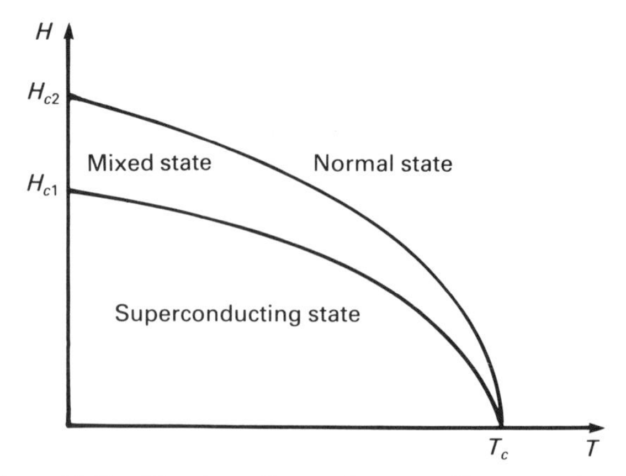

Figure 14.18. The phase diagram for a type II superconductor.

copper or bronze matrix. Because of the difficulty in preparing magnets from this compound, these magnets are generally small and use an outer magnet of Nb–Ti to supplement the field.

Nb–Ti wires are commonly produced in two forms: single-strand copper-clad, and multifilamentary. These are shown in Figure 14.21. Magnets produced from these wires as well as those of Nb_3Sn are typically of the solenoid type and, of course, do not have a ferromagnetic core.

Once a current has been established in a superconducting coil, it will circulate essentially indefinitely. Let us see how we get the current into the coil. Figure 14.22 shows the principle of energizing the magnet. A heater is wrapped around a small section of the superconducting circuit as shown in

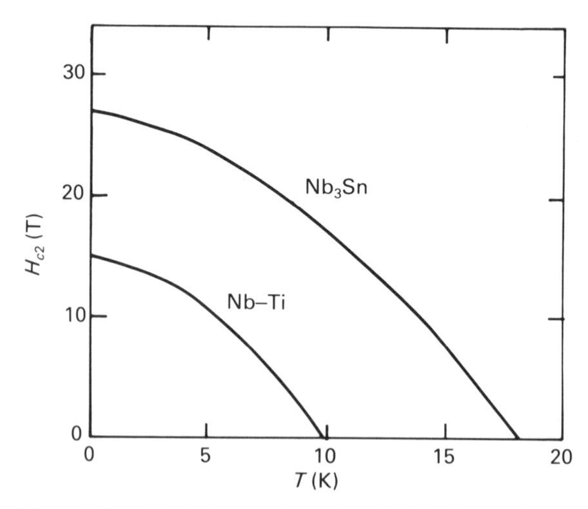

Figure 14.19. The relationship between the critical field H_{c2} and temperature for some commonly used superconducting materials.

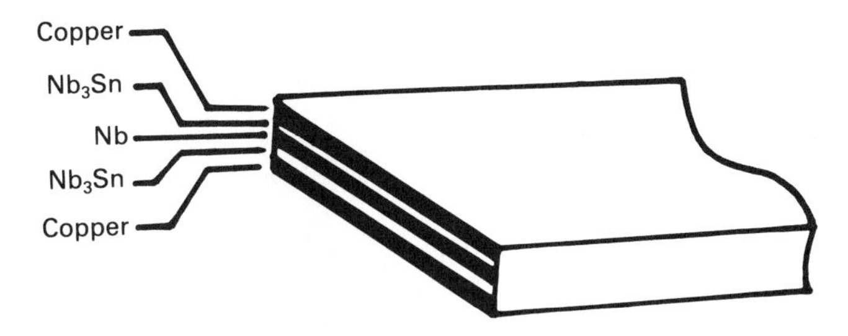

Figure 14.20. The construction of laminated Nb_3Sn ribbon for use in a superconducting magnet.

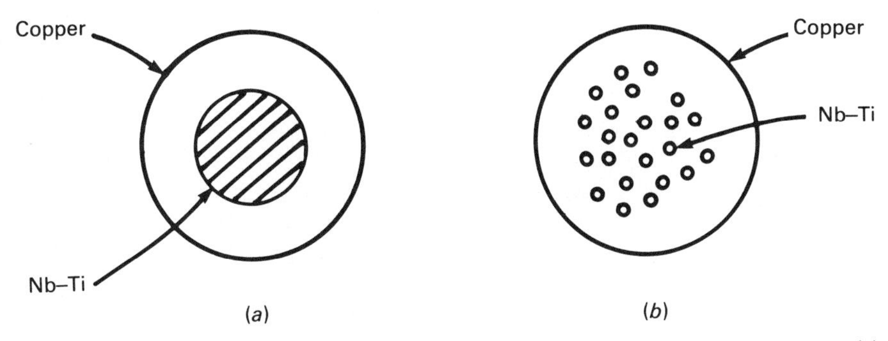

Figure 14.21. Two types of copper-clad Nb–Ti wire for use in superconducting magnets: (*a*) single-core; (*b*) multafilamentary.

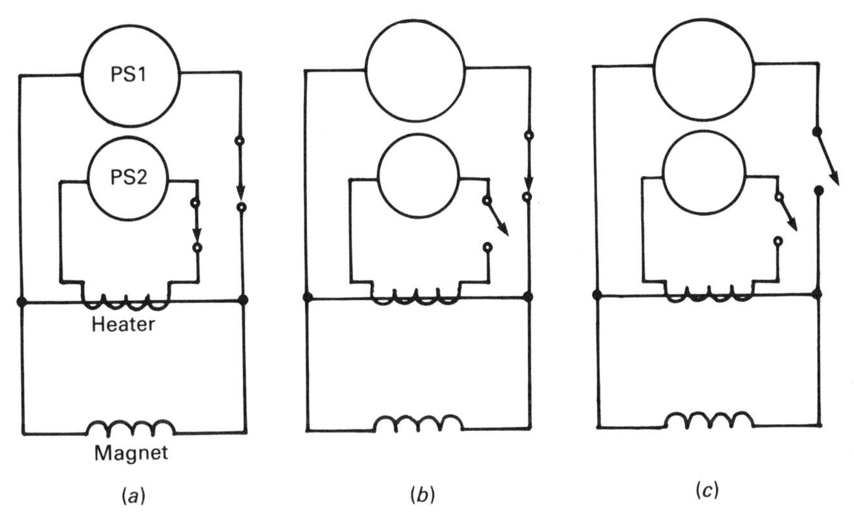

Figure 14.22. Energizing the superconducting magnet. PS1 is the main power supply and PS2 is the heater power supply: (*a*) heater on; (*b*) heater off; (*c*) main power supply off (persistent mode).

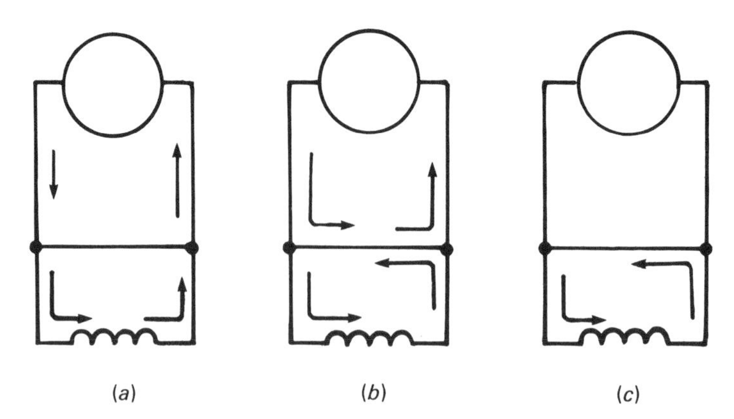

Figure 14.23. Current relationships in the states shown in Figure 14.22. Parts (*a*), (*b*) and (*c*) correspond respectively to (*a*), (*b*) and (*c*) in Figure 14.22.

Figure 14.22*a*. This is supplied by a small power supply PS2. This small link has a normal resistance of an ohm or so. Current supplied by the main power supply PS1 flows into the superconducting coil (as this is the path of least resistance). When the desired current is reached the heater is turned off as in Figure 14.22*b* and the link becomes superconducting. Figure 14.23 shows the current relationships. In this state the current flows through the power supply and the superconducting magnet. It is convenient to think of the zero current in the link as the sum of two opposite currents as in Figure 14.23*b*. The current in the PS1 loop is now reduced by turning down the power supply until this is zero. The power supply is now disconnected as in Figure 14.22*c* leaving a persistent current in the superconducting loop, Figure 14.23*c*.

The current is removed from the magnet (de-energizing) by reconnecting it to the power supply, turning the power supply current up to match the magnet current, heating the link so that it becomes normal, and using the power supply to turn down the currents.

14.3 Magnetization measurements

From measurements of magnetization versus applied field we can study the hysteresis loop of ferromagnetic materials (see Figure 14.7) and determine the susceptibility of paramagnetic and diamagnetic materials. We divide these types of measurements into four categories:

1. *Induction methods*. The sample is magnetized in an AC field and a coil is used to detect the magnetization of the sample.
2. *Vibrating-sample methods*. These are similar to induction methods except that rather than magnetize the sample with an AC field it is vibrated in a DC field.

2. *Force methods.* The force is measured on a sample in a magnetic field gradient.
4. *Indirect methods.* These measure quantities that are not the magnetization as such but that are related to the magnetization. Quantities like the electron paramagnetic resonance line width and the neutron scattering cross section are measured. These are complex methods and will not be discussed further in this work.

a. Induction methods

Induction techniques are the simplest methods and are applicable to ferromagnetic materials. These methods produce moderately accurate representations of the M–H curve. The sample is shaped as a toroid and the experimental apparatus is illustrated in Figure 14.24. The ferromagnetic toroid is magnetized by an AC magnetic field produced by a primary winding. The magnetization produced by the primary coil follows the hysteresis loop of the ferromagnetic material. The magnetic flux density B inside the ferromagnet is given by

$$B = \mu_a H = \mu_a H_0 \sin(\omega t) \tag{14.12}$$

where μ_a is the *apparent* permeability. For ferromagnets μ_a is generally determined by the *geometry* of the sample. The emf induced in the secondary coil is thus (N = number turns on secondary and A = coil cross-sectional area)

$$\begin{aligned} \text{emf} &= N\frac{\partial \phi}{\partial t} = -NA\mu_a H \\ &= -NA4\pi\frac{\partial M}{\partial t} \end{aligned} \tag{14.13}$$

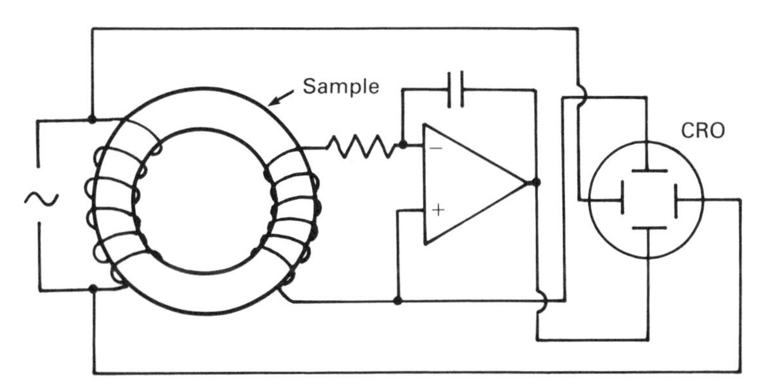

Figure 14.24. Experimental apparatus for measuring the M–H curve of a ferromagnet using the induction method.

As we are primarily interested in the magnetization itself as a function of applied field, the output of the secondary coil is integrated with the time constant of the integrator longer than the period of the driving field. The output of the integrator is proportional to the magnetization while the driving voltage is proportional to H. These are input to the Y and X inputs respectively of an oscilloscope. The image on the CRT is thus the M–H curve.

Although this is a simple and straightforward method of looking at ferromagnetic properties, it has some disadvantages. It requires samples that are sufficiently large and in an appropriate form to be made into toroids. Solids need to be cut, while foils or ribbons can be wound onto a former. Powders and liquids can generally be placed in an appropriate toroidal-shaped container. Another disadvantage is that in most cases the primary and secondary windings need to be wound on the toroidal samples, and this is generally inconvenient if a large number of different samples are to be studied. While it is applicable to large ferromagnetic samples, it is usually not sufficiently sensitive to study small samples or paramagnets.

b. Vibrating-sample methods

A simple method of measuring the magnetization of a sample is to vibrate it at a particular frequency in a constant magnetic field. The magnetization induced in the sample causes an oscillating magnetic field (because the sample is oscillating in space). This field is detected by a set of search coils. The coils do not detect the large constant applied field as this is not changing. Even when the applied field is swept, it does not change fast enough to be detected by the search coils. Figure 14.25 shows a typical system for making this kind of measurement. The sample is attached to a long rod that is connected to the cone of a loudspeaker. The speaker is

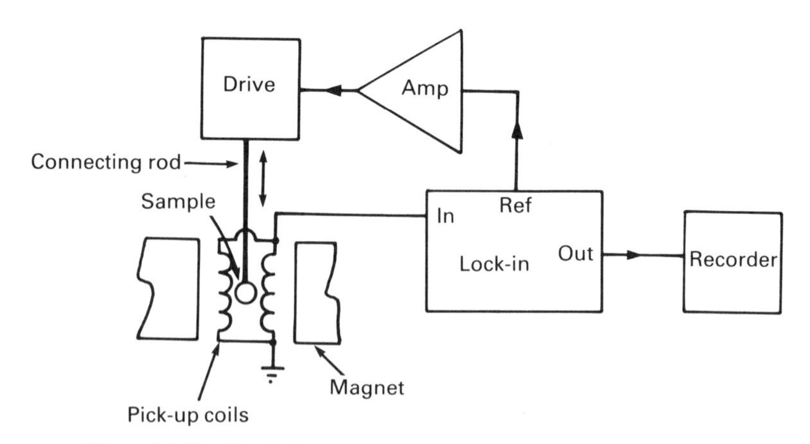

Figure 14.25. Block diagram of the vibrating-sample magnetometer.

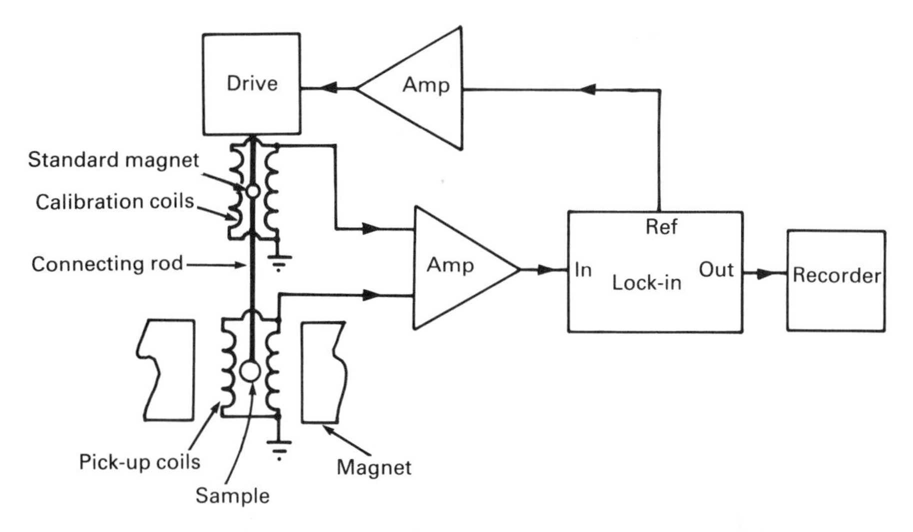

Figure 14.26. Block diagram of a vibrating-sample magnetometer using a reference magnet for calibration.

driven by an oscillator at a frequency f. The sample is suspended in the field of a large electromagnet. The search coils are suspended inside the field region of the large magnet. Here is an ideal application for the lock-in amplifier: a system is driven at a frequency f and a signal is produced at the same frequency. Figure 14.25 shows a simple method of detecting the signal using a phase-sensitive detector. The problem with this setup is that the driving frequency and amplitude will to some extent be affected by the mechanical resonance of the loudspeaker. Figure 14.26 shows a method of avoiding difficulties caused by this problem. The signal from the search coils around the sample is compared (using a differential amplifier) to the signal from a set of search coils placed around a "standard" magnet.

This kind of system is in wide use and is suitable for measuring susceptibilities of paramagnets, diamagnets, and other more unusual magnetic materials to an accuracy of about 10^{-4} A m^{-1} or better. It is also suitable for measuring the hysteresis curves of ferromagnets. In fact, its sensitivity makes this possible even if we are working with milligram quantities of sample.

c. *Force methods*

We consider several techniques under the heading of force methods. These are applicable to paramagnetic materials and, if made sufficiently sensitive, to diamagnets as well.

The Gouy balance. The Gouy method makes use of a long, cylindrical sample, one end of which is placed in a uniform magnetic field and the other

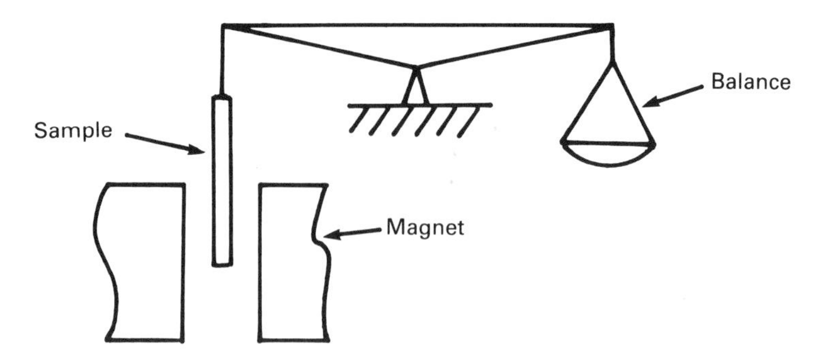

Figure 14.27. The Gouy method of measuring the magnetization of a sample.

end of which is placed in a region of zero field. The arrangement is shown in Figure 14.27. The susceptibility of the sample is determined by measuring the force on the sample as a function of applied field. The force on a sample of magnetization M (M is measured as magnetization per unit volume) in a field H is given by

$$\mathbf{F} = \nabla(\mathbf{M} \cdot \mathbf{H}) = \nabla(\chi H^2) \tag{14.14}$$

For a long sample, as shown in the figure, we integrate the force along the length of the sample (assumed to be the z direction in the figure). Thus

$$\begin{aligned} F_z &= \int dF_z = \frac{1}{2}\chi \int \frac{\partial H^2}{\partial z} dV = \frac{1}{2} S\chi \int \frac{\partial H^2}{\partial z} dz \\ &= \tfrac{1}{2} S\chi (H_1^2 - H_2^2) \end{aligned} \tag{14.15}$$

where S is the cross-sectional area of the sample in the X–Y plane. For $H_2 = 0$, this reduces to

$$F_z = \tfrac{1}{2} S\chi H_1^2 \tag{14.16}$$

Numerous apparatuses have been devised to measure this force. Some are specially designed for this kind of experiment, while others employ a standard laboratory balance to which the sample is attached by some means. The quality of the balance required depends on the size and type of sample as well as the required accuracy of the measurement. Typically, however, weight changes in the order of a few tens of milligrams are usual.

The main disadvantage of this method is that the sample needs to be both long and uniform. The larger the required magnetic field, the larger the magnet and hence the longer the sample needs to be.

Faraday method (also known as the Curie method). The Faraday method uses a small sample in a magnetic field gradient, thus eliminating the need

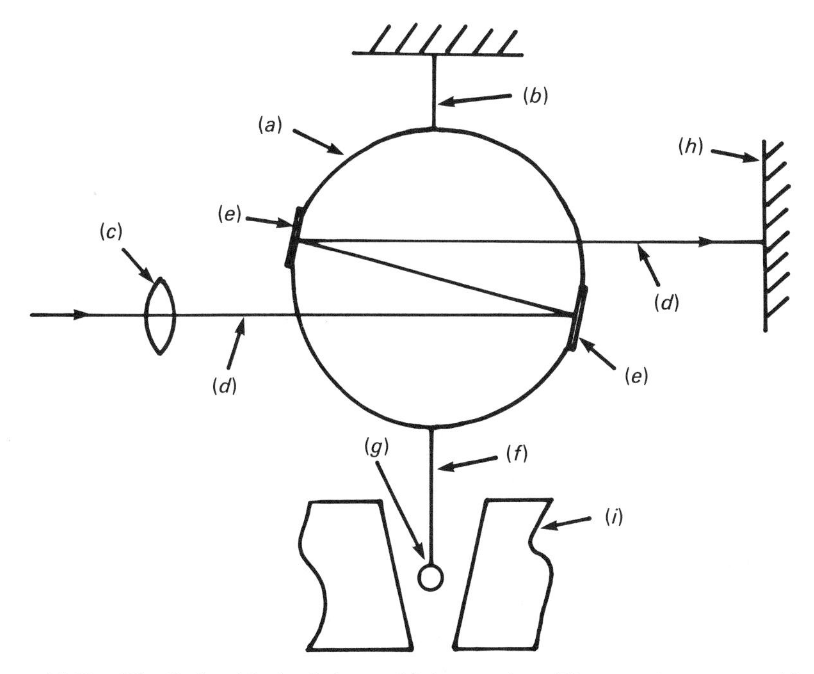

Figure 14.28. The Sucksmith ring balance: (*a*) bronze ring; (*b*) upper ring support; (*c*) lens; (*d*) light beam; (*e*) mirrors; (*f*) sample-support rod; (*g*) sample; (*h*) viewing screen; (*i*) magnet pole pieces.

for large (long) samples. From equation (14.14) we see that

$$M = F_z \left(\frac{\partial H}{\partial z} \right)^{-1} \tag{14.17}$$

A magnet provides a known field gradient and various methods have been used to measure the force on the sample. One ingenious method which is sometimes used is the Sucksmith balance, illustrated in Figure 14.28. The sample is suspended from a ring made of thin springy metal, frequently phosphor bronze. The ring has attached to it two mirrors. A beam of light is reflected from the two mirrors as illustrated. When a force is exerted on the ring it deforms slightly, the angle of the mirrors changes slightly and the path of the light beam is changed. The force, and hence the magnetization, is determined by measuring the deflection of the light beam on a screen. Once calibrated, this device is easy to use and sample changes are easy. The accuracy, however, is not as good as that of some other balances and certainly not as good as that of the vibrating-sample magnetometer.

14.4 Magnetic field measurements

a. The Hall probe

The Hall probe is the most convenient and the most commonly used method of measuring magnetic fields of ~10 A m^{-1} or larger. The accuracy for moderate fields can be better than 1 percent. Devices that consist of a Hall probe and a digital or analog meter are commercially available at prices beginning at a few hundred dollars. To understand the Hall probe we need to go back to the behavior of electrons and holes in a semiconductor.

Consider a piece of n-doped semiconductor as illustrated in Figure 14.29. We can follow a similar derivation for a p-doped semiconductor. Two applied fields are present in the problem:

1. An electric field along the x direction, providing a flow of majority electron carriers
2. A magnetic field in the z direction

Now we return to the picture of charge flow in semiconductors. From equation (1.19) we write the drift velocity for the electron carriers as

$$v_x = -\frac{eE_x\tau}{m} \tag{14.18}$$

We can write the Lorentz force acting on the electron (recall $B = \mu_0 H$) as

$$\mathbf{F} = \frac{d\mathbf{p}}{dt} = -e(\mathbf{E} + \mathbf{v} \times \mathbf{B}) \tag{14.19}$$

Consider the y component of equation (14.19),

$$\frac{\partial v_x}{\partial t} = -\frac{ev_xB}{m} \tag{14.20}$$

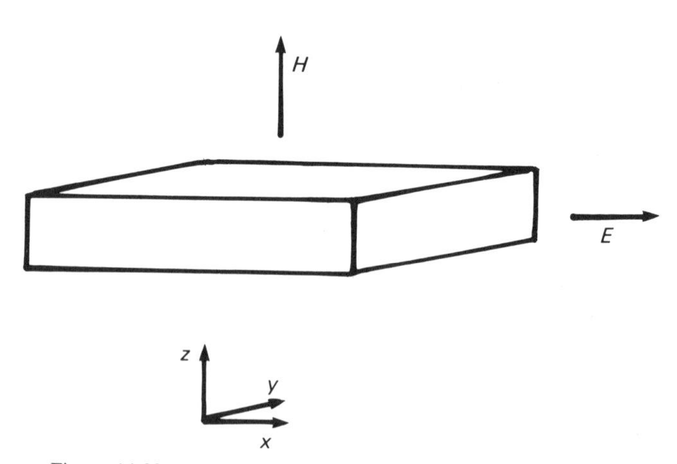

Figure 14.29. The geometry for the observation of the Hall effect.

and from (14.18),

$$\frac{\partial v_y}{\partial t} = \frac{e^2 E_x B}{m^2} \tag{14.21}$$

The drift velocity of electrons is limited by τ so that

$$\frac{\partial v_y}{\partial t} = -\frac{v_y}{\tau} \tag{14.22}$$

and (14.21) gives

$$v_y = -\frac{e^2 E_x B \tau^2}{m^2} \tag{14.23}$$

If we look at Figure 14.29, we see that charge carriers are flowing into and out of the sample along the x axis. This is because we have presumably attached wires on these two ends of the sample and have applied a voltage (electric field E_x). What about the situation along the y direction? No charges can flow this way because there is no place to go. So what about v_y? Well it turns out that an electric field E_y is set up to cancel out v_y. From equation (14.18) we can write, analogously,

$$v_y = -\frac{eE_y\tau}{m} \tag{14.24}$$

and from equation (14.23) we find that the above yields

$$E_y = -\frac{eB\tau}{m} E_x \tag{14.25}$$

E_y is known as the Hall field. We can actually measure it if we attach voltage probes across the sample in the y direction as long as we are careful not to draw any current. We see that the measured Hall field can be written as a voltage of the form

$$V_y = \frac{e\tau y_0 V_x B}{m x_0} \tag{14.26}$$

where x_0 and y_0 are the x and y dimensions of the sample, (e/m) is a constant, τ is a constant that depends on the particular sample, (y_0/x_0) is a geometric factor, and V_x is a voltage that we apply. Hence we see that our measured Hall voltage is linear in the applied magnetic field. In a more general sense, we find that V_y is proportional to the z component of B.

A Hall-probe magnetic field sensor consists of a semiconducting sample with four leads attached, two to provide V_x and two to measure V_y, a control box that includes a voltage source to provide V_x, a voltmeter to measure V_y and some electronics to convert the measured V_y into a flux density in gauss or tesla.

An example of the sensitivity of a typical Hall probe is illustrated in

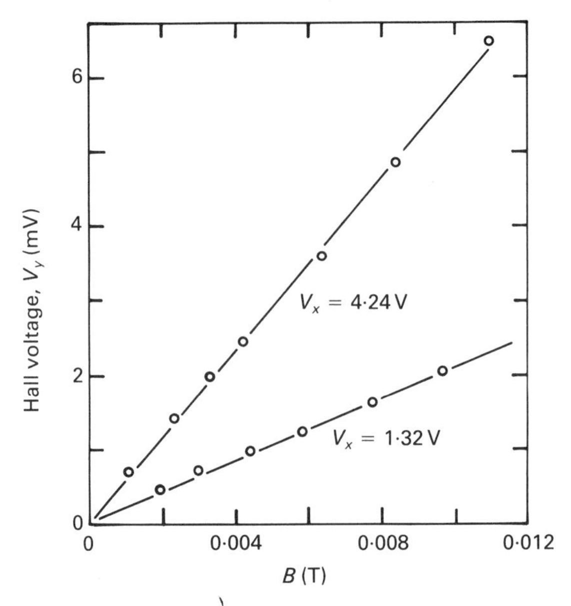

Figure 14.30. Measured Hall voltages in p-type germanium.

Figure 14.30. Values of V_y are plotted as a function of B for several different V_x. The probe geometry and characteristics are indicated in the figure.

b. The flux gate magnetometer

If we were to wind a coil around a piece of ferromagnetic material and apply a small AC voltage to the coil, then the ferromagnetic material would have a sinusoidally varying magnetization. It is best if at this point we make the assumption that the hysteresis (and M_r) of the material is small. There are many soft magnetic materials that have these features and can be used to make flux gates.

If the applied field is sufficiently large then the magnetization will not be sinusoidal in time but will saturate, as shown in Figure 14.31.

Now we consider the case where there are two coils around the ferromagnetic core that are "exactly" the same except that they are wound in opposite directions. Each coil will supply a magnetization of the form shown in Figure 14.31*c* but in opposite directions. Their total magnetization, as shown in Figure 14.32, will be zero. If a small external magnetic field (a DC field) were to be applied at the same time, the applied field, as shown in Figure 14.32*a*, would be a sinusoid with a small DC offset. As a result the saturated M curves of the two-coil system, as shown in Figure 14.32*a*, would have a DC offset—both of them in the same direction although the saturation magnetization (+ and −) would be the same. Figure

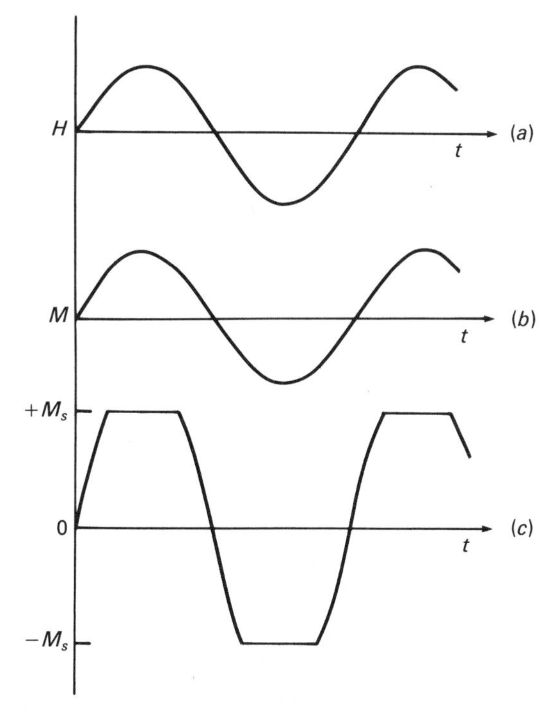

Figure 14.31. The magnetization of a ferromagnet in a sinusoidal applied magnetic field: (*a*) applied field; (*b*) magnetization; (*c*) magnetization in a sample driven to saturation.

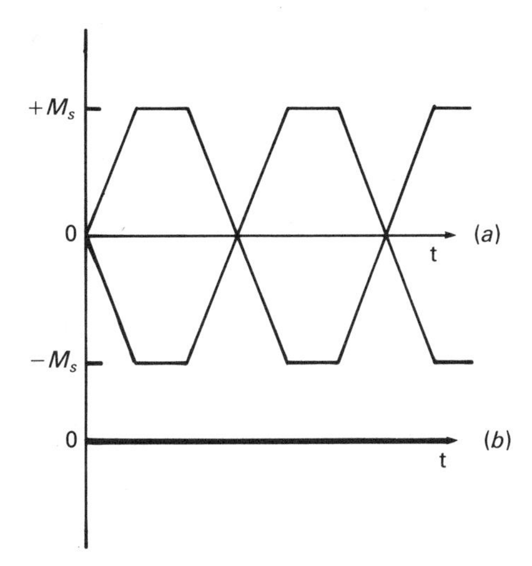

Figure 14.32. The magnetization of a ferromagnetic core inside two oppositely wound coils: (*a*) the components of M due to the two coils; (*b*) the total. The core is driven to saturation as in Figure 14.31*c*.

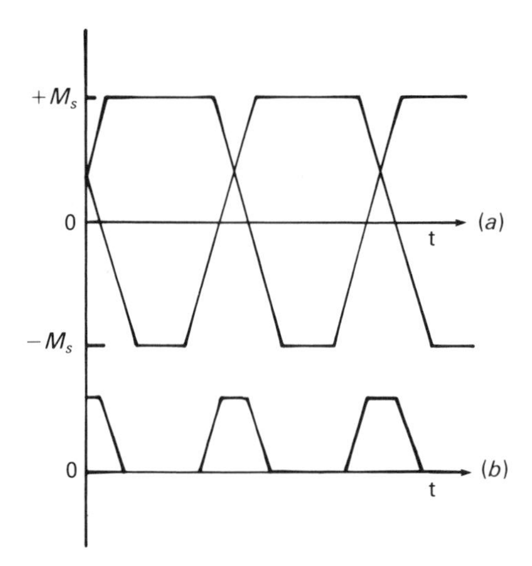

Figure 14.33. The magnetization of a ferromagnetic core inside two oppositely wound coils. As in Figure 14.32 except that there is an externally applied magnetic field.

14.33 shows this condition. Figure 14.33*b* shows that the net magnetization is no longer zero but is a small periodic quantity, the amplitude of which is proportional to the small external DC field. For this to work properly, it is essential that the two coils are exactly balanced (i.e., produce the same but oppositely directed magnetization). This is the principle on which the flux gate magnetometer works.

The flux gate magnetometer consists of a soft ferromagnetic core with two matched coils (called the primary coils or driving coils) wound in opposite directions. It is generally convenient to wind these on a former made of some nonmagnetic material. In the presence of an external DC magnetic field, the primary coils produce an AC magnetic field of the form of Figure 14.33*b*. The whole device is surrounded by an additional coil referred to as the secondary or pick-up coil. The presence of the DC field is detected by measuring the voltage produced by the secondary coil. This acts like an inductive field sensor and produces a voltage that is proportional to the time derivative of the magnetic field produced by the primary coils plus the external field. Figure 14.34 shows the relationship of these quantities. We should point out that it is the external field along the direction of the magnetization of the coil that is measured. It is important to realize that the flux gate will not work for DC fields that are larger than that required to saturate the core. For most soft magnetic materials used as flux gate cores this is in the range of a flux density of about 0.001 T.

We see that the output voltage peaks from the secondary coil are at *twice* the driving frequency. The amplitude of these peaks is related to the shape of the hysteresis loop of the core material, but most importantly it is

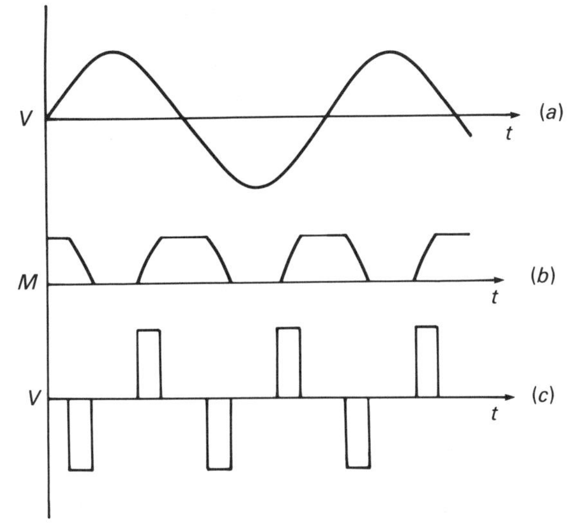

Figure 14.34. Properties of the flux gate magnetometer: (*a*) primary coil driving voltage; (*b*) core magnetization; (*c*) voltage induced in the secondary coil.

proportional to the size of the external DC field. So we now have a method, and a very sensitive one, for measuring DC magnetic fields. Figure 14.35 shows an example of the sensitivity of a typical homemade flux gate. We can see that even this simple measurement can yield values that are accurate to 10^{-6} T or better.

We can improve the flux gate by using a lock-in as illustrated in Figure 14.36. Fortunately, most lock-in amplifiers can lock on a signal twice the

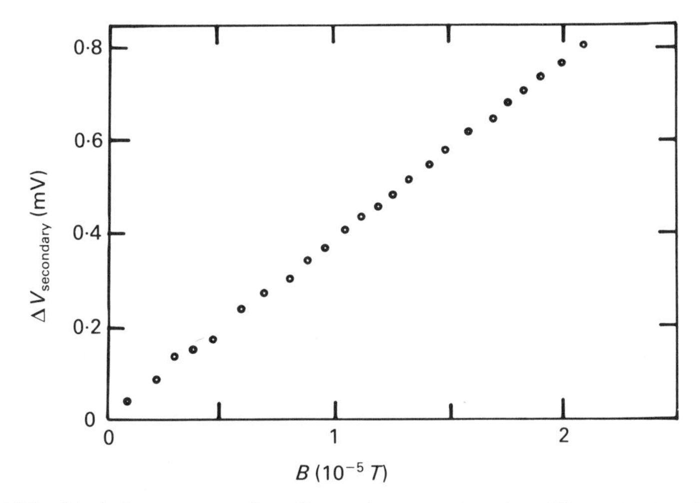

Figure 14.35. Typical response of a flux gate magnetometer. The output voltage was measured on a Hewlett–Packard 3400A RMS voltmeter.

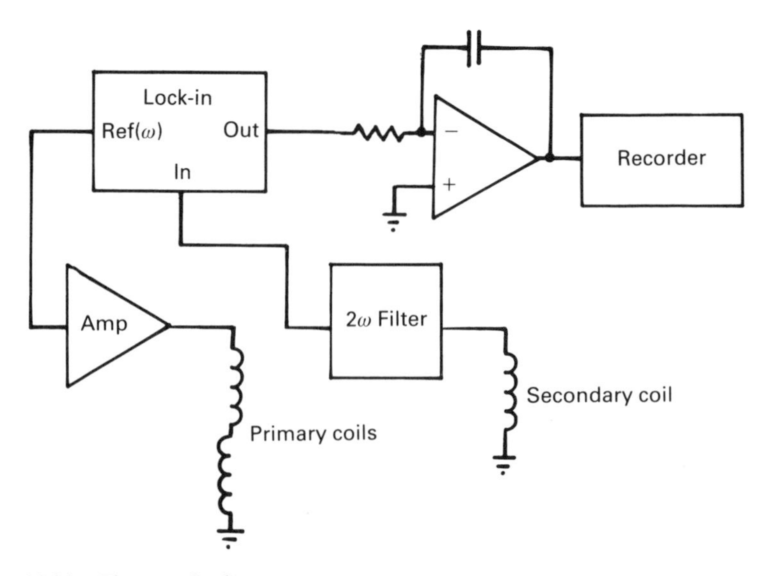

Figure 14.36. The use of a flux gate magnetometer in conjunction with a lock-in amplifier.

frequency of the reference signal as well as on the reference signal itself. It is not simple to determine the relationship of the output signal to the magnitude of the DC electric field. The details of such a calculation can be found in (for example) Primdahl or Hine (see Bibliography). With appropriate detection methods the flux gate magnetometer can be used to measure flux densities as small as $\sim 10^{-10}$ T.

Bibliography

The following references are recommended for the reader interested in obtaining more information on the topics covered in this text.

Chapter 1

A. J. Dekker. *Solid State Physics.* Englewood Cliffs, N.J.: Prentice-Hall, 1957.
C. Kittel. *Introduction to Solid State Physics,* 5th ed. New York: Wiley, 1976.
J. F. Pierce. *Transistor Circuit Theory and Design.* Columbus: Merrill, 1963.
J. N. Shive. *Physics of Solid State Electronics.* Columbus: Merrill, 1966.

Chapter 2

L. W. Anderson and W. W. Beeman. *Electric Circuits and Modern Electronics.* New York: Holt, Rinehart, Winston, 1973.
J. J. Brophy. *Basic Electronics for Scientists,* 3rd ed. New York: McGraw-Hill, 1977.
G. F. G. Delaney. *Electronics for the Physicist.* Harmondsworth: Penguin, 1969.
R. E. Simpson. *Introductory Electronics for Scientists and Engineers.* Boston: Allyn & Bacon, 1974.

Chapter 3

H. M. Berlin. *The 555 Timer, Applications Source Book with Experiments.* Derby, Conn.: E & L Instruments, 1976.
S. A. Hoenig and F. L. Payne. *How to Build and Use Electric Devices Without Frustration, Panic, Mountains of Money or an Engineering Degree.* Boston: Little, Brown, 1973.
D. Lancaster. *TTL Cookbook.* Indianapolis: Sams, 1974.
J. Millman and C. C. Halkias. *Integrated Electronics: Analog and Digital Circuits and Systems.* New York: McGraw-Hill, 1972.
D. H. Sheingold, ed. *Nonlinear Circuits Handbook,* 2nd ed. Norwood, Mass.: Analog Devices Inc., 1976.

Chapter 4

J. M. Downey and S. M. Rogers. *PET Interfacing.* Indianapolis: Sams, 1981.
D. Halliday and R. Resnick. *Physics.* New York: Wiley, 1962.

P. Horowitz and W. Hill. *The Art of Electronics.* Cambridge University Press, 1980.
R. M. Kerchner and G. F. Corcoran. *Alternating-current Circuits,* 3rd ed. New York: Wiley, 1951.
D. Lancaster. *Active Filter Cookbook.* Indianapolis: Sams, 1979.
E. M. Purcell. *Electricity and Magnetism: Berkeley Physics Course, Volume 2.* New York: McGraw-Hill, 1965.
A. L. Reimann. *Physics: Electricity, Magnetism and Optics.* New York: Barnes and Noble, 1971.
J. F. Wakerly. *Microcomputer Architecture and Programming.* New York: Wiley, 1981.
A. I. Zverev. *Handbook of Filter Synthesis.* New York: Wiley, 1967.

Chapter 5

R. P. Benedict. *Fundamentals of Temperature, Pressure and Flow Measurements.* New York: Wiley, 1969.
S. Dushman. *Vacuum Technique.* New York: Wiley, 1949.
J. F. O'Hanlon. *A User's Guide to Vacuum Technology.* New York: Wiley, 1980.

Chapter 6

American Study for Testing and Materials. *Manual on the Use of Thermocouples in Temperature Measurement* (STP470). Philadelphia: ASTM, 1970.
H. D. Baker, E. A. Ryder, and N. H. Baker. *Temperature Measurement in Engineering, Volume 1.* New York: Wiley, 1953.
T. J. Quinn. *Temperature.* London: Academic Press, 1983.

Chapter 7

C. A. Bailey, ed. *Advanced Cryogenics.* London: Plenum, 1971.
A. C. Rose-Innes. *Low Temperature Techniques.* London: The English Universities Press, 1964.
G. K. White. *Experimental Techniques in Low Temperature Physics.* Oxford University Press, 1959.

Chapter 8

F. T. Arecchi and E. O. Schultz-Dubois, eds. *Laser Handbook, Volume 1.* Amsterdam: North-Holland, 1972.
B. A. Lengyel. *Lasers,* 2nd ed. New York: Wiley, 1971.
R. M. Eisberg and R. Resnik. *Quantum Physics of Atoms, Molecules, Solids, Nuclei and Particles.* New York: Wiley, 1978.
RCA. *RCA Electro-Optics Handbook.* Lancaster, Penn.: RCA, 1974.

Chapter 9

E. A. Brown, *Modern Optics*. New York: Reinhold, 1965.
R. Kingslake. *Applied Optics and Optical Engineering, Volumes 1–4.* New York: Academic Press, 1965.
L. Levi. *Applied Optics, Volume 2.* New York: Wiley, 1980.
J. R. Meyer-Arendt. *Introduction to Classical and Modern Optics.* Englewood Cliffs, N. J.: Prentice-Hall, 1972.

Chapter 10

C. Chandler. *Modern Interferometers.* London: Hilger & Watts, 1951.
G. R. Fowles. *Introduction to Modern Optics.* New York: Holt, Rinehart, Winston, 1968.
M. Francon. *Optical Interferometry.* New York: Academic Press, 1966.
F. A. Jenkins and H. E. White. *Fundamentals of Optics,* 3rd ed. New York: McGraw-Hill, 1957.
A. Lallemand, in *Astronomical Techniques,* ed. W. A. Hiltner, p. 126. Chicago: University of Chicago, 1962.

Chapter 11

W. E. Burcham. *Nuclear Physics: An Introduction,* 2nd ed. London: Longman 1973.
H. A. Enge. *Introduction to Nuclear Physics.* Reading, Mass.: Addison-Wesley, 1966.
J. D. Jackson. *Classical Electrodynamics,* 2nd ed. New York: Wiley, 1975.
E. Serge. *Nuclei and Particles,* 2nd ed. Reading, Mass.: Benjamin/Cummings, 1977.

Chapter 12

G. F. Knoll. *Radiation Detection and Measurement.* New York: Wiley, 1979.
A. C. Melissinos. *Experiments in Modern Physics.* New York: Academic Press, 1966.
P. W. Nicholson. *Nuclear Electronics.* London: Wiley, 1974.
Ortec. *Experiments in Nuclear Science.* Oak Ridge, Tenn.: Ortec, 1971.
L. C. L. Yuan and C. S. Wu. *Methods of Experimental Physics: Volume 5, Nuclear Physics.* New York: Academic Press, 1961.

Chapter 13

E. P. Bertin. *Principles and Practice of X-Ray Spectrometric Analyses.* New York: Plenum, 1970.
B. D. Cullity. *Elements of X-Ray Diffraction.* Reading, Mass.: Addison-Wesley, 1956.

E. F. Kaelble. *Handbook of X-Rays*. New York: McGraw-Hill, 1967.
C. Kittel. *Introduction to Solid State Physics*, 5th ed. New York: Wiley, 1967.
D. E. Sands. *Introduction to Crystallography*. New York: Benjamin, 1969.
B. F. Warren. *X-Ray Diffraction*. Reading, Mass.: Addison-Wesley, 1969.
E. A. Wood. *Crystal Orientation Manual*. New York: Columbia University, 1963.

Chapter 14

R. M. Bozorth. *Ferromagnetism*. New York: Van Nostrand, 1951.
W. H. Hayt. *Engineering Electromagnetics*. New York: McGraw-Hill, 1967.
A. H. Hines. *Magnetic Compasses and Magnetometers*. Toronto: Toronto University, 1968.
E. A. Lynton. *Superconductivity*. London: Methuen, 1969.
A. H. Morrish. *The Physical Principles of Magnetism*. New York: Wiley, 1965.
E. A. Nesbitt and J. H. Wernick. *Rare Earth Permanent Magnets*. New York: Academic Press, 1975.
R. J. Parker and R. J. Studders. *Permanent Magnets and Their Applications*. New York: Wiley, 1962.
F. Primdahl, "Fluxgate magnetometers," *Journal of Physics E: Scientific Instruments* **12,** 241 (1979).
A. C. Rose-Innes and E. H. Rhodenick. *Introduction to Superconductivity*. Oxford: Pergamon, 1969.
M. N. Wilson, *Superconducting Magnets*. Oxford: Clarendon, 1983.

INDEX